PRACTICAL GEOMETRY
IN THE HIGH MIDDLE AGES

MEMOIRS OF THE
AMERICAN PHILOSOPHICAL SOCIETY
Held at Philadelphia
For Promoting Useful Knowledge
Volume 134

PRACTICAL GEOMETRY

IN THE HIGH MIDDLE AGES

ARTIS CUIUSLIBET CONSUMMATIO

AND THE

PRATIKE DE GEOMETRIE

Edited with Translation and Commentary
by
STEPHEN K. VICTOR

THE AMERICAN PHILOSOPHICAL SOCIETY
Independence Square · Philadelphia
1979

Library of Congress Catalog Card Number 78-73170
International Standard Book Number 0-87169-134-5
US ISSN 0065-9738

To Sue
and our family

CONTENTS

INTRODUCTION

   I. The Place of Practical Geometry in the Middle Ages   1
     The Nature of Practical Geometry   2
     Practical Geometry in Education   31
     Theory and Practice in Geometry   42
     Practical Geometry and Practical Concerns   53

  II. The Contents of Artis cuiuslibet consummatio and
      the Pratike de geometrie   74

 III. Procedures in the Editions, Translations, and
      Commentary   92
     Editing Artis cuiuslibet consummatio   93
     Editing the Pratike de geometrie   101
     Translating the Texts   104
     About the Commentary   105

ARTIS CUIUSLIBET CONSUMMATIO

   I. Planimetry -- Planimetria   108

  II. Altimetry -- Altimetria   220

 III. Volumetry -- Crassimetria   326

  IV. Fractions -- De minutiis   384

THE PRATIKE DE GEOMETRIE

   I. Planimetry (From ACC) -- Mesure des planeces   473

 III. Altimetry and Volumetry (From ACC) -- Mesure
      des hautesches et des perfondeces et des
      crasse mesures   507

   V. Other Geometric Material   522

  VI. Mercantile Material   550

SELECTED BIBLIOGRAPHY   602

INDEX OF LATIN TECHNICAL TERMS   615

INDEX OF OLD FRENCH TECHNICAL TERMS   623

INDEX OF ASTRONOMICAL PARAMETERS   628

GENERAL INDEX   630

PREFACE

Since beginning research toward a doctoral dissertation on practical geometry in the Middle Ages, I have changed
my views on the subject as I uncovered new evidence justi-
fying a different approach. My first assumption, and that
of most of the people I have spoken to about the topic, was
that "practical geometry" must relate somehow to architecture,
surveying and city planning, to those areas, in other words,
where geometry plays a central role in the exercise of other
professions. The study of medieval buildings, fields and
towns from extant physical evidence was not a fruitful
approach for me, and I have left it to those better trained
in the methods of archeology and art history. Since I was
working as an historian of science, I chose to concentrate
on the written tradition of treatises called "practical
geometry."

The study in depth of the two treatises presented
as part of this book, the practical geometry beginning
"Artis cuiuslibet consummatio . . ." (ACC) and the Old
French adaptation of it, the Pratike de geometrie, led me
to another view of the subject, namely that practical
geometry has its greatest importance as a popularization
of mathematics. The treatises on practical geometry were
a way of teaching some basic principles to those who would
not remain in school or university long enough to become
philosophers or theologians and would not necessarily
exercise a mathematical profession, but who might want,
or even need, some mathematics in their everyday lives.
The sampling of arithmetic and astronomy in ACC and of
commercial arithmetic and metrology in the Pratike argues
for the generally pedagogic, rather than scholastic,

ix

purpose of the treatises. The development of a vernacular
version of ACC is further evidence that the practical geometries sought their homes outside of the universities, perhaps
in the bureaucratic and commercial milieus. Nonetheless, as
the Introduction shows, the formalized structures of university education had an influence even on the non-scholastic
tradition of practical geometry. As the tradition developed,
the practical geometries acquired an increasingly theoretical
underpinning, to the point where they are sometimes considered
works on measurement rather than simply practical geometries.

ACC, whose author attempts to demonstrate most of
his geometric results, belongs fairly early in the tradition
of demonstrated practical geometry. By the time of its composition in the late twelfth century, a Latin Euclid was
widely available in the West. Thus practical geometries,
with their roots in the Roman agrimensores, Boethius, pseudo-
Boethius and Gerbert, were no longer the sole repositories
of geometric knowledge. If a practical geometry was to have
an impact, it had to offer something more than introductory
geometry to the potential reader. The author of ACC apparently broadened his audience by increasing the scope of his work.
The long sections on astronomy and fractional arithmetic
included in ACC indicate that he meant for his work to be
a general treatment of the mathematical sciences; the
treatise covers three of the four branches of the quadrivium,
music being omitted.

I have made substantial revisions and have supplemented my original doctoral dissertation study considerably
in preparing this work for publication. I have added an
edition and discussion of the thirteenth-century adaptation
of ACC, the Pratike de geometrie. Even though the Pratike
had been edited before, I felt a new edition from both manuscripts, with translation, reference to its sources in ACC,
and explicative commentary would be useful to researchers in
the history of science and the romance languages. With the
addition of the Pratike, this work acquires a further dimension.

Both ACC and the Pratike are popularizing efforts, but ACC's readers were more comfortable in Latin while those of the Pratike were more at home in French. Together the treatises give a picture of popular learning in mathematics during the High Middle Ages.

The introductory essay on The Place of Practical Geometry in the Middle Ages has been enlarged to include discussion of fourteenth-century materials. I have more fully justified the stemma of manuscripts of ACC for this edition, establishing the relationships among all known manuscripts. I have also included the marginal annotations made by a later medieval reader of ACC. The various indexes I have prepared should help the scholar find access to the various parts of the large book this work has grown into. The Latin and Old French indexes are meant as a resource for tracking down obscure technical usages. The Index of Astronomical Parameters seemed useful because of the extensive astronomical material in the book. Since ACC and the Pratike cover so many topics, the editions, with their commentaries, translations and indexes should help scholars in a variety of fields, including the history of geometry, arithmetic, astronomy, metrology and commerce. I hope my work is as helpful to them as theirs have been to me.

A very special note of appreciation goes to John E. Murdoch, who encouraged and supervised my work when the core of this book was a doctoral dissertation at Harvard and who continued to provide support as the work developed. Guy Beaujouan has shared his wealth of knowledge over the course of many years. Marshall Clagett has been very encouraging, offering advice and a place to work at the Institute for Advanced Study. Working space, time and access to libraries have also been arranged by Henry Copeland of the College of Wooster, Everett Mendelsohn of the History of Science Department at Harvard, Charles F. Montgomery of the Yale University Art Gallery and Rick Beard of Yale's Center for American Art and Material Culture.

PREFACE

The National Science Foundation supported part of my research through a doctoral dissertation grant. A Fulbright Fellowship permitted a year of study and research in Europe. Sincere thanks are due to a number of scholars who have read parts of the manuscript and saved me from many errors; Caroline Walker Bynum, Alfred Foulet, Owen Gingerich, Otto Neugebauer, Lon R. Shelby, Noel Swerdlow and Charles Zuckerman have all helped in various stages of the work.

The Publications Committee of the American Philosophical Society has been most understanding, suggesting production of this book by photo-offset from typescript as a cost-saving measure so that the text and variants, and translations and notes appear on facing pages. Although this approach has meant considerable work for my typist and me, I expect the users' convenience to justify our effort. I am very grateful to Roger Reed who did the drawings and to Donna Gold who typed this complex book in a manner suitable for publication.

To my wife, Susanne F. Roberts, whom I have exasperated for too long with problems of artificial measurement, cube root, and astronomical parameters, I owe the deepest debt of thanks. Her suggestions have been well-taken, her help is always unstinting, and her sighs of relief will echo through the house when this is done.

I.

THE PLACE OF PRACTICAL GEOMETRY
IN THE MIDDLE AGES

The history of the medieval traditions of practical geometry
is very little known; no full-length or general history of the
subject is available. Although it is not my intention to provide
such a history here, some knowledge of the place of practical
geometry in the Middle Ages is useful as a preliminary to understanding the late twelfth-century practical geometry, _Artis
cuiuslibet consummatio_, and its thirteenth-century adaptation,
the _Pratike de geometrie_, both of which I have edited below.

Information that will be useful in evaluating the significance
of the practical geometries _Artis cuiuslibet consummatio_ and the
Pratike de geometrie will also be relevant to understanding the
place of any of the medieval practical geometries. The materials
which shed light on the place of medieval practical geometry
offer insights into several aspects of practical geometry.

The first concern is the nature of practical geometry. One
might ask what was the place of practical geometry among the sciences, and what was included in the treatises themselves. In
other words, what was practical geometry thought to be by those
who composed treatises on the subject and by those who wrote on
the classification of the sciences?

The place of practical geometry in education is another area
of investigation. Its role in educational curricula, testimony
about its pedagogical effectiveness and its inclusion in works
having another educational purpose all provide information on
the place of practical geometry within medieval educational traditions.

1

A third topic to consider is the relationship between practical geometry and theoretical geometry. Both theoretical
and practical works seem to be sources of the contents of the
extant practical geometries. The question therefore arises
of the proper relationship of theory and practice. As several
of the practical geometries address themselves to the problem,
the relation indicates something of what the authors of the
treatises felt their place to be.

A last consideration on the place of practical geometry in
the Middle Ages is how it relates to practical concerns. Several treatises hint at some connection between practical geometry and what might be called the practice of geometry. Surveying and architecture are domains where one would expect
to find practical geometry applied to practical concerns.

The Nature of Practical Geometry

Practical geometry is defined by some of the treatises on
the classification of the sciences. During the twelfth and
thirteenth centuries, the nature of practical geometry is
described with increasing precision in the literature on
classification. Its place is more and more narrowly defined
as it is separated from other sciences. As its place is more
carefully circumscribed, its nature is gradually revealed.
In other words, practical geometry is not defined as an abstract concept, but in considering what it does and does not
do.

Throughout the same period and into the fourteenth century,
the practical geometries themselves vary widely. The contents
of one particular treatise can be quite different from that
of another. What is included in the various treatises does
not seem to follow a general pattern of development. The
nature of a treatise is reflected in its parts. The number
and names of the parts tell something of the author's intention for his practical geometry. Because the same name may
apply to different contents, we should establish, to some
extent at least, what is included in the various sections of
the different practical geometries.

The first use of the distinction geometria theorica et
practica seems to be that of Hugh of St. Victor in his practical geometry, the Practica geometriae.[1] Dividing the whole
of philosophy into theorica and practica is an ancient tradition, going back to Plato and Aristotle. Hugh uses the same
distinction in his work on the classification of the sciences,
the Didascalicon.[2] For the ancients, all of mathematics belongs on the theoretical side of the dichotomy. In the Didascalicon, Hugh follows the ancients, placing geometry and all of
mathematics within the theoretical sciences. In the Practica
geometriae,[3] Hugh divides geometry into theoretical and practical
branches. In other words, Hugh takes the theory-practice dyad
from the division of all of philosophy and applies it within one
of the branches of theoretical philosophy, thereby crossing the
boundaries of a domain to use the distinction in a new way.[4]

Since it is Hugh who first uses the terms geometria theorica
et practica, we might expect to see some hint of the separation
of the parts of geometry in the Didascalicon. It happens that
one of the major innovations in the Didascalicon offers to the
individual arts the distinction, analogous to that of theorica-
practica, of agere de arte and agere per artem. Hugh introduces
these terms in an effort to delineate better the areas of the
various arts and to limit the incursion of one art into the area
of another.[5] Agere de arte has to do with discussing or teaching about the art; agere per artem is the application of the
principles of the art to other endeavors.

How close the pair, de arte, per artem, is to the pair,
theorica, practica, may be established by considering any of a

_____

[1]Baron, 1955b: pp. 298-302.

[2]Ed. Buttimer, 1939. Tr. Taylor, 1961.

[3]Ed. Baron, 1966: pp. 1-64. Another MS of part of the treatise is BL Digby 166, ff.6v-8r.

[4]This practice of changing the domain of a distinction occurs
elsewhere in Hugh's work; see Baron, 1955a: p. 111.

[5]Hunt, 1948: pp. 99-100.

number of twelfth-century glosses.[6] An anonymous introduction
to astronomy, written in a blank space where a diagram was to
have been drawn, tells us

> The craftsmen (that is, the doers) of this art are two,
> one acting concerning the art (agens de arte), the other
> by means of the art (ex arte). The agens de arte is the
> teacher of the art, the agens ex arte is the practition-
> er of the art.[7]

Since ex arte is merely a reformulation of per artem, Hugh
would probably have concurred that the agens de arte, or
teacher, deals with the theoretical aspect of the art and
that the agens ex arte, the practitioner, deals with the practical aspects of the art. Still the distinction, de arte,
per artem, enunciated in Hugh's Didascalicon, does not fully
suffice to separate theoretical from practical geometry.
Nonetheless, in his Practica geometriae, Hugh does provide a
differentiation which is not only adequate but also has fuller implications. Theoretical geometry, Hugh tells us, deals
with measurements intellectually, finding dimensions by con-
templation through reason alone; practical geometry uses certain instruments and interprets by means of other proportions.[8]
The separation given in Hugh's practical geometry yields specific information about the nature of theoretical and practical geometry. However, the distinction in the practical geometry does not preclude the simpler and more general differentiation between de arte and per artem. The pair de arte-
per artem could have been used to divide geometry into two

---

[6]Ibid., pp. 100-105.

[7]Oxford, Corpus Christi College, MS 283, f.81v, as quoted
by Hunt, 1948: p. 104, nn. 1-4: "Opifices (glossed id est
factores) huius artis sunt duo, unus agens de arte, alter ex
arte. Agens de arte est doctor ipsius artis, agens ex arte
est practicus ipsius artis."

[8]Ed. Baron, 1966: 16.27-32: "Considerandum est quod omnis
geometrica disciplina aut theorica est, id est speculativa,
aut practica, id est activa. Theorica siquidem est que spacia
et intervalla dimensionum rationabilium sola rationis speculatione vestigat, practica vero est que quibusdam instrumentis
agitur et ex aliis alia proportionaliter coniciendo diiudicat."

parts, at least formally. The fuller differentiation, as
found in the _Practica geometriae_, would have completed the
treatment of geometry in the _Didascalicon_. Further, it would
have fit in well with distinctions found there. Why then does
Hugh use the distinction only in his _Practica geometriae_?
Knowing the state of geometric knowledge at Hugh's time may
help resolve the problem.

In the section of the _Didascalicon_ where Hugh describes how
he studied the quadrivium, his approach to all the quadrivial
sciences is primarily practical. Hugh's learning of geometry
seems to have involved measurement and comparison; no mention
is made of proving theorems.[9] It is obvious that Hugh did not
read Euclid as a youth; Latin translations only of an extremely
small part of the _Elements_ were available. Still there was no
shortage of works on geometry in circulation by about 1100;[10]
the two pseudo-Boethian geometries were available, as were Gerbert's geometry and other similar works.[11] These were based primarily on compilations used by the Roman _agrimensores_. Since
the _agrimensores_ were practicing surveyors, it is not surprising
that their treatises have a distinctly practical orientation.[12]

---

[9]Ed. Buttimer, 1939: pp. 114-115: "Ipso exemplo oculis
subiecto, quae ampligonii, quae orthogonii, quae oxygonii
differentia esset, patenter demonstravi. Utrumne quadratum
aequilaterum duobus in se lateribus multiplicatis embadum
impleret, utrobique procurrente podismo didici . . . Haec
puerilia quidem fuerant, sed tamen non inutilia, neque ea
scire stomachum meum onerat." See tr. Taylor, 1961: pp. 136-137.

[10]In his _Practica geometriae_, Hugh tells us he is writing his
treatise "not as though to forge a new work but to collect the
scattered old ones," ed. Baron, 1966: 15.1-2: "Practicam geometrie nostris tradere conatus sum, non quasi novum cudens opus
sed vetera colligens dissipata."

[11]The second of the geometries attributed to Boethius has been
edited by Folkerts, 1970. There is no complete edition of the
first pseudo-Boethian geometry. Books I and II may be found in
Migne, 1844-1864: vol. 63, cols. 1352-1364. Books I, III, IV
and parts of V are available in Blume, Lachmann and Rudorff,
1848-1852, v. I: pp. 377-412. Gerbert's Geometry and many related texts have been edited by Bubnov, 1899.

[12]Probably the most significant source of geometric knowledge
in the early Middle Ages is the collection of materials on sur-

The legacy of the Roman surveyors gives the works of pseudo-
Boethius and Gerbert a practical flavor as well. Thus the
geometry available in about 1100 could easily be considered
practical geometry, even though the term was not in use at
the time. The practical nature of geometry in Hugh's description accurately reflects the geometric treatises we know to
have been available at the time of Hugh's childhood.

Elsewhere in the Didascalicon, Hugh gives the parts of
geometry as planimetria, altimetria, and cosmimetria; he describes geometry as a means of measuring material things like
the sea (its depth), a tree (its height), the world, a ball,
and an egg.[13] It is quite inaccurate to consider altimetry,
planimetry, and cosmimetry to be parts of geometry in general.
Hugh seems to agree that they are part of the more restricted
subject of practical geometry, for in the Practica geometriae
he gives the same triplet as its parts. Also, Hugh's description of geometry as material measure fails to account for abstract geometry, which deals with relations besides those of
measure. We are thus led to assume that at the time Hugh
wrote the Didascalicon, he was not sufficiently familiar with
Euclid's Elements to have listed parts of theoretical geometry
or even to have realized that the geometry which he knew was
not all of geometry. If the Practica geometriae came after

---

veying, land division, etc. deriving from the practical manuals of the agrimensores. On the geometric traditions arising
from their works, see Cantor, 1875, and Thulin, 1911. On the
agrimensores in general, see Dilke, 1971.

[13]Ed. Buttimer, 1939: p. 33: "De geometria. Geometria
tres habet partes, planimetriam, altimetriam, cosmimetriam.
Planimetria planum metitur, id est, longum et latum, et extenditur ante et retro, dextrorsum et sinistrorsum. Altimetria altum metitur et extenditur sursum et deorsum. Nam
et mare altum dicitur, id est profundum, et arbor alta, id
est sublimus. Cosmus mundus interpretatur, et inde dicta
est cosmimetria, id est mensura mundi. Haec metitur sphaerica, id est, globosa et rotunda, sicut est pila et ovum,
unde etiam a sphaera mundi propter excellentiam dicta est
cosmimetria, non quia tantum de mundi mensura agat, sed
quia mundi sphaera inter omnia sphaerica dignior sit." See
tr. Taylor, 1961: p. 70.

the Didascalicon,[14] it may have been the appearance of Adelard
of Bath's version of the Elements which caused Hugh to elaborate
his definition of geometry.[15] Perhaps even Hugh's application
of the distinction practica-theorica to geometry was a way of
differentiating the geometry available in the very early twelfth
century from the "new" geometry of Euclid, which certainly de-
served to be called theoretical geometry. Thus, what had been
the extant contents of geometry before the recovery of Euclid
in a translation from the Arabic became known as practical geometry. This newly named geometry had no place among the traditional sciences. Hugh of St. Victor named it but did not place
it within his classification of the sciences.

The assimilation of Arabic learning in the West has other effects on the definition and elaboration of practical geometry.
Dominicus Gundissalinus, Archdeacon of Segovia, as translator
from the Arabic was in an excellent position to know the Arabic
traditions concerning geometry. He also was in contact with
the intellectual milieu of Northern France.[16] It is no surprise,
then, to find that Gundissalinus has an important influence on
the development of that part of the classification of the sciences that relates to practical geometry. One aspect of Gundissalinus's work concerns the relation of practical geometry
to theoretical geometry; another aspect defines the place of
practical geometry with respect to the practice of certain
crafts.

---

[14]They both seem to date from the period 1125-1130. See
Baron, 1955a: p. 116 for the date of the Practica geometriae.
Taylor, 1961, in his introduction to the Didascalicon, p. 3,
suggests a date in the late 1120's for that work.

[15]It is difficult to give a precise date for Adelard's translation. In a comment in his Astrolabe, cited in Haskins, 1924:
p. 25, Adelard tells us that the translation of the Elements
predates his Astrolabe. Haskins, pp. 23-29, dates the Astrolabe
in the period 1142-1146. It is most likely that the translation
came after 1126.

[16]See Haring, 1964: pp. 271-286.

In the work of Gundissalinus on the classification of the
sciences, the De ortu scientiarum,[17] a distinction comparable
to that between theory and practice is found in the pair
extrinsecus and intrinsecus.[18] What must be known before before beginning the practice of an art is ars extrinsecus;
what must be known in order to practice the art is the ars
intrinsecus. For Gundissalinus, the extrinsic art is very
closely related to theory; the intrinsic to practice.[19]
Gundissalinus intends the distinction to apply to all the sciences; all of them have their extrinsic and intrinsic aspects.

Even though Gundissalinus has distinguished the extrinsic
and intrinsic approaches to a subject, he in no way means for
them to be independent; they are complementary rather than
separate approaches to the same subject.[20] In fact, so close
is the reliance of the intrinsecus on the extrinsecus that
Gundissalinus warns us that

> It would be disgraceful for someone to exercise any
> art and not know what it is, and what subject matter
> it has, and the other things that are premised of it.[21]

In addition to Gundissalinus's plea for the necessity of an
extrinsic or theoretical knowledge of an art as a prerequisite
to its practice, he specifically emphasizes the merits of the
practical approach to an art. The intrinsic or practical ap-

---

[17]Ed. Baur, 1903.

[18]The distinction seems to originate in Varro; it comes to
the twelfth century from Marius Victorinus's commentary on
Cicero's De inventione; see Hunt, 1948: p. 98.

[19]Ed. Baur, 1903: p. 44: "Cum autem omnis ars dividatur
in theoricam et practicam, quoniam vel habetur in sola cog-
nicione mentis--et est theorica--, vel in exercicio operis--
et est practica--: profecto ars extrinsecus pertinere videtur
ad theoricam, ars vero intrinsecus ad practicam."

[20]Ibid., p. 43: "Non autem ista sic distinguimus quasi
artem extrinsecus et intrinsecus duas artes esse velimus, sed
quod hiis duobus modis una et eadem ars docetur."

[21]Ibid., p. 44: "Turpe enim esset alicui, si aliquam artem
exerceret, et, quid ipsa esset et cuius generis, et quam ma-
teriam haberet, et cetera que premissa sunt, ignoraret."

proach gives us both knowledge and deed, once we have the theoretical basis.[22]   Thus, both the theoretical and the practical
aspects of a science are not only mutually beneficial, but in-
terdependent in Gundissalinus's scheme. The interdependence
of theory and practice is a theme that occurs frequently in
the treatises on practical geometry, as we shall see below.

In assimilating Arabic knowledge and traditions to those of
the West, Gundissalinus was in a pivotal position for elaborating new distinctions reflecting Arabic learning in a Western
context. With respect to the divisions of practical geometry,
Gundissalinus serves as a bridge between two traditions. We
shall see that Gundissalinus, in response to the craft traditions reflected in an Arabic classification of the sciences,
distinguishes between two sorts of practical geometry.

Gundissalinus's discussion of each of the sciences is arranged into several topics. Certain topics applying to geometry
as a whole have but one answer. Seven other topic headings require two answers, one for theoretical geometry and one for
practical geometry. For a few of the topics, there are two answers for practical geometry, indicating that Gundissalinus
recognizes two types of practical geometry. The seven headings
for which the distinction between theoretical and practical
geometry is made are presented in the table following:[23]

---

[22]Ibid., p. 44: "Ars enim extrinsecus non tradit actum,
set scienciam tantum; ars vero intrinsecus et actum dat et
scienciam."

[23]Page and line numbers, separated by a period, are given
after each of the answers. Unfortunately, there is a gap in
the manuscript where the parts of practical geometry were to
have been discussed. The complete list of topics for geometry
are given in Baur's ed., 1903: p. 102: "Circa geometriam
quoque hec eadem inquirenda sunt: quid sit ipsa, quod genus
eius, que materia, que partes, que species, quis artifex,
quod instrumentum, quod officium, quis finis, quare sic dica-
tur, quo ordine sit legenda."

Topics	Geometria theorica	Geometria practica
Subject matter (materia)	Immovable magnitude outside of matter (105.19-21)	Immovable magnitude in matter (105.19-21)
Parts (partes)	Consideration of lines, surfaces solids (105.21-23)	[missing]
Species (species)	Construction, knowing, finding (106.21-107.17)	Altimetry, plani-metry, cosmimetry (108.6-14)
Purpose (finis)	To teach something (107.21)	To do something (107.21-22)
Agent (artifex)	Geometer, who renews and teaches geometry (108.15-16)	Measurers; and, in mechanical arts, makers (109.1-8)
Instrument (instrumentum)	Demonstration (108.17)	Measures; carpen-ter's, smith's and mason's tools (109.13-18)
Duty (officium)	To give reasons and dispel doubt (109.19-21)	To give measure-ments; or limits which the work should not surpass (109.24-110.2)

The two types of practical geometry implicit in Gundissalinus's scheme are evident in the last three topics in practical geometry, where he has had to provide double answers.
The separate agents, instruments, and duties of practical
geometry provide for two sorts of practical geometry; a
mechanical or fabricant's practice exists along with the measurer's practice. It may seem that Gundissalinus has added
the maker to the description in order to expand the domain of
practical geometry beyond simple measurement. It would be
revealing to know whether Gundissalinus made the addition to
take account of the actual practice of the mechanical arts,
or in response to another treatise on the classification of
the sciences. It would be most interesting if Gundissalinus
were commenting on the practice of the mechanics and artisans
of his milieu, but we have no proof that he was. Another
work on classification seems a more likely, or at least more
demonstrable, origin of Gundissalinus's elaboration.

One of Gundissalinus's major sources is al-Farabi's work
on classification, known as De scientiis in Gerard of Cremona's
translation.[24] Gundissalinus uses sources other than Farabi
in his treatment of geometry, but a partial comparison of Gundissalinus's section on practical geometry with that of Farabi
shows many similarities and an important difference. Al-Farabi
tells us that

> Practical geometry investigates lines and surfaces which
> are in a wooden body if the person who uses this science
> is a carpenter, or in an iron body if he is a blacksmith,
> or in the surface of the earth and fields if he is a
> surveyor.[25]

Gundissalinus's version reads

> The agent of practical geometry is he who employs it in
> working. There are, however, two classes who employ it
> in working, measurers and makers. The measurers are
> those who measure the height, depth or surface area of
> the earth. The makers are those who toil in working in
> the mechanical arts, as a carpenter in wood, a black-
> smith in iron, a mason in cement and stones, and simi-
> larly, every agent of the mechanical arts following
> practical geometry.[26]

The difference between the treatments lies in this: Farabi
considers the carpenter, blacksmith, and surveyor all of equal

---

[24]Baur established Gundissalinus's reliance on al-Farabi
(see Baur, 1903: p. 166), but Baur used an edition of al-Farabi
corrupted by inclusions from Gundissalinus! Nonetheless, Gundissalinus did use Farabi's work; sections are even taken whole-
sale from Farabi. Gerard of Cremona's translation of al-Farabi's
work is edited from BN lat. 9335 in al-Farabi, ed. and tr. González Palencia, 1953: pp. 143-151. Gundissalinus also made an
adaptation of Farabi's work, ed. M. Alonso Alonso, 1954.

[25]Ed. González Palencia, 1953: p. 146: "Activa igitur earum
considerat in lineis et superficiebus in corpore ligni, si illi
qui eis utitur fuerit carpentarius, aut in corpore ferri, si
fuerit ferrarius, aut in corpore parietis, si fuerit cementarius,
aut in superficiebus terrenis et cultis si fuerit mensurator."

[26]Ed. Baur, 1903: p. 109: "Artifex vero practice est, qui
eam operando exercet. Duo autem sunt, qui eam operando exer-
cent, scilicet mensores et fabri. Mensores sunt, qui terre
altitudinem vel profunditatem vel planiciem mensurant. Fabri
vero sunt, qui in fabricando sive in mechanicis artibus operando
desudant, ut carpentarius in ligno, ferrarius in ferro, cementarius in luto et lapidibus et similiter omnis artifex mechani-
carum arcium secundum geometriam practicam."

status, but Gundissalinus has made a sharp division between
the measurer and the makers. It seems that Farabi was re-
cording an Arabic view, according to which the practitioners
of the mechanical arts include the surveyor (in Farabi's
original, the word is māsiḥā). Gundissalinus has made a
sharp division between the measurer or surveyor and the
practitioners of the mechanical arts, or _fabri_ who work according to practical geometry but who do not concern themselves primarily with measure. Thus, the Arabic tradition
of the mechanical arts induced Gundissalinus to enlarge the
domain of practical geometry beyond measurement alone, which
had been its role in the West.

Certain materials which are not formally treatises on the
classification of the sciences are relevant to understanding
the classification of geometry. The practical geometries
themselves, as we shall see later, offer some information.
One of the commentaries on Euclid's _Elements_ also describes
the separate realms of practical and theoretical geometry
in a way relevant to the classification of these subjects.

Adelard of Bath wrote such an extensive commentary on the
version of the _Elements_ which he had made earlier that it
is rightly called a separate edition, the _editio specialis_
or Adelard III version.[27] In the preface of the Adelard III
version,[28] Adelard gives a classification of the sciences
which provides a place for practical geometry. Mathematics
had been a theoretical discipline in Western classifications.
The problem of finding a place for its practical aspect had
been largely disregarded. Adelard's solution to the problematic placing of practical geometry is to include it as
part of liberal practice, _practica liberalis_. In turn _prac-
tica liberalis_ is considered part of the practical side of

---

[27]The _editio specialis Alardi Bathoniensis_ is the name
Roger Bacon gave to what Clagett calls Adelard's Version III.
Clagett, 1953: pp. 23-25.

[28]Ed. Clagett, 1954: pp. 272-277. This edition replaces
Clagett's earlier one, Clagett, 1953: pp. 33-36.

the division of knowledge. Liberal practice has as its parts
the practices of each of the liberal arts.[29]

Having found a way to place practical geometry in a regular
classification, Adelard goes on to distinguish theoretical from
practical geometry. It is the master, the _artifex_, that separates the two kinds of geometry. Adelard does not explicitly
label the two types of geometry as theoretical and practical
in this part of his discussion, but that is clearly the distinction he has in mind. The geometry of the demonstrator is
clearly theoretical, that of the exercitator, practical. Adelard attributes different functions and tools to each of the
two masters of geometry.

> Indeed the master is both the demonstrator and the prac-
> titioner (exercitator). The duty of the demonstrator is
> to explain theorems for the understanding of a discipline,
> for which the seven divisions of a proposition are neces-
> sary . . . . The duty of the practitioner is to measure.
> And measurement is the assignment of a definite size.
> The instrument of the demonstrator is a stick and a table
> with sand. The instruments of the practitioner are the
> measures of geometry . . . .[30]

Adelard's practical geometer is more simply defined than
that of Gundissalinus. Perhaps Adelard preferred to remove
the incursion of craft traditions into the description of geometry. Gundissalinus's agents of practical geometry included
measurers and makers. By omitting the makers, Adelard would

---

[29]Clagett, 1954: p. 274: "Sapientia [dividitur] in theoricam et practicam. Practica in ethicam, mechanicam, et practicam liberalem . . . . Practica liberalis dividitur secundum
in artes liberales, singule enim singulas habent practicas."

[30]Ibid., p. 274: "Artifex vero est tam demonstrator quam
exercitator. Officium demonstratoris est ad intelligentiam
discipline theoremata explicare, ad quod vii sunt necessaria:
propositio, exemplum, dispositio, ratio, conclusio, ratiocina-
tio, et instantie dissolutio . . . . Officium exercitatoris est
mensurare. Est autem mensuratio certe quantitatis asignatio.
Instrumentum vero demonstratoris est radius et mensa cum pul-
vere. Exercitatoris vero instrumenta sunt mensure geometrie,
scilicet, pertica cum palma, digitus, pes, passus, et ulna.
Illis vi adde ii cum stadio miliare. Vel secundum Gerbertum,
digitus, uncia, palmus, sextalis dodrans, pes, laterculus,
cubitus, gradus, passus, pertica vel decempeda, actus minimus,
clima, porta, actus quadratus vel agripennus, iugerum vel iuger
vel iugerus, centuria, stadium, miliarium, leuca . . . ."

have returned to the Western tradition in which practical geometry concerns measure, without further definition.

The meaning and scope of Adelard's description of the practical geometer as a measurer is by no means clear. Does the
practical geometer measure for practical purposes, does he
measure natural phenomena, or does he measure as a pedagogical
device? Adelard provides no clue. We must look to other treatises on the classification of the sciences and to other sorts
of material to understand what the practical geometer was to
measure.

One aspect of thirteenth-century classifications of the
sciences makes it clear that practical geometry was not merely
concerned with measuring the physical world as an auxiliary
to the other sciences. The distinction between measurement
as an auxiliary to other sciences and as a science itself is
accomplished in the elaboration of a new category of sciences,
the scientiae mediae or middle sciences.[31] The scientiae
mediae are those sciences in which mathematics is applied in
a science of the physical world. The category seems to have
arisen because certain sciences, like astronomy and optics,
were problematic for earlier classifications. The method of
these sciences is mathematical, but their subject is the natural world. In other words, they fall between the category
of mathematics and that of natural science. In the twelfth-
century classifications, these two categories were sharply
distinguished. At the same time new sciences were being attached to the mathematical quadrivium.[32] These new mathematical sciences belong between the two categories of mathematics and natural science. In the thirteenth century the term
scientiae mediae was applied to this intermediary category.
Practical geometry, in spite of its mathematical and physical
nature, does not seem to be among the scientiae mediae.

---

[31] See Gagné, 1969: pp. 975-986, which gives an insightful
general history of the problem. It seems to be Thomas Aquinas
who first used the term "scientiae mediae."

[32] Ibid., p. 978, n. 25.

Robert Kilwardby, whose Aristotelian work on classification, De ortu scientiarum, was composed about 1250 describes
distinct sciences, e.g. perspective, that would later be known
as scientiae mediae.[33] Kilwardby, for all his distinctions,
does not list practical geometry as a separate science. He
does, however, leave a place for it and does give a reason-
able description of it as ars mensurandi or ars mensoria.
He implies that geometry, or at least ars mensoria, at its ori-
gins dealt with material things.

> . . . by assiduous study [man] found the science of mea-
> suring not only horizontal but also vertical lengths,
> and not only accessible but also inaccessible lengths.
> Similarly [the measurers] extended [the science] to ar-
> tificial measure of widths and depths and even bodies.
> They arrived at that [extension] by various relations
> and comparisons of lines and angles and various forms
> of figures, as is clear to those skilled in geometry
> and the art of measure.[34]

The extended or original form of geometry that Kilwardby calls
the art of measure is surely equivalent to practical geometry.
Although he does not use the term practica geometriae or any
closely related form, his terms ars mensurandi or ars mensoria
appear to mean practical geometry. Indeed, he gives altimetria,
planimetria, and stereometria, the three parts of the most common practical geometries, as the parts of his art of mensuration--
his ars mensurandi or mensoria.[35]

---

[33]Ed. Judy, 1976. See also Weisheipl, 1965 and Sharp, 1934,
which discuss Kilwardby's work.

[34]Ed. Judy, 1976 (cap. 11, par. 61), pp. 29-30: "
assiduo studio invenit rationem mensurandi non solum iacentes
longitudines sed etiam stantes, et hoc non solum accessibiles
sed etiam inaccessibiles. Similiter inde ad latitudinum et
profunditatum artificiosam mensurationem pertigerunt necnon
et corporum, ad quod devenerunt per varias rationes et colla-
tiones linearum et angulorum et diversarum formarum figuralium,
sicut patet peritis in geometria et in arte mensoria."

[35]Ibid., p. 30: "Unde et huius scientiae triplex est pars:
scilicet una quae mensurat longitudines et dicitur altimetria,
alia quae latitudines et dicitur planimetria, tertia quae cor-
porales dimensiones et dicitur steriometria." Perhaps Kilwardby
avoided the term practica geometriae because of the implicit
difficulty in classifying any practical subject as a branch of
the theoretical sciences.

For Kilwardby geometry is very closely related to astronomy; both geometry and astronomy deal with continuous quantity. The other members of the quadrivium, arithmetic and music, are concerned with discrete quantity. Geometry and astronomy could have connections with physics because they all deal with continuous quantity. Kilwardby further tells us that astronomy is subalternate, or subordinate, to geometry. By that he means that astronomy uses the demonstrations and conclusions of geometry.[36] By using the concept of subalternation, Kilwardby avoids the problem of sciences that seem both mathematical and natural. Astronomy is a natural science that uses mathematics.

Kilwardby considers astronomy to be an art of measuring; like geometry, astronomy measures continuous quantity. Indeed he uses the term ars mensurandi, which describes practical geometry, in his description of astronomy. Astronomy does after all concern itself with measuring the earth, the planets and celestial orbs. Astronomy also relies on geometry in another way; it uses geometric instruments. Observations made with instruments are the foundation of the demonstrations of astronomy, which demonstrations are themselves geometric.[37] Geometry's important role in astronomy

---

[36]The role of continuity is discussed ibid. (cap. 14), pp. 36-39. Kilwardby describes the subalternation of astronomy to geometry ibid. (cap. 16), pp. 41-48.

[37]Ibid. (cap. 12, par. 66), p. 31: "Tandem artem mensurandi corpora infima applicuerunt summis corporibus, scilicet caelestibus, ex illa comperientes eorum magnitudines et distantias ab invicem et a terra." Ibid. (cap. 16, par. 115), pp. 47-48: "Ad hoc enim quod applicetur demonstratio geometrica astronomicae scientiae cadunt quaedam instrumenta media, ut praedixi, et haec instrumenta arte geometrica mechanice fiunt, ut astrolabium et instrumentum considerationis et sphaera et consimilia, et continentur sub machinativa sicut caementaria et carpentaria et omnino architectonica. Et ita per machinativam potest subintelligi cum quibusdam aliis etiam astronomia ratione instrumentorum ei servientium super quorum consideratione fundatur demonstratio geometrica descendens in astronomiam." See also ibid. (cap. 12, par. 67), pp. 31-32.

explains why in some of the practical geometries, especially in Artis cuiuslibet consummatio, so much space is devoted to astronomy.

Kilwardby does not have a very long or explicit definition of the art of mensuration. He seems to take for granted the place of ars mensurandi as part of geometry and even of astronomy. In describing the other sciences, he indirectly defined practical geometry. His definition reflects more than the tradition of practical geometry; it reflects the other sciences as well. That is, Kilwardby gives practical geometry its place among several sciences that have developed or acquired a larger scope since practical geometry first appeared in a classification of the sciences.

Nevertheless, the treatises on practical geometry reflect anything but a static tradition. The practical geometries themselves grow and change, elaborating and enlarging their own domain. Some of the ways in which practical geometries developed during the twelfth, thirteenth, and fourteenth centuries may be discerned by considering the explicit definitions of practical geometry given in the treatises. Another, less direct, way of getting at what the practical geometries were is by comparing how they are divided into sections and what each section contains.

It was in Hugh of St. Victor's Practica geometriae that the distinction between theoretical and practical geometry was first drawn. Hugh offers a definition of practical geometry in his treatise, "practical [geometry] is that which is done by means of certain instruments and which judges by inferring proportionally some [distance] from some other ones."[38] Hugh's definition of practical geometry describes his own treatise quite well. Most of Hugh's treatise does concern itself with using instruments in the indirect and proportional measurement of some distance. As we shall see below, not all practical geometries limit themselves to the handling of instruments. The

---

[38]Ed. Baron, 1966: 16.31-32: "[geometria] practica vero est que quibusdam instrumentis agitur et ex aliis alia proportionaliter coniciendo diiudicat."

instruments used by Hugh include the astrolabe, a right triangle, several staffs and reeds, a mirror, and a gnomon.
None of these instruments is unusual; most of the practical
geometries use the same ones. Most notable, perhaps, is
Hugh's omission of the quadrant; however, the astrolabe may
be used for all of the measurements performed by the quadrant.

The other major aspect of Hugh's definition of practical
geometry involves the use of ratio. Most of the measuring
techniques Hugh describes rely on a proportion. The distance
to be measured has a certain ratio to a known distance. That
ratio is the same as one determined using an instrument. The
resulting proportionality is used to compute the unknown distance. The ratios most used are those of the sides of right
triangles sharing the same hypotenuse. The equivalence of
ratios serves Hugh both as a measuring technique and as a
justification for his procedures. Thus the use of ratio is
as important an aspect of Hugh's work as is the use of instruments. Indeed, they are inseparable for Hugh. In other
practical geometries a distinction between the use of instruments and the application of ratios will appear.

For the parts of practical geometry, Hugh offers the triad
altimetria, planimetria, cosmimetria.[39] Hugh's chapter on
the measurement of heights, De altimetria,[40] follows most
closely both aspects of his definition of practical geometry;
it shows how to measure heights using instruments and proportional relations in almost every proposition.

Hugh's desire to limit his treatise to those aspects of
practical geometry using both instruments and proportion is
probably the reason for the extreme brevity of the section
on the measurement of surface areas, De planimetria.[41] Measuring areas usually involves multiplying two lengths, an
arithmetical procedure not requiring the use of instruments.

---

[39]Ibid., 17.33-52. As discussed above, the same group is
given as the parts of geometry in the Didascalicon, ed. Buttimer, 1939: p. 33. Cf. tr. Taylor, 1961: p. 70.

[40]Ibid., pp. 22-46.

[41]Ibid., pp. 47-48.

The last section of Hugh's _Practica geometriae_ deals with what Hugh calls _cosmimetria_, measurement of the cosmos.[42] The term _cosmimetria_ also applies to measuring anything disposed in a circle, according to Hugh.[43]   Other practical geometries, as we shall see below, use different terms for their sections on volumes, e.g., _steriometria_ or _crassimetria_.[44] Hugh's _cosmimetria_ does, however, concern itself primarily with the dimensions of the cosmos; it treats the sizes and proportions of the sun and earth. As an aid in determining the sizes of the sun and earth, it describes the use of the gnomon in finding solar altitudes. Even though Hugh uses proportionality and even certain instruments in his section on _cosmimetria_, he does not provide much in the way of measuring volumes. We might have expected some material on measuring solids if only because of the analogy of the three parts of the treatise with the three types of measure, linear, surface and solid, which Hugh describes in the _Prenotanda_ to his treatise.[45]

The aspects of Hugh's practical geometry that we have chosen to discuss are developed in another twelfth-century practical geometry, known by its opening as _Geometrie due sunt partes principales_.[46] Where Hugh had spoken of the use of instruments and ratios as both being necessary for practical geometry, the author of _Geometrie due sunt partes principales_

---

[42]_Ibid._, pp. 49-64.

[43]_Ibid._, 17.47-52.

[44]See Baron, 1957: pp. 30-32 and Thorndike, 1957: p. 458.

[45]Ed. Baron, 1966: pp. 15-16.

[46]Nan Long Britt has provided a critical edition of the treatise in her unpublished doctoral dissertation, Britt, 1972: pp. 208-255. She calls the treatise by its generic name _Practica Geometrie_. It is sometimes difficult to distinguish the practical geometry from the quadrant treatise having a very similar incipit (see below). In addition to the fourteen used by Dr. Britt, the following seem to be manuscripts of _Geometrie due sunt partes principales_: Vat. lat. 3114, 13 c, ff. 81v-83r; Basel F. IV. 18, 13-14 c, ff. 26v-30r; BN 7381, 15 c, ff. 199r-209r; CLM 27105, 15 c, ff. 36r-46r; Krakow 1968, 15 c, pp. 1-20; VI 5184, a. 1482, ff. 55r-58v; BN 7433, 16 c, 83v; Yale MS bound with item 42, ff. 102v-107v.

defines two separate kinds of practical geometry. One kind,
artificial measurement, relies on the use of instruments;
the other, non-artificial measurement, uses ratios and proportions. As the author tells us in the introduction to his
treatise:

> There are two principal parts of geometry: the
> theoretical and the practical. The theoretical is
> when we contemplate the proportions of quantities
> to one another by speculation of the mind alone.
> The practical is when we obtain knowledge of a less
> known quantity from some [knowledge] of known quan-
> tities by means of the senses . . . . This investi-
> gation of quantity is called measurement. One [kind
> of] measurement is artificial, the other non-artifi-
> cial. The non-artificial is when one or many names
> of the ratio of a known quantity to an unknown one
> is substituted for the knowledge of the unknown quan-
> tity. Artificial measurement is when we arrive at
> knowledge of the unknown quantity through consider-
> ing instruments and working with numbers.[47]

By elaborating the distinction between artificial and non-
artificial measurement, the author establishes a new kind
of practical geometry. This new kind of practical geometry
does not require the use of instruments at all but instead
performs computations using ratios and proportions. Even
where the treatise explains how to compute surface area,
proportion, rather than multiplication, serves as the basis
of the technique.[48] The existence of this new kind of prac-

---

[47]Ed. Britt, 1972: pp. 208.2-209.8: "Geometrie due sunt
partes principales: theorica et practica. Theorica est quando
sola mentis speculatione quantitatum proportiones adinvicem
intuemur. Practica est quando per aliquam notarum quantitatum notitiam quantitatis minus note sensitive comparamus
. . . . Huius vero quantitatis investigatio, mensuratio
appellatur. Mensurationum alia artificialis, alia inartificialis. Inartificialis est quando per unicam vel multiplicem
appellationem note quantitatis ad ignotam notitia ignote quantitatis supponitur. Artificialis mensuratio est quando per
considerationem in instrumentis et opus numerorum ad notitiam
quantitatis ignote venimus."

[48]For example, ibid., p. 243.1-7: "Quod si volueris
scire quantitatem alicuius quadrate superficiei, considera
proportionem unius lateris ad mensuram per quam mensurare
volueris. Et duc illam proportionem in se; productum erit
area quadrati. Verbi gratia, latus quadrati sit 12 pedum,
mensura 4 pedum, proportio 12 ad 4 est tripla. Duc ergo

tical geometry necessitates some knowledge of the rules for
proportional computation. Right after the introduction, the
treatise explains the rules of ratio and proportion needed
for non-artificial measurement.

Geometrie due sunt partes principales gives as the three
species of artificial measure the three parts that Hugh had
given for the whole of practical geometry, "The three species
of artificial measure are altimetry, planimetry, stereometry
or cosmimetry."[49] The author of Geometrie due sunt partes
principales seems to feel that the triad which Hugh had elaborated belongs to the more restricted domain of instrumental
or artificial measurement. Even though the author makes cosmimetry synonymous with stereometry, he does not deal with measuring the cosmos. In fact his section on stereometry is
quite short, giving a few simple propositions on the ratios of
sides and volumes of solid figures. In other words, Hugh's
title for the third branch of measurement appears in Geometrie
due sunt partes principales even though its contents are totally
different.

Another treatise quite similar to the practical geometries
is the treatise known as Tractatus quadrantis or Quadrans vetus,
probably by Johannes Anglicus of Montpellier.[50] The treatise
had a very wide distribution; some seventy-five manuscripts are
extant today.[51] Quadrans vetus deals with the construction and

---

tria in se et habes novem, et sic quadratum nonuplum erit ad
quadratum mensure."

[49]Ibid., p. 210.1-2: "Artificialis mensurationis tres sunt
species: altimetria, planimetria, steriometria vel cosmimetria."

[50]I concur with Britt's choice of Johannes Anglicus of Montpellier as author. She takes issue (Britt, 1972: pp. 13-27)
with Paul Tannery's attribution of the work to Robertus Anglicus. Tannery's discussion of authorship accompanies his edition, Tannery, 1897. The question of authorship is also discussed in L. Thorndike's articles, Thorndike, 1943 and Thorndike, 1947.

[51]Britt uses 63 manuscripts in her edition. Because of the
similarity of Quadrans vetus and the practical geometry of which
the manuscripts are listed in n. 46, it is sometimes difficult
to tell which treatise is meant in a catalogue entry. The fol-

use of an instrument having the form of a quarter of a circle.
This instrument, the old quadrant or quadrant with cursor, is
marked with scales which permit its use for determining time
and measuring altitudes of celestial bodies and heights of
terrestrial objects. The precise date of the treatise is unknown. Nan Britt shows it was written before 1284 and suggests it is from the thirteenth century; Emmanuel Poulle be-
lieves it dates from the twelfth century.[52]

Quadrans vetus has as its concern the use of an instrument
in measuring. That alone would make us suspect a relation
to a practical geometry. But a much stronger connection
exists; it seems to be an adaptation of the practical geometry Geometrie due sunt partes principales. Part of the
introduction of Quadrans vetus is very close to the introduction to Geometrie due sunt partes principales.

> There are two parts of geometry: the theoretical and
> the practical. The theoretical is that which contem-
> plates proportions, quantities and their measures by
> the speculation of the mind alone. The practical is
> when we measure the unknown quantity of some thing by
> the experience of the senses. There are, however,
> three species of artificial measure, which is called
> practical, namely, altimetry, planimetry and stereo-
> metry.[53]

---

lowing seem to be manuscripts of Quadrans vetus not used by
Britt in her edition: Boncompagni 323, 13 c, ff. 4-9; Escor-
ial O. II. 10, 13 c, ff. 64r-68v; BL Tanner 192, 14 c, ff.
67-74; Boncompagni 51, 14 c; Milan H 76 sup., 14-15 c, ff.
52r-56r; VI 5239 B, 14-15 c, ff. 19v-23v; BN lat. 7294, a.
1434, ff. 49-63, 45-48; CLM 10662, a. 1436, ff. 206r-217v;
CLM 27105, 15 c, ff. 36r-(45v); BN lat. 16649, 16 c, ff.
50v-56r.

[52]See Britt, 1972: pp. 1, 13-14, and Poulle, 1972.
Poulle's article gives a good general history of the quadrant
and astrolabe treatises and a concise description of the instruments.

[53]Ed. Britt, 1972: 100.2-103.4: "Geometrie due sunt
partes theorica et practica. Theorica est que sola mentis
speculatione quantitates, proportiones et earum mensuras
intuetur. Practica est quando alicuius rei quantitatem ignotam experimento sensibili mensuramus. Mensurationis autem
artificialis, que practica dicitur, tres sunt species, scilicet altimetria, planimetria et steriometria."

In spite of the similarity of this text to that of Geometrie
due sunt partes principales, certain differences are to be
noted. The differences are significant, indicating the relationship between the two treatises. The definition of practical geometry in Quadrans vetus is like the definition of
artificial measurement in Geometrie due sunt partes principales. That is, Quadrans vetus limits practical geometry to
the use of instruments in measuring. Another difference is
in the parts of artificial measurement. Johannes Anglicus
does not include cosmimetria as a synonym for steriometria as
had the author of Geometrie due sunt partes principales. It
is almost as if Johannes Anglicus recognized that there was
nothing in either treatise dealing with measuring the cosmos.

The differences and similarities in the introductions to
the two treatises indicate at least that the treatises are related. The nature of the relation can be determined with some
degree of assurance. It seems most likely that Quadrans vetus
was written later, taking those parts of Geometrie due sunt
partes principales that would be useful.[54] The proof that
Quadrans vetus comes later hinges on those aspects of the introduction that we have considered above. Johannes Anglicus does
not need the distinction between artificial and non-artificial
measurement; indeed, he does not use the distinction. Artificial measurement adequately provides for measurement using an
instrument. In fact, even Hugh of St. Victor's definition of
practical geometry would have been sufficient as a description
of the aims of Quadrans vetus. Would Johannes Anglicus have
used only one branch of the division of practical geometry into
artificial and non-artificial measurement if he had invented the
term? Of course not. Clearly then, Johannes Anglicus shortened
the introduction to a treatise in which the division was fully
elaborated. That treatise would have been Geometrie due sunt
partes principales.

---

[54]Tannery had assumed the practical geometry was based on
Quadrans vetus. Britt uses different arguments from mine to
show that Geometrie due sunt partes principales is one of the
sources of Quadrans vetus; see Britt, 1972: pp. 50, 64, 73-79.

Further evidence for the priority of Geometrie due sunt partes principales may be adduced. The author of the earlier treatise had as the third part of artificial measurement either cosmimetry or stereometry. Johannes Anglicus leaves out cosmimetry even though the term would have more properly applied to the astronomical concerns of Quadrans vetus than to the terrestrial measurements of Geometrie due sunt partes principales. The author of Geometrie due sunt partes principales probably took over Hugh of St. Victor's name for measurement in three dimensions somewhat uncritically. In revising the introduction, Johannes Anglicus dropped the superfluous and confusing term, leaving only the clearer steriometria.

In a very real way, then, Quadrans vetus may be considered a practical geometry. We have seen how much of its material comes from a practical geometry. Even if we knew nothing of its source, we could call Quadrans vetus a practical geometry. Instruction in using instruments and performing computations is probably the defining characteristic of a practical geometry; instruments and computation are essential to Quadrans vetus as well. Thus Johannes Anglicus's treatise is a practical geometry and perhaps the most widely diffused of them all.

Chronologically, the next practical geometry to have been composed is the anonymous treatise beginning "Artis cuiuslibet consummatio . . . ," edited below.[55] The author of Artis cuiuslibet consummatio, or ACC, lets us know when he composed the text by an almost incidental citation of an astronomical parameter. In giving the values for the solar and planetary apogees, he tells us that the values, which are computed, not observed, are for his own time. "But at the present time, in the year of the Lord 1193, the apogee of the sun is . . . ."[56] By virtue of another astronomical value, the equinoctial al-

---

[55]Reference will be made to my edition using book, proposition and line numbers. Thus II, 26.6-8 means Book II, Proposition 26, lines 6-8.

[56]ACC, II, 26.6-8: "Est autem hodierno die ipse altus locus solis 2 signa 27 gradus 40 minuta, scil. anno domini 1193."

titude of the sun from his point of observation, we know that
he was probably writing in Paris. "The result will be the
equinoctial altitude, which is 42 degrees in the Parisian zone."[57]

Since ACC is analyzed rather fully below,[58] we will limit
our considerations to the division of the treatise into sections.
The parts of ACC are different from those of the other practical geometries considered here. The author divides his work
into four parts instead of the more common three. The first
three parts, similar to those of other practical geometries,
are planismetria, altimetria, and crassimetria. The added part
concerns computation with fractions, both common and sexagesimal,
or as the author calls them, geometrical and astronomical fractions. This section on fractions is added because it is necessary for the other parts, according to the author.[59]

Besides adding a fourth part, ACC is notably different from
other practical geometries in the contents of its section on
altimetry. Most of the practical geometries limit altimetry
to the measurement of terrestrial heights. ACC has an adequate
section on using instruments to find the usual heights of towers
in different circumstances, but there is much more. Book II,
altimetry, has extensive material on astronomy.[60] It is difficult to determine why so much astronomy is included. The traditional practical geometries scarcely consider astronomy, giving a proposition or so on finding the altitude of the Sun. If

---

[57]ACC, II, 5.9-11: "Et erit equinoctialis altitudo, que in
climate parisiensi est 42 graduum." Cf. II, 7.5-6; II, 9.8;
II, 10.6 and II, 18.6.

[58]See chapter II.

[59]ACC, I, 0.24-30: "Opus autem nostrum in 4 distinguimus
particulas. In prima, planismetriam instruimus, superficierum
quantitates investigando. In secunda, altimetriam, mensurare
alta, docebimus. In tercia, capacitates corporum et crassitudines invenire docemus. In quarta, geometricas et astronomicas minutias ad predicta necessarias docere promittimus."
Elsewhere, in III, 21.13, the author names the third part more
briefly as crassimetria.

[60]The contents of all four parts of the treatise will be
discussed more fully in chapter II below.

the astronomical material was added following some plan on
the part of the author, it may have been the precedent of
the material on solar altitudes that induced the author to
give a long section on astronomy. Or perhaps it was one of
the quadrant treatises, like Quadrans vetus, with their combination of practical geometry and practical astronomy that
served as a model to the author. In any case, he acknow-
ledges the importance of astronomy in his treatise. As he
tells us in the preface, "this work serves principally geometry, secondarily astronomy."[61]

Having been given an explicit acknowledgment that the
treatise is meant to be useful in astronomy, we have a better understanding of the role of Book IV, the section on
fractional computation. Much of Book IV deals with sexagesimal fractions; it is these fractions that are of primary
utility in astronomy. Even if sexagesimal computations are
not needed in the other parts of ACC, they would be necessary for further work in astronomy. Together, the astronomical material in Book II and the sexagesimal arithmetic in
Book IV make ACC as much an introduction to astronomy as to
geometry.

In 1220 Leonardo Pisano or Fibonacci wrote his Practica
geometriae.[62] He is the same Leonardo Pisano who in his commercial arithmetic, the Liber abaci, developed what is now
known as the Fibonacci sequence.[63] Leonardo's geometry is
considerably longer than any of the earlier practical geometries. Some of the greater length results from Leonardo's

----

[61] ACC, I, 0.30-32: "principaliter geometrie, secundario
astronomie, hoc opus deserviat."

[62] Ed. Boncompagni, 1857-62, v. 2: pp. 1-224. Boncompagni's
edition is based on a single MS, Vat. Urb. lat. 292. Other
MSS containing the treatise include BN 7223, 15 c, ff. 1r-
188r; BN 10258, 17 c, pp. 1-349; BN nal 1207, 19 c, ff. 1v-
509v (a copy of Ugh. 259); FN II.iii.22, 15 c, ff. 2r-241v
(missing Dist. 8); FN II.iii.23, 16 c, ff. 1r-191v; FN
II.iii.24, 14 c, ff. 1r-147r; Vat. Ottob. lat. 1545/1546,
17 c, ff. 1r-341v/1r-375v.

[63] In determining the size of a population of rabbits re-
producing at a regular rate.

adding new sorts of material. For example, besides the usual
heights, areas, and volumes measured with various instruments,
Leonardo also provides sections on square and cube roots, on
the division of fields among partners, on arcs and chords, and
on some algebraic solutions to geometrical problems. Much of
the added material in Leonardo's work comes from the Islamico-
Judaic tradition, which had a broader scope at the time. Indeed most of Leonardo's added topics were traditional to the
Arabic and Hebrew science of measure. For that matter, a siz-
able part of the Practica geometriae is borrowed from Savasorda's Liber embadorum.[64] The added material in Leonardo's
Practica geometriae gives it a much broader scope than the
other practical geometries we have considered. It could be
called a general text-book in geometry instead of a practical
geometry. The extent and nature of the theoretical parts of
the treatise as well as the relation between the theoretical
and practical aspects of Leonardo's work will be treated below
in the section on theory and practice.

The oldest known treatise on geometry in the French language
is a practical geometry.[65] It is largely a translation and
and adaptation of Artis cuiuslibet consummatio, edited below,
rather than an original composition in old French; the notes
to my edition of the French text, which I have called Le Pratike de geometrie from the subject indicated in its first sen-
tence, below, should make its connection with ACC abundantly
clear. Le Pratike de geometrie was written in Picard dialect
before 1276.[66] It begins with a list of its topics,

---

[64]See Vogel, 1971: pp. 609, 611 and Levey, 1970: p. 22.
Savasorda is the name by which Abraham bar Ḥiyya ha-Nasi is
usually known. His geometry was translated into Latin in
1145 by Plato of Tivoli. See Curtze, 1902a.

[65]The treatise has been edited by Charles Henry, 1882.
Henry's edition is based on a single thirteenth-century MS
from the Bibliothèque Sainte Geneviève in Paris. The treatise
in the MS currently bearing the number 2200 occupies ff. 151v-
161r. Another copy of the geometry is found in BN franç. 2021,
ff. 155r-169r, also from the thirteenth century.

[66]Henry, 1882: p. 51.

> We begin a work on the practice of geometry which
> we shall divide into three parts: in the first part
> we shall teach how to find the measure of surfaces,
> in the second to find the measure of heights and
> depths and of solid measures, in the third to find
> the geometric and astronomical fractions appropriate
> to the two preceding parts.[67]

The measurement of surfaces, heights, depths, and volumes,
announced in the introduction, are entirely to be expected
in a practical geometry. Consideration of fractional computation is somewhat rare; fractions do, however, make up a
significant part of ACC. Le Pratike de geometrie gives standard procedures for finding simple surface areas and volumes.
The methods it gives are word for word translations of parts
of Books I and III of ACC.[68] Certain corruptions make it
clear that ACC is not a translation and enlargement of the
French geometry.

The material on measuring heights and depths and the section on fractions, promised in the introduction, are not
present in the extant manuscripts. We cannot tell whether
it was the translator or some later scribe or editor who
left out the material on fractions and altimetry. As evidenced in the original Latin of ACC, those sections are relevant more to astronomy than to practical geometry and perhaps
too difficult for an introductory text. The nature of the
omissions and literal nature of the translation make it clear
that the oldest French geometry is a translation of ACC.
In spite of the literalism of the translation, the text, as
edited by Henry from a single manuscript, is quite corrupt.
A new edition making reference to the source of the translation and using both of the extant manuscripts is included in
the present work.

---

[67]Pratike, I, 0.2-8, edited below. Victor Mortet, 1904:
p. 938, has made some improvements on Henry's text. Mortet
announced a new edition of the treatise that seems never to
have appeared, ibid., p. 937 and in Mortet, 1909: p. 55.

[68]After the brief introduction, the treatise contains
translations of at least parts of the following propositions
of ACC: I,1 - I,9; I,11 - I,17; I,21 - I,24; I,26 - I,36;
III,1 - III,9; III,11 - III,12; III,14; III,13; III,15 -
III,17.

The author of the French translation, or perhaps some later
editor, has added some material to the translation and adaptation of ACC. Some of the added contents is geometric, treating
areas and volumes in a way similar to that of Books I and III
of ACC.[69] The rest of the additions in the Pratike de geometrie
concern mercantile or commercial arithmetic; some of it is purely
arithmetical--tables of squares and algorithms for handling large
numbers, but the more interesting part explains calculations of
various coinages and alloys.[70]

Even in the fourteenth century, Dominicus de Clavasio uses
altimetria, planimetria, and stereometria as the principal topics
of his Practica geometriae, composed in 1346.[71] Dominicus is
concerned with both artificial and non-artificial measurement.
The emphasis on ratio and proportion and its use in non-artificial
measurement is significant. Indeed, for Dominicus, to measure
something is defined as finding its ratio to a known or standard
measurement.[72] Because of the importance of ratio in his treatise, Dominicus devotes most of his introduction to explaining
several rules for dealing with ratios, giving abstract rules
and numerical examples for computing with many types of ratios,
e.g. multiple, superparticular, superpartient, multiple superparticular, and multiple superpartient. The rules deal with
finding the ratio between two known quantities, finding one quantity when another quantity and their ratio are known, finding a
whole quantity and its part when the ratio and difference between
the whole and part are known, and finding the fourth proportional
when the other three are known. Thus the introduction of Dominicus's work provides not only the techniques needed for carrying
out measurements but also a general introduction to computing
with ratios.

---

[69]I have separated these geometric additions as Book V of
the Pratike.

[70]The mercantile parts of the Pratike I have edited as Book
VI.

[71]Ed. Busard, 1965: p. 520.

[72]Ibid.

The section of Dominicus's _Practica geometriae_ on altimetry
explains a large number of indirect measuring techniques using
instruments such as the astrolabe, quadrant and some others
as well as an _instrumentum gnomonicum_, which was described in
the introduction. In other words, its subject seems to be
artificial measurement. Nonetheless, each of them is justified by reference to some proportion, especially as related
to similar triangles. Thus, non-artificial measurement provides justification for the techniques of artificial measurement.

In dealing with areas, Dominicus has less recourse to measuring instruments; still, manipulating ratios and proportions
plays an important role in his computations of areas, both
in computing areas and in finding altitudes of triangles in
order to compute their areas. Dominicus appeals to ratios
fairly often in his consideration of solid figures. Both
ratios of one kind of solid to another, of known volume, and
ratios of similar solids of different dimensions are treated
in the section on solids. In each section of his _Practica
geometriae_, Dominicus de Clavasio makes extensive use of
ratios and proportions to supplement the traditional techniques of practical measurement. In fact, the role of ratios
and proportions is so great that one is tempted to conclude
that Dominicus has written a practical treatise on proportion
or at least that he used practical geometry as a vehicle for
studying the rules and applications of ratios and proportions.

Our considerations of the parts of some of the practical
geometries composed in the twelfth through fourteenth centuries show little uniformity of content. Some constant ten-
dencies are, however, evident. All of the treatises on practical geometry concern themselves with measurement. Measurement in practical geometry is carried out either using instruments or proportional relations, or both. That the treatises all deal with measure in some form is not surprising;
after all, the science of measure is as close as we came to
finding a definition of practical geometry in the treatises
on classification. All the practical geometries intend to

treat measurement in one, two, and three dimensions. Several
of the practical geometries touch on astronomical matters;
some of them contain extensive treatments of astronomy. All
of the practical geometries describe the use of some measuring
instrument; some explain several measuring devices. Many of
the treatises contain rules for proportional computation. In
sum, then, a practical geometry may be simply described as a
treatise concerned with measurement, in one, two, and three
dimensions, using either instruments or proportional relations
or both. A practical geometry may contain further material
to develop or extend its scope, following the author's particular interests.

Practical Geometry in Education

Since the practical geometries we have considered all seem
to have some instructional purpose, we would do well to enquire
about the role of practical geometry in medieval education.
One may refer to a variety of sources to learn about the place
of practical geometry in pedagogical traditions. Indeed, the
incomplete information about such matters obligates us to use
a range of materials. University statutes yield almost no information about the place of mathematics in the curriculum in
the twelfth and thirteenth centuries.[73] Despite the silence of
the regulations of the universities in this matter, something
of the place of practical geometry may be understood.

In some sense, the treatises on the classification of the
sciences are reflections of a curriculum. That is, the treatises
offer a program in which the relations between the various sciences are spelled out. It is not for us to decide whether the
philosophical considerations of the classifications find expression in the university curriculum. The lack of regulations about
the curriculum makes such an evaluation impossible. Nonetheless,
the treatises on classification offer an idealized program, one
which may have provided a model or goal for teaching. Indeed,
the subtitle of Hugh of St. Victor's Didascalicon, De studio

---

[73]See Kibre, 1969: p. 175.

legendi is a certain indication of pedagogical concerns.
Thus we may generally expect to see the concerns of the
classification treatises reflected in education. Remember-
ing that the university was not the only seat of education
in the twelfth and thirteenth centuries, we shall look for
evidence of the role of practical geometry in statements
about what education should be and in material which arose
in an educational milieu.

The study of the liberal arts was considered a basis for
the study of Scripture during the earlier Middle Ages. The
concept of using secular learning as propaedeutic to the
study of the Bible is not new to the twelfth and thirteenth
centuries; it goes back to Augustine's On Christian Doctrine[74]
and continues through the Middle Ages. Whether or not the
study of the liberal arts had any effect on Biblical exegesis
in the earlier Middle Ages is not at issue here. However,
by the twelfth century, evidence of training in the liberal
arts is manifest in Biblical commentaries.

Hugh of St. Victor makes the arts and sciences a basis
for the study of Scripture.[75] In the Didascalicon Hugh describes how he studied the trivium and quadrivium as a child.[76]
His description of learning geometry makes it clear that it
was practical geometry he studied.[77] Thus we have evidence
both that Hugh knew some practical geometry and that he considered the arts and sciences useful in studying Scripture.
We are led to look for the effects of his knowledge of practical geometry in his Biblical exegesis. In fact, the use of
geometry in explaining Scripture is a reality for Hugh, not
an idle homage to an unachieved goal.[78]

---

[74]De doctrina christiana, II, 40-42.

[75]De sacramentis, prol., Ch. 6, Migne, 1844-1864, v. 176:
col. 185C.

[76]VI, 3, tr. Taylor, 1961: pp. 136-137; ed. Buttimer, 1939:
pp. 114-115.

[77]Ibid., tr. Taylor, 1961: p. 136; ed. Buttimer, 1939:
pp. 114-115.

[78]See Smalley, 1964: pp. 83-106.

In Hugh's literal exposition of Noah's ark, De arca Noe morali, he offers us a geometric description of the ark, complete with comments on the size of the cubit used, the way of finding the area and even the diagonal of the oblong which comprises the ark.[79] The description of the size of the ark has the exegetical purpose of showing that the ark in Genesis was big enough to hold all the animals and their required food.[80] Hugh was by no means the first Biblical commentator to concern himself with the size of the ark. Origen had previously solved the problem by claiming that Moses had used the geometer's cubit, equal to six ordinary cubits, in giving the dimensions of the ark.[81] After all, Moses grew up in Egypt, the birthplace of geometry; naturally he would use the geometer's cubit.

Still, Hugh explains much more geometry than he needs strictly for exegesis. Besides describing the geometer's cubit, Hugh gives almost a brief course in metrology.[82] He explains that an inch is equal to two thumbnails placed end to end, that there are four inches to a palm, four palms to a foot, one and one-half feet to a small cubit and six small cubits to a large cubit. Then he shows the ark to be some forty thousand inches long. Next, as if to make the ark more real, he finds that it is more than half a mile long.[83] In

---

[79]Migne, 1844-1864, v. 176: cols. 628-629.

[80]Ibid., col. 628A: "Sed sunt qui dicant, quod haec magnitudo ad tot genera animalium, et ad eorum cibos, quibus per annum vescerentur, capiendos non sufficeret."

[81]Augustine reports Origen's solution to the size of the ark in his Quaestiones in Heptateuchum, lib. I, q. 4 (Migne, 1844-1864, v. 34: col. 549): "Quam quaestionem cubito geometrico solvit Origines, asserens non frustra Scripturam dixisse quod Moyses omni sapientia AEgyptiorum fuerit eruditus, qui geometricam dilexerunt. Cubitum autem geometricum dicit tantum valere quantum nostra cubita sexvalent. Si ergo tam magna cubita intelligamus, nulla quaestio est, tantae capacitatis arcam fuisse, ut posset illa omnia continere."

[82]One of Hugh's minor works, "De ponderibus et mensuris," ed. Baron, 1956: pp. 132-137, is metrological. The definitions of measures are very like those given in Hugh's De arca Noe morali.

[83]Migne, 1844-1864, v. 176: col. 628C-D.

this way Hugh has explained both the ark and metrology. The
ark is also the occasion for a lesson in geometry. Hugh describes how to compute the area of a rectangle and its diagonal.[84] The methods of computation are very like those of
a practical geometry; the procedure is explained without a
proof. As in a practical geometry, numerical examples are used
to make the technique clear. Here, the examples relate to the
dimensions of the ark.[85] We must conclude that Hugh was concerned with teaching geometry in a general way. Had he given
merely the results of his calculations, his students might not
have been able to follow his example in carrying out their own
literal interpretations of the Bible.

Hugh's concern for education and his interest in the literal
sense of Scripture provide an explanation for his having composed a practical geometry. Perhaps he perceived a lack of
geometric treatises suiting the needs of fledgling exegetes
and decided to fill that gap. Both the geometry in Hugh's
exposition of Noah's ark and in his Practica geometriae would
aid in interpreting the Bible. The geometry in the De arca
Noe morali is suited to the problems of the ark, but goes
beyond to teach in a general way; the Practica geometriae is
general, but may be applied to particular problems.

The use of geometry in explaining Scripture is not found
only in Hugh. Richard of St. Victor also uses geometry in
literal exegesis. Richard was a Scot who became prior of
the Abbey of St. Victor in 1162. He developed some of Hugh's
teachings on the classification of the sciences and intended
to create an encyclopedia as an aid in exegesis, following
the plan of Hugh's Didascalicon.[86] As far as I know, nothing
original on the place and division of geometry in the curri-

---

[84]Ibid., cols. 628D-629A.

[85]For example, area is explained, ibid., cols. 628D-629A:
"Quod si etiam scire delectat, quod quadratorum, sive pedum,
sive cubitorum fuerit embadion, id est area, multiplica
majus latus per minus, et quot a summa inde excreverit, tot
quadratorum erit area. Verbi gratia:[Duc]quinquagies trecen-
ti fiunt quindecim millia, tot cubitorum quadratorum est area."
I have corrected to area what is obviously an error, arca.

[86]Smalley, 1964: p. 106.

culum of studies can be found in Richard's work. Geometry has
a role in exegesis for Richard, but perhaps a different one
than for Hugh. Hugh used practical geometry; in some sense at
least, Richard's geometry is also practical geometry.

Richard was attracted to the luxurious and sensual aspects
of literal exegesis.[87] His literal interpretation seems to
have found fullest play in his descriptions of the architecture
described in the Bible. He wrote on the tabernacle, on the temple, and on the visions of Ezechiel.[88] The only Biblical architecture he neglected is Noah's ark, so extensively treated by
Hugh. To give a literal interpretation of architecture means
to describe it. Architectural description would seem to require
some use of geometry. It is precisely to geometry that Richard
turns as an aid in understanding Ezechiel's second vision. Richard announces he will make coherent Ezechiel's vision that even
St. Gregory had declared to have no literal sense.[89] In the
vision Ezechiel is confronted by a celestial being who uses a
measuring reed and cord to give the dimensions of the buildings
of the visionary city. It is the inconsistency of the measurements of the vision that makes interpretation difficult. But
since the angel may be considered a heavenly geometer, the vision
ought to be understood. Richard undertakes to investigate the
measurements revealed by the angel as diligently as possible,
using the authority of Scripture and reason.[90]

---

[87]Smalley, 1964: p. 107, cites his work Benjamin minor, cap.
15 and 24; Migne, 1844-1864, v. 196: cols. 10-11, 17, where
Richard claims that Scripture "raises the spirit by pleasing
the sense."

[88]Expositio difficultatum suborientium in expositione taber-
naculi foederis, Migne, 1844-1864, v. 196: cols. 211-222; De
templo Salomonis, ibid., cols. 223-242; In visionem Ezechielis,
ibid., cols. 527-600.

[89]In visionem Ezechielis, ibid., cols. 527-528.

[90]Ibid., col. 539C-D: "Cum nos viderit singulorum aedifi-
corum situm, qualitatem et mensuram subtiliter quaerere et investigare, et cum tanta diligentia assignare et singula quaeque
prout possumus, vel Scripturae auctoritate, vel rationis attes-
tatione probare."

Some of the passages are genuinely problematic. Richard hopes to make sense of them by interpreting them in a different way.[91] For example, one of the measurements of a large hall as described to Ezechiel is eighty cubits; that same dimension, seen from its other end, is given as one hundred cubits.[92] Richard's solution to this apparent inconsistency is simple and clever. He tells us that if the building is considered as being on the side of a hill, the dimensions may be understood by anyone skilled in geometry.[93] The geometry Richard uses is very simple indeed. He tells us that one end of the building must be sixty cubits higher than the other.[94] Thus, the dimensions comprise a right triangle having sides of sixty, eighty, and one hundred cubits. Richard's demonstration of how those dimensions are disposed is far more extensive than it need be. In effect, he goes through the construction of the three-four-five right triangle which represents the needed dimensions of sixty, eighty, and one hundred cubits.[95] Thus Richard is able to make consistent the puzzling dimensions of Ezechiel's vision. Without some knowledge of geometry, he would have been hard pressed to explain the architectural vision of Ezechiel. Richard's geometry is not

---

[91] Ibid., cols. 527D-528A: "Nam si eamdem de qua hoc dicit litteram velimus secundum aliam acceptionem discutere, fortassis etiam juxta historicum sensum, ex ipsa poterimus congruum intellectum eruere."

[92] Ibid., col. 574D: "Hoc tamen inter ista et orientale et pavimentum fuit, quod illud secundum unam dimensionem octoginta cubitos, ut dictum est, habuerit, secundum aliam vero dimensionem centum cubitos, etiam secundum planum possedit."

[93] Ibid., col. 575A: "Elevantur itaque a gemino latere aquilonare, et australe pavimentum, sed orientale quidem et interius atrium ab uno latere tantum. Si autem quaeritur de interiori atrio quantum ab una parte elevabatur, facile hoc et absque scrupulo a geometricae disciplinae perito deprehenditur."

[94] Ibid., col. 575A: "Absque dubio atrium, quod octoginta cubitos habuit in base, et centum in superficie, sexaginta cubitis excelsius fuit in eminentiori parte."

[95] Ibid., cols. 575B-576A.

as obviously drawn from the tradition of practical geometry as
was Hugh's. Still, because of the importance of measurement
in Richard's work, his use of geometry may legitimately be
called practical geometry.

The importance of geometry as an aid to understanding Scripture is also pointed out by Roger Bacon. In the Opus majus he
describes the utility of geometry for Biblical studies. Geometry is to be used to derive the literal sense of Scripture;
the literal sense, in turn, is a prerequisite to the spiritual
sense. Geometry should be applied particularly where something
made by man is to be understood.

> Since, therefore, artificial works, like the ark of Noah,
> and the tabernacle with its vessels, etc., and the temple
> of Solomon and of Ezechiel and of Esdra and other things
> of this kind almost without number are placed in Scripture,
> it is not possible for the literal sense to be known un-
> less a man have these works depicted to his sense, but more
> so when they are pictured in their physical forms.[96]

In this Bacon is hardly original; he seems to be merely reflecting the Victorine tradition. If we had an example of Bacon's use
of geometry in Biblical exegesis, we might determine whether it
is practical geometry to which he appeals. However, because of
Bacon's ideas on teaching geometry, we can be fairly sure he
would turn to practical geometry to aid in understanding the
Bible.

In Bacon's ideas about teaching mathematics, practical geometry plays an important role. Bacon wants to rebuild mathematical education by emphasizing the practicality of mathematics
and particularly of geometry.

> I want to show something about the utility and worth of
> mathematics so that we may be aroused to its sweet study.
> For everything is neglected, howsoever distinguished it
> may be, as long as its worth and utility are not known.[97]

---

[96]Tr. Burke, 1928: pp. 232-233, corrected following ed.
Bridges, 1897-1900: p. 210: "Cum igitur opera artificilia,
ut arca Noae, et tabernaculum cum vasis suis et omnibus, atque
templum Salomonis et Ezechielis et Esdrae et huiusmodi alia
pene innumerabilia ponantur in scriptura, non est possibile
ut literalis sensus sciatur, nisi homo ad sensum habeat haec
opera depicta, sed magis figurata corporaliter."

[97]Communia mathematica, ed. Steele, 1940: p. 6: "Volo aliquid invenire circa utilitatem et laudem mathematice ut ad ejus

A restoration of mathematics seems necessary to Bacon because
few students were learning mathematics in his time. He attributes the scarcity of mathematicians to the prolixity of
mathematics.

> Since, however, the books and teachers of mathematics
> dwell upon the proliferation of conclusions and demon-
> strations, no one can arrive at an acquaintance of
> this science in the usual way unless he puts in nearly
> 30 or 40 years, as is clear of those who were eminent
> in these sciences, like Robert of blessed memory, re-
> cently bishop of the church at Lincoln, and brother
> Adam Marsh and master John Bandoun and others of this
> sort. And therefore few study this science. Yet,
> without it, the other sciences cannot be known, as I
> have shown above. Wherefore Latin students have the
> greatest impediments to wisdom because of this multi-
> plication of mathematical conclusions and demonstra-
> tions, and chiefly because to this multitude of con-
> clusions is annexed the cruel and awesome difficulty
> of the techniques of proof. It is in this way then
> that students reject this science not only because
> the insurmountable proliferation of conclusions and
> proofs is heaped on them, but because the difficulty
> is infinitely increased.[98]

Bacon may have had the tradition of practical geometry in
mind when he outlined his program for the restitution of
mathematical study. His program would reduce the number of
theoretical propositions as far as possible, leaving only

---

studium sauvis excitemur. Omnis enim res negligitur quantumcumque sit eximia dum ejus laus et utilitas ignorantur."

[98]Ibid., pp. 117-118: "Quoniam autem libri et doctores
mathematice insistunt multiplicationi conclusionum et demonstracionum, ideo nullus potest pervenire ad noticiam illius
sciencie secundum modum vulgatum nisi quasi ponat .30. vel
.40. annos, ut planum est in eis qui floruerunt in hiis sci-
enciis, sicut dominus Robertus felicis memorie nuper epis-
copus Lincolniensis ecclesie, et frater Adam de Marisco, et
magister Johannes Bandoun, et huiusmodi. Et ideo pauci student in hac sciencia, et sine hac sciri non possent alie,
ut superius demonstravi. Quapropter studentes latini habent
maxima inpedimenta sapiencie propter multiplicationem conclusionum et demonstracionum mathematice, et precipue quia huic
multitudini est annexa crudelis et horenda difficultas in
modo demonstrandi, ita quod studentes spernunt hanc scienciam,
non solum quia eis ingeritur multiplicacio conclusionum et
demonstracionum inpertransibilis, set quia difficultas adi-
citur infinite."

those of obvious utility.[99]   In addition, Bacon would have the
remaining propositions shown not by rigorous proof, but by numerical example, much as is done in practical geometries.[100]
Bacon even maintains that the student may acquire a better knowledge of mathematics through numerical example than through
formal demonstration, as his discussion of Euclid V, 8 shows.

> I could prove the same proposition among the others
> which proceed, but this proposition is pleasing be-
> cause of its great utility, so that it is clear that
> it can be known better and more surely without a dem-
> onstration than with it.[101]

Utility, simplicity and demonstration by example are thus the
requirements for the more teachable mathematics in Bacon's
scheme. The tradition of practical geometry was well known
to Bacon; he had great confidence in its utility.[102] Thus he
probably felt that practical geometry would serve at least some
of the needs of his pedagogical program.

Restoring the learning of mathematics leads another thirteenth-
century teacher to turn to practical geometry. This teacher, a
certain P., makes explicit the instructional purposes of his practical geometry, the Canones in triangulum pictagoricum de mensuris
practica geometrie.[103] P. makes the instructional purposes of

---

[99] Bacon gives a list of the few propositions of theoretical
mathematics necessary in his program, ibid., pp. 118-120 and
140-143.

[100] See chapter II below.

[101] Ed. Steele, 1940: p. 122: "Possem quidem in aliis que pro-
cedunt hoc idem ostendere, set illa placet propter magnam ejus
utilitatem, ut pateat quod sciri potest sine demonstracione meli-
us et cercius quam cum ea." The proposition under consideration
is obviously Euclid V, 8; misabbreviation or misreading would account for "5^{is} 8" as "communis 8."

[102] See the section "Practical Geometry and Practical Concerns,"
below.

[103] BL Digby 166, ff. 8v-12v. The same manuscript contains
part of Hugh of St. Victor's Practica geometriae with a preface
added by the same P. He intends the selection from Hugh's practical geometry to serve as a basic introduction to the subject.
In fact P. calls his version of Hugh's geometry "Una bona generalis introductio in practicam geometrie." P.'s preface is found
on f. 6v. The part of Hugh's Practica geometriae on ff. 6v-8r
comprises Baron's edition, 1966: p. 16, line 1 to p. 25, line 54.

the <u>Canones</u> quite explicit. Perceiving a neglect of the quadrivium, he proposes to dissipate "the smoke of ignorance
[which] befogs the natural ability of clerks who ought to be
clear about science."[104] The particular clerks about whom P.
is concerned are the friars; he was assigned to write a treatise for the use of "paupers."[105]

In spite of the audience specified, there is nothing in
the treatise that is not of general utility. Indeed the
treatise, which describes the workings of a particular geometric instrument, is more general than many of the other
practical geometries. P. approaches the subject from general
philosophical principles. He cites Aristotle to justify beginning from first principles, "A general cognition, granted,
however, that it be a universal and indeterminate cognition,
naturally is always to be preferred to particulars, according
to Aristotle."[106] P. does not merely pay lip service to generality; he offers a general introduction to geometry, describing the types of lines and figures, the division of a
circle, the utility of sexagesimal fractions, the theoretical
basis for his instrument, and the arithmetic needed for its
use. Even where he discusses the use of his "Pythagorean triangle," an instrument in the form of a 45-degree right triangle with movable pinnules through which to sight, he gives
general principles rather than specific instructions. These

---

[104]BL Digby 166, ff. 9r-9v: "Sed modernis temporis ignor-
ate mathematice in studiis mirabilium rerum maxime arcium
liberalium et maxime quadruvialium tantus ignorancie fumus
obvubilat ingenia clericorum ut qui clari esse deberunt in
sciencia . . . ."

[105]Ibid., f. 8v: "Per instrumentum trigonum pictagoricum
pro pauperibus ad hoc specialiter deputatum ut temporibus,
laboribus et valoribus de quorum custodia curam rationabili-
ter applicare." It is not particularly surprising for the
treatise to be directed to the mendicants. Because of a
ruling of their order, during the thirteenth century the friars studied profane knowledge only within their own houses.
See Leff, 1968: p. 104.

[106]BL Digby 166, f. 8v: "Cognicio generalis licet sit
universalis et confusa tamen specialibus secundum Aristoti-
lem naturaliter semper preferenda."

principles belong more to the realm of philosophy than to the
realm of theoretical mathematics. For example, rather than
basing practical geometry on considerations of abstract dimension, he appeals to the geometry visible in the cosmos, relating dimension to the diameters crossing the circle of the horizon and perpendicular to its plane.[107]

Knowing the intended audience of P.'s treatise helps put its
very general and philosophic approach in better perspective.
P. seems to have felt that a philosophic treatment of geometry
would be of greater relevance to the student of philosophy.
After all, by the thirteenth century, instruction in the liberal arts was intended to be an introduction to philosophy. In
P.'s Canones we have a practical geometry related to the needs
of university education. Perhaps we can even see P.'s instructional goal as related to that of Bacon; they both perceive a
neglect of mathematical studies. While Bacon's solution was to
turn from the theoretical to the more particular and concrete,
P. moved in the direction of greater generality, providing a
more general philosophical basis even while working in practical
geometry.

Practical geometry, therefore, had a definite place in medieval education. The treatises on the classification of the sciences give a theoretical explanation of its role. The realiza-
tion of those theoretical concerns is evidenced in the educational practice of the Middle Ages. Treatises arising in several
educational milieux reflect a knowledge of practical geometry and
a concern for teaching it. Practical geometry served as a propaedeutic to at least two educational traditions. Both the Scrip-

---

[107] Ibid., f. 10r: "Fundamentum ergo altimetrie et planimetrie
necnon et steriometrie sit circulus orizontis cuius diameter ori-
entis et occidentis, meridionalis vel borealis se invicem ortogonaliter minime mediocres et maxime istarum semper sint per
centrum ortogonaliter transeuntes minime circa fines eorumdem
secundum mediocres erunt inter hec duo puncta sibi invicem ortogonaliter obviantes se tangentes non secantes secundum diametrum
secantes sursum vel subtus, ante vel retro, a dextris vel sinistris ad modum sinuum et bardagarum."

tural and philosophical traditions in the High Middle Ages
were claimed to rest on a basis in the liberal arts. At
least part of that foundation for both traditions was provided by the study of practical geometry.

Theory and Practice in Geometry

A place was found for practical geometry among the sciences as it was distinguished from theoretical geometry. Defining practical geometry meant dividing geometry into two
parts.[108] The separation seemed more important than the connections between the theoretical and the practical. Nonetheless, the relation between the two parts of geometry merited
the attention of some of the authors of geometric treatises.
The introductions to the various treatises sometimes specified the relationship between the theoretical and practical
aspects of geometry, as did some treatises on the classification of the sciences. The introductions and classifications,
however, offered rather laconic definitions of the two branches
of geometry.

More about the connections between theoretical and practical geometry may be learned from the treatises themselves.
Some of the practical geometries show the influence of the
theoretical tradition; some of the theoretical geometries
are affected by the practical tradition. The interconnections between the traditions appear not so much in the subject matter of the treatises but rather in their techniques.
Consideration of a few treatises will indicate the nature and
purpose of the relation between the two parts of geometry.
The interconnections sometimes appear to have been developed
for pedagogical reasons.

The extent to which the two traditions occur varies in the
different treatises. Some of them show greater influence of
the practical aspects of geometry; theory plays a larger role
in others. Since the proportion of theoretical and practical
geometry changes, one is led to seek a pattern of development.

---

[108]See the section on the Nature of Practical Geometry,
above.

The relative roles of the two traditions do appear to evolve, even when a few treatises are considered. In this section I shall deal with only a few geometric works of the High Middle Ages. Therefore, any conclusions about the evolving proportions of theory and practice are necessarily tentative. Nonetheless, from the Adelard III version of the Elements to Artis cuiuslibet consummatio on to Leonardo Pisano's Practica geometriae, Johannes de Muris's De arte mensurandi, and Dominicus de Clavasio's Practica geometriae, there seems to be a developing theoreticism in geometry.

Adelard of Bath's special edition of the Elements of Euclid, the Adelard III version, was important to the elaboration of geometric traditions in the Middle Ages. It was in the introduction to Adelard's edition that a satisfactory way of placing both the theoretical and practical branches of geometry within the liberal arts was found. Adelard also made clear the distinction between the two subjects by specifying the duties of the masters of theoretical and practical geometry.[109] Thus it is clear that Adelard knew both the geometric traditions of his time.

Since Euclid's Elements is virtually a paradigm of theoretical geometry, one might not expect to find much concern with practical geometry in Adelard's edition. In this case, however, our expectations prove wrong. Adelard's knowledge of practical geometry has marked effects on his edition of Euclid, effects which may account for some of the special characteristics of Adelard's edition. One of these singular features is an emphasis on the didactic.[110]

As John Murdoch has pointed out, throughout the Adelard III version, Adelard tries "to relate the whole geometry, as it were, of commensurable magnitudes to the science of practical measure."[111] One manifestation of Adelard's program is the long list of practical measures and their equivalences in the preface

---

[109]See above, pp. 12-14.

[110]Murdoch, 1968: p. 93.

[111]Ibid., p. 83.

to Adelard's special edition.[112] Practical measure, as we
have seen above, is one of the characteristics of practical
geometry. Several of the practical geometries include similar lists of measures.[113] Indeed, Adelard provided his list
as an illustration of the tools of the practical geometer.

Adelard's familiarity with the techniques of practical
geometry has another result. In propositions treating continuous magnitudes, Adelard frequently uses numerical proofs
instead of general geometric ones.[114] This use of numerical
proof amounts to a weakening of the distinction between
arithmetic and geometry. Murdoch considers the erosion of
the distinction to be one of the hallmarks of the medieval
Euclid.[115] He does not explain, however, the possible source
of procedures like Adelard's. Adelard may have been moved
to disregard the difference because of his knowledge of practical geometry. The practical geometries of the twelfth
century and their predecessors often give only a numerical
example by way of proof. Sometimes, as in the case of _Artis_
cuiuslibet _consummatio_, no numerical example is given, but
appeal is made to proof "by argument from number," _ratione_
numeri.[116] Adelard may have preferred the more obvious numerical proofs of the tradition of practical geometry. He
might have used them because he found their simplicity pedagogically useful. Breaking down the separation between
arithmetic and geometry would then be an unintended result
of Adelard's having espoused the methods of practical geometry. In other words, one of the characteristics of the
medieval Euclid arose from the use of a pedagogical aspect
of practical geometry within the realm of theoretical geometry.

---

[112]Clagett, 1954: pp. 274-275.

[113]See _ACC_, I. 32.11-15 and IV, 1.3-12 and notes thereto.

[114]Murdoch, 1968: p. 87.

[115]_Ibid_., p. 86.

[116]See chapter II, below.

Connections between theory and practice may be perceived in
treatises on practical geometry as well. One practical geometry
in which the relation between the two geometric traditions is
particularly marked is Artis cuiuslibet consummatio. From the
beginning of the treatise, the author makes it clear that both
theory and practice are important. In the prologue to treatise,
the author describes the special benefits to be gained by those
who know both aspects of an art. He considers theory without
its practice devoid of much interest. The practice adds immea-
surably to the pleasure one might derive from the theory. As
he puts it, those who would study only the subtleties of theory
are "failing to reap where they sowed the richest fruits as if
picking a spring flower without fruit."[117] The author goes on
to apostrophize about the pleasure one may derive from practicing each of the arts of the quadrivium. The practical aspect
of geometry that he praises consists of investigating the quantities of assorted geometric entities, once the relationships
between these things have been proven.[118] Thus, practical geometry serves to complete the study of geometry.

The prologue to ACC almost makes the practical side of geometry seem subservient or secondary to the theoretical. The
contents of ACC show that theoretical concerns play an important role within this practical geometry. Again the techniques
of proof point out the nature of the relations between the two
traditions of geometry. ACC has some proofs that grow out of
the tradition of practical geometry. The appeal to proof "by
argument from number" is an aspect of ACC closely related to earlier practical geometries. Other methods of proof in ACC are
connected more with those in the theoretical tradition. A fair
number of the proofs in ACC are acceptably rigorous. They would
fulfill the canons of proof in Euclidean geometry, even though

---

[117]ACC, I, 0.7-9: "eius theorice invigilant practicam postponentes fructus uberrimos ubi seminaverunt non metentes quasi
florem vernum sine fructu legentes."

[118]ACC, I, 0.15-18: "Quid magnificentius quam superficierum
et corporum, lateribus et angulis probatis per geometriam, eorum
quantitates scire examussim et investigare."

they are not as difficult as many of those in the Elements.
For example, most of the propositions in ACC giving the areas
of triangles rely on a single proof technique.[119] The proof
amounts to constructing a rectangle of which the triangle is
half.[120] The proof is perfectly general; the result could
be found once and applied to any triangle. However, ACC re-
turns to the same proof technique for each case, probably to
help the reader understand the general applicability of the
technique. Thus, the author of ACC seems to be teaching how
to apply a rigorous, general proof to individual cases.

The author of ACC knew some of the major works of theoretical geometry. It would, however, be extremely difficult to
show that the author used specific works of geometry as a
model for his theoretical proofs. Knowledge of any single
theoretical treatise would probably have sufficed; familiarity with several members of the tradition would not have made
a great difference. We do, however, know that the author of
ACC was acquainted with several works which qualify as theoretical geometry. He appeals to them for proofs which he
does not care to give. The authors to whom he refers include
Euclid, Archimedes, and Ptolemy.[121] Works of these authors
would have served as excellent models of theoretical geometry.
Whether or not the original proofs of ACC were inspired by a
particular theoretical work would be very difficult to show.
However, it is not necessary to single out a model; the simple fact that ACC refers to them points out substantial connection between the two geometric traditions. Since the
author of ACC refers to proofs found in theoretical works,
and since he frequently offers his own proofs, it is likely

---

[119]The technique is that of ACC, I, 2.32-49; it is also
applied in I, 3.6-9; I, 4.6-8; I, 5.3-4; I, 7.3-4; I, 8.3-4;
I, 11.3-4; I, 13.4-8; and I, 15.8-16.

[120]One of the sides is constructed parallel to the base
of the triangle through its vertex. The two remaining sides
of the rectangle are drawn parallel to the altitude of the
triangle through the ends of the base. The rectangle so constructed has an area that is twice that of the triangle.

[121]Euclid in III, 5 and III, 6; Archimedes in I, 22;
Ptolemy in II, 1; II, 11; II, 14; II, 17; II, 25; II, 26;
II, 27; and III, 4.

that he was inspired by the tradition of theoretical geometry.
His practical geometry is, indeed, rather theoretical. By
using theory in his practical geometry, he is able to teach
not only the practice, but something of the theory of geometry.

For both Adelard of Bath and the author of ACC the two
branches of geometry were mutually complementary. Proof techniques derived from one served the purposes of the other. In
another practical geometry, separating the procedures of the
two kinds of geometry seems more important than integrating
them. Leonardo Pisano offers two paths through his Practica
geometriae. The two approaches to his work reflect the two
traditions of geometry. Each path is meant for a different
group of readers.

In introducing his work, Leonardo differentiates between
the two types of reader he hopes to serve, "those who would work
following geometric demonstrations and those who would proceed
following common usage or, as it were, lay custom."[122] To proceed according to geometric demonstration means to appeal to
the theoretical tradition. Leonardo's practical geometry makes
extensive use of theoretical proof. Extended geometric demonstrations accompany virtually every proposition, making the
treatise perhaps a more theoretical than practical geometry.

The other approach to his work, following "common usage or
lay custom" is clearly different from rigorously demonstrated
geometry. The less theoretical path through the work is intended
for those readers not looking forward to "geometric subtlety"
like that of Euclid.[123] After giving a series of enunciations
from Book I of the Elements, Leonardo encourages the less theoretically oriented reader. "Without all of these propositions,
and without the extraction of roots, those who might want to proceed adequately with those things which I will point out below

---

[122]Ed. Boncompagni, 1857-1862, v. 2: p. 1: "ut hi qui secundum demonstrationes geometricas; et hi qui secundum vulgarem
consuetudinem, quasi laicali more, in dimensionibus voluerint
operari."

[123]Ibid., p. 2: "Multa enim sunt que oportet scire eos,
qui in mensuratione, et divisione corporum, secundum subtili-
tatem geometricam procedere volunt, que in Euclide aperte monstratur."

for convenience in place of those extractions and propositions."[124] Thus, the alternative path is a simpler approach,
requiring fewer proofs and less theory.

Even though we know that "common usage" involves a non-
theoretical approach, we cannot assume it to be identical
with practical geometry. Nonetheless, "common usage" does
have some connections with the tradition of practical geometry. By considering what Leonardo offers as an alternative to the extractions and propositions, we may determine
just what he meant by "common usage or lay custom." Discussing the measurement of triangles, Leonardo describes "the
common way which surveyors (agrimensores) ought to use and
which is sufficient for the measurement of all triangles."[125]
The method described is simply a way of determining the altitude of the triangle to be measured; the agrimensor is to use
a cord as a compass in constructing the line from one vertex
of the triangle perpendicular to the opposite side.[126] The
altitude and base are used to compute the area of the triangle. In this case, the measuring technique belongs to what
some practical geometries call artificial measurement.[127]
Another example of artificial measurement that Leonardo
recommends to the agrimensores "who want to proceed following
the common way" is a method for measuring the length of an
arc without extensive computations.[128] He suggests using a
flexible rope or a series of reeds placed along the curve of

---

[124]Ibid., pp. 2-3: "Sine his omnibus, et sine radicum inventione, ipsi qui secundum vulgarem modum procedere volue-
rint:cum his que ostendam inferius, convenienti loco poterunt
sufficienter procedere."

[125]Ibid., p. 43: "Modus vulgaris quo uti debent agrimensores, et est sufficiens in mensuratione omnium trigonorum."

[126]Ibid.

[127]See above, pp. 19-21.

[128]Ed. Boncompagni, 1857-1862, v. 2: p. 95: "Sed hec
talis investigatio non est operanda ab agri mensoribus, qui
secundum vulgarem modum procedere volunt."

the arc to be measured. The length of the arc is approximated
by the length of the rope or by the sum of the lengths of the
reeds.[129] The "common way" of measurement for Leonardo, then,
is that used by _agrimensores_ or surveyors. The techniques of
the _agrimensor_, however, are those of artificial measurement,
that is, of part of practical geometry.

That Leonardo's work deals with techniques of measurement is
hardly surprising; it is after all called _Practica geometriae_.
What is somewhat unusual is the concern it shows for theoretical
geometry. Most of the treatise is occupied with proofs, geometric demonstrations intended for those who would take the more
rigorous path through the book. The theoretical and practical
are not explicitly connected in Leonardo's geometry; nonetheless,
some relation is implicit, since theoretical and practical techniques of geometry are both present. There is much more theory,
but the work has as its subject and title practical geometry.
Leonardo has, therefore, produced a rather theoretical practical
geometry.

Johannes de Muris, one of the most important fourteenth-century mathematicians, also concerned himself with practical geometry. The form of his work in practical geometry is rather
unusual; Johannes chose to complete the unfinished practical geometry of an author anonymous even to him.[130] The original author
of the treatise, known as _De arte mensurandi_, completed only as
far as the first part of the fifth chapter of his projected eleven
chapters. Johannes de Muris completed the remaining chapters, as
foreseen in the author's description of the contents of the treatise. Furthermore, Johannes added several chapters of his own
at the end of the work.

The author of the first part of the _De arte mensurandi_ seems
to have been interested in astronomy; his parts of the treatise
contain material on trigonometry and on sexagesimal fractions,

---

[129]_Ibid._

[130]For a discussion of the role of Johannes and the unknown
author, see my article, Victor, 1970. Prof. Marshall Clagett
discusses the issue in greater detail in his forthcoming _Archi-
medes in the Middle Ages_, v. 3, pt. I, ch. 2.

both matters of importance for an astronomer. Even though
Johannes de Muris was fulfilling the prescriptions of the
original author, his work shows considerable creativity. In
effect, Johannes is writing a new kind of geometry; his geometry, though having measurement as a major concern, does not
deal merely with the art of practical measurement. Techniques
of practical measurement occupy a place in the De arte mensurandi but a small place. Most of the practical measuring or
"artificial geometry" occurs in the part composed by the original author. Johannes's part deals almost exclusively with
"non-artificial" measurement, measurement based on notions of
proportion. Because non-artificial measurement is, in a
sense, the study of proportion, it lends itself better to
theoretical treatment than does artificial measurement,
which has instruments rather than theorems as its subject.
Johannes's treatment is quite theoretical; he often offers
proofs of his theorems, and at the very least he indicates
where proofs may be found. In those places where he does not
give his own proofs, Johannes usually directs the reader to
Euclid or to Archimedes. Because of his concern for theory,
Johannes has produced a comprehensive theoretical treatment
of the art of measuring.

The De arte mensurandi is so extensive that it includes a
good bit of material on what might be called geometric algebra, like the contents of Books II and VI of Euclid's Elements,
which contain geometric means of solving problems for which
we would use algebraic equations. The De arte mensurandi is
not the first practical geometry to deal with geometric algebra. Geometric algebra is not present in the early Latin
agrimensorial material, the source of many of the later practical geometries. It does, however, play an important part
in the Islamic tradition of practical geometry[131] and in
Latin works based on Islamic materials, like Leonardo Pisano's
Practica geometriae. Johannes de Muris's treatment of geometric algebra is extensive and theoretical. He appeals to
proofs, either his own or Euclid's, to justify his propositions.

---

[131]See above, p. 27.

Johannes carried his treatise beyond the eleven chapters intended by the original author. He added other chapters to complete his work on the art of measuring. What he chose to add
is further material on ratios and proportions. Chapter 12 deals
with what Johannes calls binomia; a binomium, in our modern
terms, would be a linear magnitude composed of an integer and a
surd. The remainder of the book treats proportions and quantities in plane figures and solids. Thus, Johannes's additions to
the scope of the original author's plan comprise parts of the
contents of Books VII, VIII, IX, and especially X of Euclid.
All in all, the De arte mensurandi is a compendious treatment
of practical geometry, full of both proofs and numerical examples. It is of broader scope and more fully theoretical than
any of the other practical geometries up to its time.

Dominicus de Clavasio's Practica geometriae is as theoretical as that of Johannes de Muris, though its scope is much more
limited. Dominicus does not include nearly as much Euclidean
material as the De arte mensurandi. Nonetheless, the influence
of Euclidean theoreticism is clearly felt. Dominicus, as we
have seen above, bases his geometry on the concept of ratios
and proportions. It is not surprising, therefore, that after
defining measurement of a quantity, he begins an extensive discussion of ratio and proportion. It is interesting to note that
the propositions concerning ratio and proportion are called sup-
positiones, almost as if he were trying to axiomatize his practical geometry. The propositions of his geometry, which he calls
constructiones, do indeed have a highly theoretical tone, even
though they deal largely with the use of instruments in measuring. Each measurement, at least all those dealing with rational
quantities, is justified by reference to propositions both about
ratio and proportion and about geometric relations. The geometric
justifications come directly from Euclid: Dominicus cites them
by proposition and book number. Many of them, of course, deal
with proportions in similar triangles. He even cites from Witelo's Perspectiva the proposition about the equality of angles
of incidence and reflection as justification for a measurement

using a mirror.[132] Dominicus also provides definitions where he deems them necessary. In other words, Dominicus manages to turn practical geometry into an almost purely theoretical enterprise. It remains practical in the sense that the measuring techniques could be applied to the measurement of real objects.

Dominicus makes an interesting distinction between the mensor geometrie and the mensor laicus; this separation is similar to that of Leonardo Pisano between those who proceed theoretically and those who follow "common usage or lay custom." Dominicus makes his distinction after having explained how to find the altitude of a known triangle both by measuring the altitude of a triangle similar to a given one but of a more convenient size, and by calculation using the Pythagorean theorem. He explains:

> I have put down this and the two preceding constructions in order to give a way of measuring areas that the non-geometric measurers do not know, as will be seen below, since this ought to be the difference between the geometric and lay measurer: that what the lay measurer knows how to measure by going and dashing around the sides of a field with his rods and cords, the geometric measurer will know standing still by mental reflection alone or by drawing lines.[133]

Thus Dominicus distinguishes between those who measure by going to the site where the measurement is to be made and those who measure by reflecting and calculating while staying still.[134] The distinction may seem comparable to that between

---

[132]Ed. Busard, 1965: p. 539.

[133]Ibid., pp. 559-560: "Istam constructionem et duas alias precedentes posui, ut darem unum modum mensurandi areas qualiter nescirent mensores non geometrie mensurare, sicut videbitur inferius, quia ista debet esse differencia inter mensorem geometrie et laicum, quia illud, quod laicus scit mensurare cum perticis et cordis eundo et discurrendo per latera campi, mensor geometrie sciet stando ex sola consideracione mentis vel linearum protractione . . . ."

[134]Ibid., p. 560: "quot esset quodlibet latus, ipse [mensor laicus] nesciret et quanta esset area nisi iret supra locum. Sed ista debet mensor geometrie scire in considerando et quiescendo."

artificial and non-artificial measurement. However, given Dominicus's extensive use of examples employing instruments and his very theoretical treatment of the geometry of instrumental measurements, it is likely that he intends to separate a practitioner's or surveyor's ordinary procedures from the geometer's understanding of those procedures, an understanding which provides surer and more efficient results.

Dominicus does not provide two alternative paths through his Practica geometriae as Leonardo Pisano had. Dominicus's geometry seems not to be intended for the "lay measurer" in any way. For Dominicus, the rigorous and contemplative approach is the one worthy of concern. Thus, it is yet a further stage in the growth of theoreticism in geometry. By the time of Dominicus de Clavasio in the mid-fourteenth century, even the geometry dealing with practical measurement had a fully theoretical basis.

Both theory and practice are present in the geometric treatises considered in this section. Each aspect of geometry lent support to the other. For the most part, the interconnection took the form of borrowed proof techniques. Theoretical geometry used the pedagogically useful proofs of practical geometry. Practical geometry adopted some of the rigorous aspects of theoretical geometry, perhaps in order to teach about the nature of proof. The use of theoretical methods in practical geometry seems to have increased between the twelfth and fourteenth centuries. At first their role was ancillary to the purposes of practical geometry. Once proofs had found a place in practical geometry, their role increased and changed. Theoretical proof became the goal even of practical geometry.

Practical Geometry and Practical Concerns

Knowing that a series of treatises called practical geometries exists, one may naturally enquire about the way in which these works are practical. Are these geometries practical in the sense of being applicable? And if applicable, then to what? Thus we are led to investigate the type and extent of connection which might be found between the tradition of practical geometry and various practical concerns. The problem is a difficult one; it

amounts to trying to establish links between a literary tradition, that of the practical geometries, and an almost entirely
non-written one, that of certain craftsmen or artisans. Any
connections that might be found, however, are to be prized,
for they would tell something about both the training of some
groups of artisans and the possible audience for the treatises
on practical geometry.[135]

The contents of the practical geometries hint strongly at
some connections with practical concerns. The treatises contain material to suggest that instrument-makers and surveyors
are some of those to whom a practical geometry might have
been useful. Analysis of the treatises themselves, however,
can show only that they might have been usable. Other written
materials support the hope of demonstrating a connection between the treatises and the practice of certain crafts, like
architecture and surveying. Still other documents provide
some indication of how certain artisans proceeded in their
work. But it is only when written sources are compared with
one another and with other less verbal evidence that connections may be established. These connections result from
analogies between the specific techniques of certain craftsmen and those given in some of the practical geometries.
Even then, the relation is tenuous. All we can have is a
possibility or, at best, a likelihood that medieval craftsmen learned something they might have applied from the treatises on practical geometry.

Some relation to practical concerns is suggested in the
practical geometries themselves. Surveyors and instrument-
makers could have found the practical geometries useful. The

---

[135]These problems were considered at the First International Conference on Philosophy, Science, and Theology in the
Middle Ages in September 1973. The proceedings, ed. Murdoch
and Sylla, 1975, include several papers and discussions of
the relation of theory and practice in the Middle Ages. Most
immediately relevant to my concerns here is Guy Beaujouan's
"Réflexions sur les rapports entre théorie et pratique au
Moyen Age" and the discussion following it, pp. 437-484.

treatises contain material about artificial measurement which
could have been useful to surveyors.[136] The procedures for measuring distances are given in sufficient detail to provide elementary instruction in surveying.[137] Knowledge of metrology
would also be necessary to the practicing surveyor; this too
is given in many practical geometries.[138] Methods for computing
the areas of fields, another requirement of the surveyor, appear
in the treatises.[139] Anyone interested in making certain scientific instruments might also have turned to the treatises on
practical geometry. Artificial measurement required the use of
instruments. Instructions for making at least a few scientific
instruments are found in most of the practical geometries. The
most common of these instruments is the quadrant.[140] The treatises also describe the construction of other simple instruments.[141]
Some of the treatises include material of an intrinsically practical nature. Some of it is not geometric and may indeed seem
to be unrelated to practical geometry. For example, one of the
treatises offers, in addition to the usual geometric content,
material on mercantile practice, money changing and alloying.[142]
All of this material appears to be usable, but it may not have

---

[136]See above, pp. 20-21.

[137]For example, see ACC, I, 1.

[138]See ACC, I, 32.11-15 and IV, 1.3-12, and the French Pratike, V, 2; V, 6-8, and notes thereto.

[139]See ACC, I, 32 and 33 and the notes thereto. In a sense,
all the problems about measuring area are about measuring fields.

[140]The larger part of Quadrans vetus deals with making a
quadrant. See above, pp. 21-22 and n. 52.

[141]For example, ACC gives instructions for adding simple devices to an astrolabe (II, 36) and a quadrant (II, 37). It also
tells how to make a gnomon with a scale for measuring the length
of its shadow (II, 38) and a circle to be used to determine
which way is south (II, 3). A construction necessary for making
an astrolabe is also explained in ACC (I, 25).

[142]What I have edited below as the Pratike de geometrie, VI
seems mercantile in tone and content. Thus VI, 1, 3, 4, 6, 9
and 10 concern exchange rates or prices for commodities; VI, 5
and 11 treat alloys.

been applied. The presence of material on surveying, instru-
ment-making, money changing, and alloying does not prove that
the authors were writing for surveyors, instrument-makers,
merchants, or mint-masters. Examples which appear practical
may have been included to illustrate treatises whose only
function was to teach geometry or mathematics in general.
Other support is required to show a practical intent in the
practical geometries.

A few discussions of practical geometry exist outside of
the treatises themselves. These writings, contemporary with
the practical geometries, argue for the actual utility of
practical geometry. One might say that these works are almost commentaries on the role of practical geometry in practical concerns. They are, therefore, related to the treatises
on the classification of the sciences.

In fact, one of the works on the classification of the sciences offers evidence that practical geometry included certain
practical concerns. Gundissalinus's treatise on classification, as we have seen above,[143] describes as part of practical geometry a mechanical or fabricants' practice. Gundissalinus thereby indicates that a real connection exists between practical geometry and the crafts of the mason, carpenter, blacksmith, and surveyor. Gundissalinus was, however,
echoing an Arabic rather than a Western tradition. Nonetheless, he would not have included this aspect of applied geometry if there was nothing like it in the Latin world. Thus
there is strong likelihood that the connections that Gundissalinus describes in his treatise, connections between certain
crafts and practical geometry, actually existed during his
time.

Roger Bacon attributes a very broad utility to practical
geometry. In his Communia mathematica, Bacon gives an elaborate description of the parts of practical geometry. For
each of the parts, Bacon gives a brief description of its

---

[143]See above, pp. 10-12.

utility. The parts are as follows:[144]

A. For the use of men in governing families and cities
    (42.32-44.15)
    1. Knowledge of isoperimetric and space-filling
        figures (43.13-18)
    2. How to measure lines, surfaces and solids
        (43.19-33)
        a. altimetry
        b. planimetry
        c. superiometry (for stereometry?)
    3. How to draw cities, castles, houses, etc.
        (43.33-34)
    4. How to make canals and conduits, bridges and
        ships, etc. (43.34-37)
    5. How to construct machines that add to human powers
        and abilities (43.37-44.13)
    6. How to construct machines of war (44.13-15)
B. For the construction of scientific instruments
    (44.34-47.12)
    1. Astronomical instruments (45.1-19)
    2. Musical instruments (45.19-26)
    3. Optical instruments (45.26-35)
    4. Instruments of the science of weights (45.35-37)
    5. Instruments of experimental science (45.37-46.33)
    6. Surgical and medical instruments (46.33-35)
    7. Alchemical instruments (46.35-47.1)

Bacon certainly knew the tradition of practical geometry. The
parts of Bacon's schema closest to the treatises called practical geometries are those dealing with the measurement of lines,
surfaces and solids, part A. 2. above, and with the construction
of astronomical instruments, part B. 1. Bacon even calls altimetria, planimetria, and superiometria the "three famous parts"
of practical geometry.[145] Even though he uses "superiometria,"
he probably means "steriometria," one of the standard names for
the third part of the most common practical geometries.[146] There
is thus no doubt that Bacon was familiar with some of the treatises on practical geometry.

---

[144]Ed. Steele, 1940 : pp. 42-47. Bacon's division into parts
is not at all clear. Steele's text, p. 42, tells us that practical geometry has eight major parts; the text is either corrupt
or incomplete, for Bacon gives only two major parts. I have
emended the text in A. 3., changing ista to tertia. The numbers
in parentheses refer to page and line numbers in Steele's edition.

[145]Ibid., p. 43.

[146]Misreading an abbreviation could account for the error,
which might be either the editor's or the scribe's.

Some question remains about the extent to which the parts
of Bacon's scheme may be called parts or applications of practical geometry. If practical geometry could be applied in all
those areas, then it would be a very practical subject indeed.
Most of Bacon's applications, however, seem to require the use
of almost no geometry in any explicit or systematic way. A
few of Bacon's parts may very well reflect areas in which geometry may be applied. Part A. 1. is clearly geometric; the
study of figures having the same perimeters and figures which
will fill up space is an aspect of geometry rather than a
practical application of it. The traditional practical geometries belong in A. 2., as we have seen. Except for B. 1.,
astronomical instruments, most of part B. would not demand
recourse to treatises on practical geometry. The technology
of parts A. 4., 5., and 6. could proceed very well with very
little formal knowledge of geometry. In A. 3., the drawing
or laying-out of cities, castles, houses, etc., Bacon seems
to be referring to surveying or city planning and architecture. Thus, one might most profitably look for applications
of practical geometry in these crafts. The possibility does
remain that Bacon overstated the utility of practical geometry in all areas. But unless we appeal to other evidence,
we cannot properly determine whether Bacon was describing the
actual practice of geometry in his time or merely expressing
a hope for a broader utility for geometry.

Fortunately, we possess materials to help evaluate the extent to which practical geometry may have been applied in surveying, city planning, and architecture. This evidence consists of reports about how surveying and building were carried out. If the methods described by these reports or implicit in them are similar to methods given in some of the
practical geometries, then some connection might be assumed.
Of course any interrelation derived in this way is conjec-
tural. We cannot show a very definite relation between the
practice of certain crafts and the works on practical geometry; we can establish only the likelihood of the connection.
Nor can we determine, in this way, whether the application

was developed from material in a practical geometry or whether
the example in a practical geometry is a reflection of surveying or architectural practice. In spite of all these cautions,
parallel procedures in the practice of crafts and practical
geometries seem the only way of showing any sort of relation
at all.

By the twelfth and thirteenth centuries, surveyors were employed as professionals. What we can learn of their techniques
shows parallels to some of the propositions of practical geometries. For example, Antoine De Smet has compiled a list of
surveyors for the major towns of Belgium. He holds the profession to be a widespread one, "Il y avait dans la plupart
des villes au moins un mesureur (geometricus, agrimessor,
mensurator ou landmeter) qui faisait spécialement les mesurages
de biens contestés, de routes, de cours d'eau, etc."[147]

The accounts of the city of Bruges record payments to at
least three different surveyors before the end of the thirteenth
century. One Johannes Troebel, in records of payment for work
he performed is called, respectively in 1282, 1284, and 1285,
lantmetere, geometricus, and agrimessor.[148] In 1297 two agri-
messores worked on fortifications at Bruges.[149]

De Smet also gives a testimony that a certain Walterus de
Kirchem, generalis mensurator terrarum, in 1253 made a survey
of lands of the abbey of Saint-Trond. This done, Walter, in
the presence of various officials, swore to the size of the
property "secundum legitimam mensurationem."[150] Since a virga

---

[147]De Smet, 1949: p. 783.

[148]Ibid., p. 785 and n. 3.

[149]Ibid., p. 785 and nn. 4-7.

[150]Ibid., p. 792 and n. 4: "Hoc facto, predictus Walterus,
mensurator, requisitus a supradicto Lamberto de Porta, scultetus,
coram scabinis et multis aliis ibi presentibus, sub fidelitate
prestita, respondit quod predicta terra, secundum legitimam
mensurationem, habebat 50 bonuaria et 3 virgas magnas cum virga
de 20 pedibus: hoc est 4 mansos et 43 virgas magnas, cum virga
de 20 pedibus. Dixit etiam dictus mensurator, sub fidelitate
prestita, quod ipse acciperet dictam terram pro tanto et daret

of a defined length, 20 feet, was used and an oath was required, it seems certain that this professional surveyor
would have made his measurements as accurately as possible.
If Walter's survey had been careless, the eventuality of a
more accurate survey would have entailed the unwilling breaking of his oath. Therefore, Walter would probably have used
all the skill and knowledge available to him.

Some of the knowledge possessed by surveyors in the Middle
Ages may have come from practical geometries. Most of the
treatises on practical geometry contain material which would
have been useful to a professional surveyor like Walterus de
Kirchem. Computations of the areas of fields of different
shapes are taught in all of the practical geometries.[151]
Since a surveyor would probably need to give the area of the
lands surveyed, knowledge of practical geometry would provide
him with applicable techniques. If the field were anything
but rectangular, he would often have to divide it into triangles, the areas of which might be more easily determined.
To find the area of a triangle, however, one needs to know
its altitude. A method used by surveyors for finding the
altitude of a triangle is reported by Leonardo Pisano in his
Practica geometriae.[152] Leonardo's method belongs to the
tradition of artificial measurement. Another method, a computational one, for finding the altitude of a triangle is
found in Artis cuiuslibet consummatio.[153]

---

eam pro tanto, si ad eum pertineret . . . ." De Smet cites
the passage from Henri Pirenne, Le livre de l'abbé Guillaume
de Ryckel (1249-1272): Polyptyque et comptes de l'abbaye de
Saint-Trond au milieu du XIIIe siècle (Bruxelles, 1896),
p. 184.

[151]For example, ACC explains how to find the areas of
fields in the shapes of a trapezoid (I, 32) and an irregular
quadrangle (I, 33). See the notes in ACC for other practical
geometries. All procedures for finding areas may, of course,
be applied to fields.

[152]Ed. Boncompagni, 1857-1862, v. 2: p. 43. I have described the method in the preceding section of this chapter.

[153]ACC, I, 4.

A surveyor also has to know the size of various units of
measure and how those units are related to one another. In
the example above, the size of Walter's measure is specified
in the report of the survey. The report also gives the measurements of the lands of the abbey in two different, but
equivalent, sets of units.[154] Thus, knowing the relations
among the different systems of units was very important to a
surveyor. Again, a surveyor could have turned to most of the
practical geometries for instruction in metrology. ACC, for
example, gives two lists of measures and their equivalences.[155]
A metrological discussion also occurs as part of a proposition
giving methods for finding areas in ACC and for finding volumes
in the Pratike.[156] Leonardo Pisano gives a more extensive treatment of measure in his Practica geometriae. He explains in great
detail the various measures used in his native Pisa. Measuring
cultivated lands and open fields seems to have required different systems of measure in Pisa.[157] In addition, the various
systems of measure are quite complicated in themselves and in
their interrelations. Dominicus de Clavasio's Practica geometriae also gives several lists of measures;[158] indeed, he uses
ratio with a common measure (famosa quantitas) as the key to his
whole treatment of practical geometry.

Connections like those between the craft of surveying and
the treatises on practical geometry also exist for the field of
city-planning. Many new towns were created in the High Middle

---

[154]See note 150 above.

[155]ACC, I, 32.11-15 and IV, 1.3-12. See the notes to those
propositions for other treatises.

[156]ACC, I, 32.5-10; V, 2 and V, 6-8.

[157]Ed. Boncompagni, 1857-1862, v. 2: p. 3: "Mensurantur
quidem agri, et spatia domorum cum perticis, et pedibus, et
uncijs linealibus. Sed camporum embada colliguntur in stari-
oris, et panoris, et soldis, et denariis, vel in partes unius
panori: domorum quoque spatia colliguntur in scalis, et in
partibus scale, et in soldis, et denarijs."

[158]Ed. Busard, 1965: pp. 524, 555, 569.

Ages.[159] Someone seems to have been responsible for planning
or laying out each of these new towns. We know, for example,
that in the thirteenth century a "black clerk" called Richard de Escham was paid for laying out the streets of Baa, a
bastide or new town set up near Bordeaux in the reign of
Edward I. Besides measuring out the streets, Richard seems
to have been general superintendent of the enterprise, for
all of which work he was quite well paid.[160]

Maurice Beresford in his exhaustive study of the new towns
of England, Wales, and Gascony shows that the rectangular
grid was well adapted to a simple setting out of building
plots, placeae, and streets, and that such a grid was fairly
common. A grid for streets and building plots would have
been easily established, for it required only the ability to
divide a line into equal lengths and to set other lines perpendicular to the first.[161] Even though we have few written
testimonies of this process, Beresford feels it was widely
applied, as evidenced by the "dumb witness of the sites in
England, Wales, and Gascony that have the simple rectangular
grid of streets."[162]

---

[159]See Beresford, 1967.

[160]Trabut-Cussac, 1961: pp. 142-143. The entry in the accounts for June 11, 1287 records the payment to "Ricardo de
Escham, clerico nigro, deputato per Thesaurium Garderobe ad
expensas bastide de Baa juxta Burdegalam de novo constructe,
pro stipendiis diversorum operariorum circa bastidam predictam, videlicet quorumdam querencium fontes in bastida et men-
surancium stratas ville et aliorum mundancium vicos in bastida et viam versus eandem de spinis et opturancium vicos
antiquos, et pro calce carianda emptis, cum cariagio rerum
predictarum, cordis ad mensurandum villam, . . . et pro aliis
minutis expensis factis per manus dicti Ricardi circa bastidam predictam, prout patet per particulas inde liberatas in
Garderoba: .xlj. li. xij. s. ix. den. chip., qui valent in
sterlingis: .vij. li. xij. s. vj. den."

[161]Beresford, 1967: p. 146.

[162]Ibid., pp. 146-147.

Material from the tradition of practical geometry reflects the use of a grid system in laying out a town. Three propositions from the practical geometry _Artis cuiuslibet consummatio_ would have been of immediate use to a medieval city planner.[163] These theorems involve finding the number of small plots in a trapezoidal, in a triangular or in a circular area. The last of these propositions is even explicit in its object, "To infer the number of houses in a round city."[164] Several of the new towns of the Middle Ages were in fact established with circular boundaries.[165] Of course it would be difficult to establish with any generality the number of plots of a given size that would fit into an area of a specified shape. The techniques of a practical geometry would give a rough estimate.

In the propositions from _ACC_ mentioned above, all the houses or plots are assumed to be the same size. Although certainly not in all of them, in some of the new towns all the building plots were indeed of the same size.[166] For example, at Monségur (Gironde) the royal charter specifies the size of the building plots: 24 by 72 feet. The same rental was to be paid for each plot.[167]

In some of the new towns, different rents were assessed for different sections of the town, and the plots were of various sizes.[168] This situation made the assessment and bookkeeping more difficult, but because of it we have a particularly valu-

---

[163] _ACC_, I, 34-36; cf. I, 31.

[164] _ACC_, I, 36.1: "Civitatis rotunde numerum domuum colligere."

[165] Beresford, 1967: pp. 155, 589 gives the example of a Gascon town, Fourcès (Gers). There is an aerial photograph of a round city in Italy, Cittadella, in Hyde, 1973: pl. 1B.

[166] Ibid., p. 147.

[167] Mortet and Deschamps, 1911-1929, vol. 2: pp. 291-293, give parts of the charters of Eleanor of Provence (July 26, 1265) and of her son Prince Edward (June 26, 1267).

[168] On different rents, see Beresford, 1967: pp. 24-25; on various sizes, _ibid._, pp. 16-18.

able rent roll for New Winchelsea, Sussex.[169] The list of
assessed rents gives next to the name of the tenant, the size
of his plot in _virgae_ and the rent due.[170] These areas were
most easily computed for a rectangular plot, but many of the
areas between two pairs of streets, or _insulae_, were not rectangular but oblique or trapezoidal.[171] At least some of the
plots within these "city blocks" would have had corners which
were not right angles. Just as for the surveyors discussed
above, whoever measured and assessed these unsquare plots
would have had little trouble computing areas if he had even
some very rudimentary knowledge of practical geometry.

Practical geometries would therefore have provided useful
information for the practicing surveyor or city planner. However, all of this material on surveying does not prove that
the authors of the practical geometries meant them to be applied. Indeed, the reasons for including such material may
have been traditional rather than practical. After all, the
tradition of practical geometry had its origin in the works
of the Roman _agrimensores_.[172] Since the _agrimensores_ were
practicing surveyors and city planners, it is natural to expect a practical flavor in works deriving from their hand-
books. On the other hand, the antiquity of the handbook tradition does not prevent the medieval successors to that tradition from being practical guides as well. Thus, Leonardo
Pisano's description of surveying techniques and measures reflects the special practices of Pisa. In other words, Leonardo's treatise is more closely tied to contemporary surveying methods than with those of the ancient Roman field-
measurers. Therefore, the manuals of practical geometry may
indeed have been useful to the medieval surveyor or town
planner. He could have learned various techniques of mea-

---

[169]_Ibid._, pp. 14-28.

[170]This _virga_ is a square 5 1/2 yards per side or 1/160
part of an acre. _Ibid._, p. 18.

[171]_Ibid._, fig. 1, facing p. 16.

[172]See above, p. 5.

surement from them. The practical geometries could have also
served as references in metrology. The fact that the practical
geometries contain material so like that needed for surveying
and city planning suggests that they may have been used by the
practitioners of those professions in the Middle Ages.

The historian who would try to show that written treatises
influenced the practice of architecture has an even more difficult problem. The few literary testimonies about the practice
of architecture during the twelfth and thirteenth centuries
show that the architects knew some geometry of a practical kind.
And certainly well-aligned towers, straight rows of columns,
and overall geometrical regularity hardly indicate an ignorance
of geometry. What the source of the geometric knowledge of medieval masons or architects may have been is rather difficult to
determine. Lon Shelby in a pertinent article has come to some
interesting conclusions about the problem.[173] He argues per-
suasively that no single written work or secret key to Gothic
architecture is the basis of the geometry of medieval masons
or architects. Rather, the geometry of medieval masons involved
the construction or manipulation of a few geometric figures.
He does, however, acknowledge that one can perhaps find the
"particular procedures used by particular masons at particular
times and places."[174] Thus, some hope remains of finding parallels between the techniques of medieval architects and materials
in some of the practical geometries. The conclusions that may
be reached about the relation between treatises on practical
geometry and architectural practice will not be very strong.
Sometimes a first consideration leads us to expect stronger connections than actually hold.

A case of this sort is the work done by an architect for the
abbey church of St. Denis before July 14, 1140, requiring the
use of certain geometric instruments. Abbot Suger, writing in
1144, describes the work of "equalizing" a new apse with an exis-
tent structure,

---

[173]Shelby, 1972: pp. 395-421.

[174]Ibid., p. 421.

> Moreover, it was cunningly provided that--through
> the upper columns and central arches which were to
> be placed upon the lower ones built in the crypt--
> the central nave of the old [church] should be
> equalized, by means of geometrical and arithmetical
> instruments, with the central nave of the new addi-
> tion; and, likewise, that the dimensions of the old
> side-aisles should be equalized with the dimensions
> of the new side-aisles . . . .[175]

The equalization, perhaps alignment is closer to the meaning, would probably not have inspired Suger to make special
note of the instruments unless the new apse presented some
special problems. Frankl's analysis of the situation points
out what the problems were.[176] The new choir was to extend
beyond the old Carolingian crypt, which itself extended beyond the old apse and was to be the floor of Suger's choir.
The piers of the new crypt, under the planned choir, had to
be set so the new choir would line up with the nave; the
same piers also had to be placed so the new choir would have
the same width as the recently completed west end of the nave.
In order that services could continue, the old apse had to
be left in place while the new one was being built. Thus
the architect's work was hindered by two obstacles, the old
apse and the crypt extending behind it. Because of those obstacles it was not possible to sight straight down the nave;
therefore, the architect must have used techniques of indirect measurement.[177] To carry out these indirect measurements, the architect would have had to use the "geometrical
and arithmetical instruments" emphasized by Suger.

---

[175]Panofsky, 1946: pp. 100-101, gives both the translation
above and the text: "Provisum est etiam sagaciter ut superioribus columnis et arcubus mediis, qui in inferioribus in
cripta fundatis superponerentur, geometricis et aritmeticis
instrumentis medium antiquae testudinis ecclesiae augmenti
novi medio aequaretur, nec minus antiquarum quantitas alarum
novarum quantitati adaptaretur . . . ." The first stones for
the foundation of the new apse were laid ceremoniously using
mortar mixed with water from the dedication (on June 9) of
the just-completed narthex. See ibid. and notes.

[176]Frankl, 1960: pp. 8-10.

[177]Ibid., p. 9.

The use of instruments brings to mind the practice of artificial measurement found in practical geometries. However,
the instruments used at the church of St. Denis need not be
those of practical geometry. It is difficult to try to determine the techniques Suger's architect would have employed to
align the new apse. Perhaps a base line established parallel
to the axis of the nave would provide the simplest solution.
Then the distance from this line to the new rows of columns in
the apse would be measured out equal to the distance from the
base line to the rows of columns of the narthex and west front.
The new columns were, in fact, set so their centers would lie
on the line of centers of the narthex columns. The procedures
both of centering and aligning would have required the architect to lay out a series of right angles or a rectangle. Thus,
some knowledge of the geometry of parallels and perpendiculars
was needed for aligning the columns.

Even though some geometry was used by the architect at St.
Denis, it need not have had any relation to the treatises on
practical geometry. In that it was applied to a practical purpose, it could be thought to be practical geometry. However,
the geometry required for this aspect of the church of St. Denis
is so simple that an oral tradition, taught perhaps during ap-
prenticeship, would have sufficed. In my conjecture about how
the architect would have proceeded, I noted that the architect
would have had to know how to set out parallel and perpendicular
lines in order to establish his rectangles. The technique would
probably have consisted of applying a carpenter's square at the
appropriate places. The architect could have then measured off
the required distances. Thus, no appeal to a treatise on practical geometry was required to complete this part of the work.
Nonetheless, the architect may not have worked in the way I have
supposed. And even if he did, we cannot assume that he did not
know some treatise on practical geometry. If some more positive
conclusion is required, it would be that there were a variety of
instruments available for the "practice of geometry" in the Middle Ages; only some of them were related to the techniques of
artificial measurement as found in a practical geometry. In

this case, even promising material did not yield up the hoped-
for results.

Another architectural document of the High Middle Ages
suggests that connections between architecture and treatises
on practical geometry may have indeed existed. Villard de
Honnecourt, an architect who flourished around 1225-1250 in
Picardy has left us his sketchbook, a valuable testimony of
the practice of medieval architecture.[178] The work of a man
known as Magister 2, a close follower and compatriot of Villard, made important contributions to the sketchbook begun
by Villard. It is Magister 2 who wrote about the practice of
masonry and geometry and described some of Villard's machines
and inventions.[179] In a note accompanying the material on
practical architectural geometry, Magister 2 tells us that
"All these figures are taken from geometry."[180] One might
hope to find that "geometry" here means some treatise on
practical geometry. Others have assumed it did.[181] However,
the material was probably not taken directly from a practical
geometry.[182] It may, nonetheless, have some relation to the
tradition of practical geometry.

---

[178]Ed. Hahnloser, 1935. References to the plates will be
to Hahnloser's plate numbers, which are standard. A recent
reproduction of the plates renumbers them and makes no substantial editorial additions: ed. Bowie, 1959.

[179]The masonry and geometric techniques are found in plates
39-41, the inventions in plates 44-45. Hahnloser on pp. 195-
198 gives the few ideas we have of Magister 2's personality.

[180]Ed. Hahnloser, 1935: pl. 39 s) and p. 114: "Totes ces
figures sunt estraites de geometrie."

[181]Shelby, 1970: p. 399, n. 11 cites Hahnloser, 1935: pp.
196-197, 240-242, 254-259 and Victor Mortet, "La mesure de la
figure humaine et le canon des proportions d'après les dessins de Villard de Honnecourt, d'Albert Durer et de Léonard
de Vinci," Mélanges offerts à M. Emile Chatelain (Paris, 1910),
pp. 367-371 as having taken "geometrie" to mean a treatise on
practical geometry.

[182]Shelby considers the issue an important one too. He
argues that "geometrie" in this context means constructive
geometry or the manipulation of geometrical forms. See
Shelby, 1970: pp. 398-411, especially pp. 410-411.

If we look into what the figures "taken from geometry" contain,[183] we might discover what connection, if any, they have
with treatises on practical geometry. A wide variety of problems is contained in this section of the sketchbook. Each consists of a sketch plus a label. Neither part is rightly considered without its complement. In order to make sense of the
problems, we might divide them into several classes.

The first group includes problems which might have been taken
from a practical geometry. The point of these problems is to
measure an inaccessible object. Magister 2 gives us three of
these problems: (1) "This is how you find the height of a
tower." (2) "This is how you find the width of a river without
crossing it." (3) "This is how you find the width of a distant
window."[184] The first of these is perhaps the most common of
all problems in practical geometry. The accompanying diagram
shows a man sighting the top of a tower along the hypotenuse of
a right isosceles triangle. The height of the tower is the same
as the distance from the man's eye to the base of the tower.
Measuring the height of a tower occurs as a problem in virtually
every practical geometry.[185] Many different techniques are used
to solve the problem, but sighting along the hypotenuse of a triangle, especially a right isosceles triangle, is probably most
frequent.

A second type of problem presupposes knowledge of a technique
of geometry for the completion of a practical problem. Of this
sort is the method for finding the diameter of an engaged column.[186] The procedure is explained by a pair of drawings. The

---

[183] Ed. Hahnloser, 1935: pl. 39-41.

[184] Ed. Hahnloser, 1935: pl. 40 k) and pp. 119-120: "Pa[r]
chu prent om le hautece d'one toor." Plate 39 l) and pp. 108-
109: "Par chu prent om la largece d'one aive, sens paseir."
Plate 39 m) and pp. 109-110: "Par chu prent om la largece d'one
fenestre ki est lons."

[185] See the notes to ACC, II, 30-II, 33.

[186] Ed. Hahnloser, 1935: pl. 39 a) and p. 104: "Par cue
pren[t] um la grosse d'one colonbe que on ne voit mie tote."

first gives a method of transferring to a plane three points
taken from the surface of a column at the same horizontal
section. Then the second drawing is supposed to show how to
find the center from which to draw a circle passing through
those points.[187] The distance from the center to one of the
points is the diameter of the column. The picture given is
not an adequate explanation; the architect would also have to
know how to construct a circle to pass through three points.
An implication of a Euclidean theorem would give the construction,[188] but _Artis cuiuslibet consummatio_ has as one of its
propositions "Given three points not in a straight line, to
find the center."[189] Thus, a practical geometry completes
the solution.

The doubling of a circle will serve as an example of how
a theorem of practical geometry is simplified by Magister 2's
techniques. He explains how to "make two vessels so that one
holds twice as much as the other."[190] The accompanying picture shows a cylindrical object with a mason's square placed
against it so that both legs touch it. The circumference of
the larger vessel is traced by placing a pencil or other trac-
ing instrument in the corner of the square and, keeping the
legs touching the cylinder, rotating the square to trace out
the larger circle. The area of the large circle is twice
that of the smaller one. If two vessels of equal height have
these circles as bases, then the volume of one equals twice
the volume of the other. In effect, Magister 2's technique
amounts to finding a circle circumscribed around a square
tangent to the smaller circle. The method is simple and direct. _ACC_ gives computations in which the ratio of a square

---

[187]_Ibid._, pl. 39 b) and pp. 104-105: "Par chu trov'om le
point en mi on caupe a conpas."

[188]Euclid IV, 5 gives the construction of a circle circumscribing a triangle; the construction of a circle through
three points is mathematically identical.

[189]_ACC_, I, 25.

[190]Ed. Hahnloser, 1935: pl. 39 q) and pp. 112-113: "Par
chu fait om ii vassias, que li ons tient ii tans que li
autres."

to an inscribed circle and an inscribing circle are used.[191]
The results are only numerical; therefore, the ratio of the
areas of the circles is not pointed out.[192]

Most of the other problems in this section of the sketchbook
deal with questions of pure masonry or of building practice:
how to lay out a stone to fit a particular place or how to en-
sure that parts of the structure line up at the same level.
Interesting though these techniques may be, they do not seem
to touch on practical geometry to a degree that would merit
their discussion here.

The types of problems that Magister 2 confronted show that
he was familiar with a kind of geometry very similar to that
of a treatise on practical geometry. And of course he knew
architectural practice and its particular problems. The drawings, even if only a shorthand, show an understanding of practical application which surpasses that of the practical geometries. But his notes would not, in themselves, suffice to
teach an apprentice the geometric procedures of architecture;
an oral explanation and manual demonstration would have been
required.

The geometric problems in the sketchbook may very well have
been "taken from geometry." In that case, Magister 2's comment
probably means something very general, perhaps something like
"all these problems are geometric." They could have arisen from
a variety of sources, including an oral tradition of the architect's craft, a particular treatise on geometry, or general knowledge arising from a tradition of several treatises. Even if
identification of the source is not possible, the connections
with practical geometry are certainly implicit. In the most specific sense, Magister 2's notes and figures are practical geometry.

That we cannot demonstrate the historical patrimony of the
techniques of the sketchbook is perhaps no misfortune at all.

---

[191] ACC, I, 28 and 29.

[192] Problems concerning the ratios of volumes of cylindrical
vessels having lengths and diameters in various ratios are
given in ACC, III, 9-III, 11. The case given by Magister 2
is not among them.

What we have instead may be more valuable. In the sketchbook
we have evidence of a fully assimilated geometric tradition,
a suggestion of how geometry appeared after having been
learned and passed on by non-literary means. The fact that
certain techniques of the treatises on practical geometry appear in this written record of an oral tradition demonstrates
a connection, even if not a filiation. The traditions are
similar and parallel; the resemblance makes it difficult to
decide which is parent and which is offspring.

Even if I have been unable to show that the practical geometries were applied to practical concerns, the connections
between the practices of architecture and surveying and the
practical geometries are significant. The treatises may
never have been applied by architects and surveyors during
the Middle Ages; they were, however, written to appeal to
the practically oriented reader.

We know the treatises grew out of a practical tradition,
that of the _agrimensores_. That they continued to circulate
even and especially after a more complete geometry, that of
Euclid, was available must mean that they filled at least a
pedagogical need.

The inclusion of non-geometric material in some of the
practical geometries, especially in _ACC_ and the _Pratike_,[193]
also indicates a wide pedagogic scope. Our discussion of
the use of the practical geometries has been limited largely
to "geometric" crafts and professions--surveying, architecture, city planning. Once we consider the non-geometric
material--fractions, astronomy, and even mercantile practice--
included in some of the practical geometries,[194] our inter-

---

[193]E.g., material on astronomy, fractions, monies.

[194]Although among the works considered here only the _Pra-
tike_ contains mercantile problems, most of the Italian practical geometries of the Middle Ages and Renaissance are either
part of or appended to a mercantile treatise, a _libro d'abbaco_.
See A. Fanfani, 1951. Warren Van Egmond in 1976 completed a
catalog and description of many _libri d'abbaco_ as a doctoral
dissertation at Indiana University in the Department of History and Philosophy of Science.

pretation must become broader. The wide range of material in
the practical geometries is an indication that the treatises
were possibly intended for two groups. The first group is the
one considered up to now: those craftsmen who would make direct use of geometry in their work. The non-geometric material
might have been included because these users or their teachers
wanted an exposure to some of the other uses of mathematics.
The other group comprises non-professionals, people who would
have wanted to learn the rudiments of geometry as part of a general education, those who had no intention of becoming mathematical professionals. Thus the treatises may have been pedagogic
in the broader sense, that is they may have taught about geometry to all who might wish to learn it and not only to those
who would use it professionally.

Even if the practical geometries were used to teach geometry
to all comers, the practical feel of the treatises is still important. The practical examples from geometry and from the
other branches of mathematics found in the treatises indicate
strong concern for practicality and a belief in the pedagogic
value of the practical. The practical geometries were an effective and simple introduction to geometry. Their tangible
examples contributed to their effectiveness, an effectiveness
demonstrated by the age and continuity of the tradition.

INTRODUCTION

II.

THE CONTENTS OF ARTIS CUIUSLIBET CONSUMMATIO
AND THE PRATIKE DE GEOMETRIE

A general discussion of the contents of Artis cuiuslibet
consummatio and its French adaptation, the Pratike de geometrie can offer certain insights that the running commentary, accompanying the edition, cannot. The running commentary is analytic; its information is specific to the techniques and sources of individual propositions or parts of
propositions. Another approach will yield a broader view of
the treatises and their authors. Once the material in the
treatises is classified and considered in categories different from those of their presentation, some idea of the relation of the treatises to their sources and some notion of
the authors' originality may be had. The reason for using
another classification than that of the authors is that,
even in terms of medieval categories, a different ordering
is acceptable. Further, this other grouping reveals more
about the author's use of his sources.

The author of ACC divided his work into four sections:
measurement of areas, heights, volumes, and computation with
fractions. As we have seen, the first three are standard to
the tradition of practical geometry; including fractions
seems to be an innovation of the author.[1] Much of the section on altimetry relates more to astronomy than to geometry;
the part on fractional computations, minutimetria, belongs
to arithmetic. Thus, the contents of ACC may be divided
among geometry, arithmetic and astronomy. In other words,
all the branches of the quadrivium but music appear in the

---

[1]Above, p. 25.

74

treatise. The inclusion of arithmetic and astronomy is related,
to some extent at least, to the purposes of practical geometry.[2]
Only parts of the subjects of arithmetic and astronomy appear
in the treatise; it is not the author's intention to offer basic
instruction in the quadrivium. Nonetheless, division of the
propositions of ACC according to the branches of the quadrivium
is in accord with medieval classifications of the sciences.

The propositions that appear most strictly geometric are those
dealing with measurement. Following a distinction of other practical geometries, they fall into two types, artificial and non-
artificial measurement.[3] Artificial measurement requires the
use of some instrument in measuring some object. Our concern
here is with the measurement of terrestrial objects, since we
are treating the geometric problems of ACC.

There are six propositions in the treatise that belong to artificial measure in geometry; they all deal with measuring
lengths. All but one of them appear in the section on altimetry;
the lengths they measure are disposed vertically.[4] The remaining
proposition uses an instrument to measure the length of a horizontal line. It is the first problem in the section on planimetry.[5] Its inclusion there gives an interesting bit of information on what the author conceives plane measure to be; it is the
measurement of things disposed horizontally. In other words,
not only dimension but also position affect how the author class-
ifies a measurement.

Another group of propositions, not strictly part of artificial
measurement, is important to its purposes. These are the propositions explaining how to make certain instruments.[6] Two of the

---

[2]See above, pp. 25-26.

[3]The distinction appears in Geometrie due sunt partes principales; see pp. 19-21, above.

[4]ACC, II, 30-34.

[5]ACC, I, 1.

[6]ACC, II, 36-38. I have not included those instruments of a
temporary nature.

three instruments could be used for measuring lengths or
heights in both astronomy or geometry.[7] The other is purely
astronomical.[8]

The largest number of propositions in ACC explain non-
artificial measurement in geometry. In fact, there are so
many that a further division into types seems justified.
The first type of non-artificial measure includes procedures
for computing lengths, areas, and volumes. The other types
are also computational, but result in numbers, not measures.
These last propositions of non-artificial measure concern
ratios of areas and of volumes and the number of areas of
one sort that fit into an area of another sort.

The computations of lengths are related to the material
on areas; they make up part of the section on planimetry. In
this category we find how to determine the lengths of altitudes and diagonals of various straight-sided figures and of
the diameter and circumference of circles. The altitudes
are treated as part of the procedure for finding the area of
a triangle or simply as an investigation of one of the properties of a figure.[9] Other propositions give methods for
finding the diagonal of a rectangle or the diameter and circumference of a circle.[10]

Areas are the greatest concern of the author in non-
artificial measure. Not only are the areas of a great variety of figures discussed, but the material on lengths is related to considerations of area. In addition, there is ma-

---

[7]ACC, II, 36 and 37 describe how to add shadow squares
to an astrolabe and a quadrant.

[8]ACC, II, 38 explains how to make a measuring scale for
use with a gnomon.

[9]ACC, I, 4 computes the altitude of a triangle auxiliary
to finding its area. I, 11 finds the altitude of half a
rhombus, again as an explanation of the area.

[10]ACC, I, 14 gives the diagonal of a rectangle. I, 6 computes the diameter of a circle inscribed in a right triangle;
I, 21 gives the circumference of a circle in terms of its
diameter.

terial on the ratios and division of areas. Areas of an as-
sortment of triangles and quadrilaterals comprise much of the
section on planimetry.[11] The determination of the area of a
circle is also given.[12] The author also includes material on
polygonal numbers as if they were areas.[13] The propositions
dealing with what I have called ratios of areas describe the
effect of doubling the diameter of a circle on its area,[14]
the method of finding a square with an area equal to that of a
circle and vice versa,[15] and the difference between the area of
a circle and a square inscribing it or inscribed by it.[16] The
remaining area problems give the number of rectangles that will
fit into areas of a variety of shapes.[17]

Volumes are treated almost exclusively in the section on
solid measure, crassimetria.[18] Methods for computing the volumes or capacities of different solids are treated, as are
ratios of some solids. Volumes are computed for spheres, cubes,
cylinders, cones and their frusta, and rectangular solids.[19]

---

[11]Triangles are treated in ACC, I, 2-5, 7 and 8; quadrilaterals in I, 9-13, 32 and 33.

[12]ACC, I, 22 computes the area of a circle; I, 23 gives the
procedure for finding the surface of a sphere.

[13]ACC, I, 15-20 will be discussed with arithmetic, below.

[14]ACC, I, 24.

[15]ACC, I, 26 and 27.

[16]ACC, I, 28-30.

[17]A larger rectangle in ACC, I, 31; a trapezoid in I, 34;
a triangle in I, 35; and a circle in I, 36.

[18]The single exception is ACC, II, 35 which shows how to
find the capacity of a well. It occurs just after a proposition on measuring the depth of a well.

[19]Spheres are computed in ACC, III, 3 and 4; a cube in III,
1; a cylinder in III, 2; a cone in III, 6; the frustum of a
cone in III, 19; a solid made of two frusta in III, 18; and a
rectangular solid in III, 12. In addition, some of the same
solids appear in everyday dress as a barrel in III, 7; a tun
in III, 8; a well in II, 35; and a chest in III, 13.

Just as in the section on plane measure, the relation between
changes in various dimensions and the corresponding changes
in volume are treated.[20] Pyramidal numbers are treated as if
they were volumes.[21]

Measurement, then, is the primary concern of the section
on geometry. Most of the measurement is computational rather
than artificial. Euclidean geometry contains many other kinds
of material that ACC and most of the practical geometries
omit. In the realm of plane geometry, for example, the Euclidean doctrine of congruent triangles is very important.
However, it does not appear in the practical geometries. As
pointed out in the discussion of the nature of practical geometry, measurement is the defining characteristic of the subject.[22] The definition is borne out in the geometric material of ACC, almost all of which concerns measurement, either
artificial or non-artificial.

The distinction between artificial and non-artificial measurement seems to apply to the astronomical propositions of
ACC as well. To use this division of practical geometry in
astronomy is, of course, to extend its domain. However, a
justification for the broader use is found in a standard
medieval description of the quadrivium. Both geometry and
astronomy have as their subject continuous magnitude; the
difference between them is that geometry studies magnitude
at rest while astronomy deals with magnitude in motion.
Once the astronomical contents of ACC have been considered,
how well the distinction applies may be seen.

---

[20]ACC, III, 9-11 discuss changes in the volume of a tun
resulting from changes in diameter and length. III, 14 shows
the effect of doubling the sides of a chest. III, 15 and 16
convert the volume of a square chest to a tun and conversely.

[21]ACC, III, 20 gives a method for finding a pyramidal number having a triangular base; III, 21 has a square number as
base. The propositions will be considered along with the
other arithmetic material.

[22]Above, p. 30.

Measurement in astronomy is somewhat different from that in geometry. Astronomy needs to know the angles between lines sighting heavenly bodies; distances are less important, at least to medieval astronomy. Thus, most of the astronomical propositions of ACC involve measuring angles. The direct measurement of angles using instruments of some kind is what I have decided to call artificial measurement in astronomy. The astronomical analogue to finding areas and volumes is computing angles from other measurements.

Finding altitudes is one of the uses of artificial measurement in astronomy. Indeed, finding altitudes is probably the author's justification for including the astronomical material in the section on altimetry.[23] ACC explains how to measure the altitude of the sun, a star, the sun at equinox and the pole with various instruments that give the shadow of the altitude.[24] A rather simple instrument is used to determine which way is south.[25] All the rest of the artificial measurements are performed using an astrolabe. The astrolabe is called on in measurements of the time of day, solar and planetary longitude, the direction and latitude of a planet and the degree of a star.[26]

Determining angles is also the purpose of the computational or non-artificial part of astronomical measurement. Computations based on shadows or angles are used to derive other angles. A

---

[23]See above, pp. 25-26.

[24]ACC, II, 1 uses the shadow found, I imagine, by means of a gnomon to compute the altitude of the sun. Thus, II, 1 is both artificial and computational. II, 2 explains a method of using a pair of sticks to find a stellar altitude. The equinoctial altitude of the sun is determined using a gnomon in part of II, 5. The pair of sticks is used to measure the altitude of the pole in II, 8.

[25]The instrument described in ACC, II, 3 consists of a circle and a gnomon.

[26]ACC, II, 13 explains how to find the time. Solar longitude is found in II, 19. Planetary longitude is determined in II, 22. The direction of a planet is given by II, 23. Planetary latitude is measured in II, 24. The method for finding the degree of a star is taught in II, 21.

few of the problems amount to what may be called plane trigonometry; others require merely an addition or subtraction.
Still others deal with results related to spherical trigonometry.

Plane trigonometry is involved in the computation of an
altitude from a shadow and vice versa.[27] The time of day
is derived approximately by means of trigonometric relations.[28] Simple arithmetic is performed on angles to find
the altitude of the sun at equinox and solstice, to determine the latitude of a region from the equinoctial altitude
of the sun, to deduce the zenith distance of the pole, and
to derive the ascendent from the altitude.[29] Right and
oblique ascensions, important parameters in spherical
astronomy, are also found.[30]

The author also offers brief discussions of some other
astronomical material in the section of ACC dealing with
altimetry.[31] The purpose of this last group of propositions
is not obvious; it may have been included to complete the
discussion of astronomy. In any case, as it is presented in
ACC, it has little relation to the measurement of heights or
angles and therefore little relation to practical geometry,
even in the larger sense of ACC.

---

[27]ACC, II, 1 computes an altitude from the shadow; II, 14
performs the inverse operation.

[28]ACC, II, 11 uses the sine of the altitude of the sun to
establish the time.

[29]Equinoctial altitude is found in ACC, II, 5; solsticial
altitude in II, 6; latitude in II, 7; zenith distance in II,
10; and ascendent in II, 12.

[30]Right ascensions are treated in ACC, II, 15-17; oblique
ascensions in II, 18.

[31]Declination of the sun and stars appears in II, 20;
planetary equations in II, 25; planetary apogees in II, 26;
the length of the year in II, 27; the nodes of the planets
in II, 28; and seasonal variations in II, 29.

Since some of the astronomical contents and a large part of
the geometry of ACC use computation to arrive at measurements,
it is not surprising to find a part of ACC devoted purely to
techniques of computation. The arithmetic in ACC appears to
have an ancillary role; it is useful for the computations needed
in the geometric and astronomical sections, especially in their
non-artificial or computational aspects. The division into artificial and non-artificial branches does not apply to arithmetic
because arithmetic has little to do with measure. Another division must be found for the part of ACC that deals with arithmetic.

.The author of ACC separates the parts of the section on fractional computation, minutimetria, in a way that will be of use
to us. He divides the section on fractions into parts on common fractions and sexagesimal fractions--geometric and astronomical minutes, as he calls them.[32] The whole section also
contains another division: into multiplication, division, and
the extraction of roots.[33] The part on roots splits into sections on cube and square roots, each treating integral and fractional roots.[34] Most of minutimetria is concerned with the extraction of roots, but it begins with a few propositions on
simple computation with fractions.[35] The first proposition in
this group is also metrological, perhaps providing a justification for treating common fractions in this practical geometry.[36]
The actual extractions of cube and square roots make up only a
small part of the remaining propositions.[37] The rest of them

---

[32]ACC, IV, 1.1-2 has a statement of the author's intent in
the chapter.

[33]See ACC, IV, 4.1-4.

[34]See ACC, IV, 9.1-4; IV, 12.1-2; and IV, 15.1-2.

[35]ACC, IV, 1-3 treat aliquot and "whole" parts; IV, 4-8 deal
with products and quotients of sexagesimal fractions.

[36]ACC, IV, 1 explains that units of measure are to be treated
as numbers and square units as square numbers.

[37]Cube roots are taught in ACC, IV, 12 and 13; square roots
in IV, 20 and 26-28.

give identities determining the kinds of quantity that will
have roots, and procedures for converting numbers to a kind
having a root.[38] Part of some of the techniques for finding
square roots consists of converting the root found in deci-
mals, either in part or wholly, into sexagesimal fractions.

The other material falling into the category of arithmetic is scattered through three of the parts of ACC. Number
theory is perhaps a better title for this material than
arithmetic. Even though it is presented as treating areas
and volumes, its subject is really figurate numbers, that
is, polygonal and pyramidal numbers.[39] Only in the section
on roots are the figurate numbers treated as numbers.[40]

The catalog of the contents of ACC has classified its
materials following a scheme legitimate to the High Middle
Ages, even though different from that of the author of the
treatise. Having done so, we have a better idea of what
kinds of problems the author included in ACC. How the different types of material are related to one another has been
explained, at least in a general way. Thus, we have a partial explanation of why this somewhat curious mixture of material occurs in ACC.

Some of the author's reasons for including material from
outside the tradition of practical geometry may relate to
his particular sources. A summary of these sources should
thus illuminate his way of using sources and give some indication of what was available to him. As a result of these

---

[38]ACC, IV, 14-19 and 21-24 explain which forms have which
kind of root. Few of the identities given in these propositions are interesting; most are trivial. IV, 25 converts
fractions with odd sexagesimal powers into ones with even
powers.

[39]Polygonal numbers appear in ACC, I, 15-20; pyramidal
numbers in III, 20 and 21.

[40]ACC, IV, 9-11 deal with producing square numbers from
triangular, square and pentagonal numbers. The same propositions explain how to find the side of a given polygonal
number.

considerations, we may come to evaluate the extent of the author's originality and his relation to his intellectual milieu.

The sources of ACC belong to several traditions. Rather
different materials are used for the geometric, astronomic,
and arithmetic sections of ACC. As we have noted, material
from the Roman agrimensores is at the origin of the tradition
of practical geometry.[41] Medieval geometries growing out of
the agrimensores texts are the primary source of much of the
geometric contents of ACC. The author gives numerous problems
almost certainly derived from the geometries falsely attributed
to Boethius and Gerbert as well as from genuine Gerbertian material.[42] The greater part of these works is based on the material of the agrimensores. In his use of these early medieval traditions, the author is rather conservative. Only occasionally
does he appeal to Euclid, whose works were readily available at
the time. On the other hand, if he had strayed too far from the
early geometric traditions, his work might have ceased to be a
practical geometry.

Even where he has borrowed from earlier material, the author
makes his own contributions. He usually adds proofs to the propositions he borrows from his sources.[43] At times, rather than
giving the proof in detail, he simply indicates how to proceed.
Several of the propositions, the sources of which are not apparent, seem to be the author's own. They are primarily propositions dealing with ratios of areas and volumes. They are not
difficult conceptually, but they provide practice in non-
artificial measurement. Thus, in spite of the author's extensive
use of his sources in the geometric material, much of it is ori-

---

[41]See above, pp. 5-6, 64.

[42]Because the sources have been indicated as far as possible
in the running commentary to ACC, I will not cite specific propositions in discussing sources. It should be enough to refer
to the types of material, as specified in the general discussion
of contents, above. References to specific propositions are
given there.

[43]For the proofs of geometric propositions he relies largely
on two techniques: the creation of a rectangle double the area
of a triangle and proof "by argument from number." See above,
pp. 45-46.

ginal. Indeed, he added enough to the pastiche of borrowed
material to make the geometric part his own creation.

The reliance on sources may be more thorough in the astronomical and arithmetical contents. However, I have not been
able to identify the author's sources for those sections with
much precision at all. The sources for the geometric material were readily apparent; at times I was even able to determine which family of manuscripts the author had at hand.[44]
The astronomical and arithmetical texts that might have been
the sources of ACC are much more diffuse. There seem to be
many more possible sources, few of which have been edited in
modern times. Thus, the closest I can come to identifying
the texts on which ACC was based is to determine to which
traditions the materials belong.

The two types of astronomical measurement in ACC, artificial and non-artificial, appear to have as sources two different kinds of treatises. The large part of the artificial
measurements of ACC are performed using an astrolabe. All
the propositions using astronomical instruments are analogous
to propositions in some treatise on the astrolabe. Thus, it
is very likely that a work on the astrolabe was the source
of ACC's material on artificial measurement. We have not
established which of the many astrolabe tracts served as the
author's source, but that is not absolutely necessary. The
treatises on the use of the astrolabe are derived from Arabic
sources. They all seem to cover the same topics, many in very
similar ways, making up what could be called a single tradition. Thus, the tradition of works on the astrolabe may be
called the source of that part of ACC that treats artificial
measurement in astronomy.

The non-artificial measurement in the astronomical part of
ACC is concerned with computing a variety of astronomical
parameters. Almost all of the propositions of this type are
similar to theorems found in any of a large number of astronomical works. Again, it has been impossible for me to tell

---

[44]For example, see ACC, II, 34.

which of the many analogous propositions might have been used
by the author of ACC. Here too, as for the material on artificial measurement, one type of work encompasses nearly all
the propositions on astronomical computation. It is apparent
that these propositions derive from some Arabic source, at
either first hand, or in some more westernized form. The technical terms from Arabic astronomy reveal the Arabic rather
than classic origin of this section.[45] There is an Arabic tradition in which the method of computing the various parameters
is explained. That tradition comprises the commentaries on a
set of astronomical tables. Such a commentary might give the
methods used to compute the values appearing in it, as well as
instruction in using the tables.[46]

In order to understand the astronomical computations explained in ACC, I turned to some of the available commentaries
on astronomical tables. Propositions analogous to those of
ACC appeared with such frequency that it was impossible to decide whether any one of them was an immediate source. Even the
particular values for some of the astronomical constants of ACC
did not permit a determination. The constants were similar to
those of several of the commentaries but identical with none of
them. Once again, then, I can identify the source of the material in ACC only as being a member of a tradition, that of the
commentaries on astronomical tables.

The author of ACC seems to have used his source rather uncritically. Several of the computations of astronomical parameters make almost no sense at all. Sometimes he has given only
the numerical procedure for a computation with no reasons for
the procedure.[47] I am tempted to assume that the author sometimes did not understand the material he borrowed from some commentary on astronomical tables. The author's material on non-
artificial measurement in astronomy does not readily reveal a

---

[45]For example, altifa in ACC, II, 1; azimuth and almucantarath in II, 9; and genzahar in II, 28.

[46]See Kennedy, 1956.

[47]See ACC, II, 18, for example.

principle of choice. That is, much of the astronomical content of <u>ACC</u> seems to have little relation to measurement.
The choice of material may have been dictated by the source.
The author seems to have taken over the large part of some
commentary on astronomical tables, adapting it where he could
but transmitting some material he did not understand. The
unassimilated material in <u>ACC</u> reveals both that the author
failed to understand the subtleties of astronomical computation and that he did not have a very original contribution to
make.

Thus the two kinds of astronomical material in <u>ACC</u> arose
from two different traditions, that of the astrolabe treatises and that of the astronomical table commentaries. There
was greater justification for including the astrolabe material: it is close to the tradition of artificial measurement.
The rest of the astronomical content is somewhat anomalous.
Its connections to practical geometry are so tenuous that
even the author of <u>ACC</u> was not able to assimilate it.

The main source of the arithmetic part of <u>ACC</u> was probably
one of the treatises on practical arithmetic. In fact, much
of the fractional computation is quite close to that in John
of Seville's <u>Liber</u> <u>algorismi</u> <u>de</u> <u>practica</u> <u>arismetrice</u>.[48] Even
the specific numerical examples of <u>ACC</u> are often the same as
those of John of Seville. The title of John's work reveals
its Arabic origin, <u>algorismi</u> being a latinized form of al-
Khwarizmi. If John's work was not the source of the section
on fractional computation, then some other treatise belonging
to the same tradition certainly was.

The author of <u>ACC</u> may have been somewhat original in the
section on fractional arithmetic. He even indicates that
some of the material is his own, making a plea for the reader's approbation.[49] The author seems to have fully assimilated the material he borrowed. That is, he understood it
well enough to give examples and elaborations of his own.
Thus, his claim to originality may be believed.

---

[48]Ed. Boncompagni, 1857.

[49]<u>ACC</u>, IV, 14.39-41.

The other sort of arithmetic present in ACC is what I have
called number theory because its subject is figurate numbers.
This material all could have come from some of the early medieval treatises on geometry, perhaps those attributed to Boethius or Gerbert. Thus, it belongs to the same tradition as the
geometric part of ACC. It is really part of number theory, but
historically it belongs to the geometric tradition.

In sum then, ACC was more original in some parts than in
others. The geometric material arose out of what is certainly
the most traditional source of the contents of practical geometry. Perhaps the author's greater familiarity with the early
medieval geometric traditions led him to make his most original
contributions in the geometric parts of ACC. The inclusion of
material on arithmetic and astronomy was a departure from the
tradition of practical geometry. That material, however, was
not absorbed in an original way. The section on arithmetic
shows a little originality, but much of the astronomical material
was scarcely understood. In other words, the author found himself overextended where he tried to include new material. His
originality had greatest play in his own subject of practical
geometry, where the sources were the most traditional.

The contents of the Pratike de geometrie deserve separate discussion. Most of the comments that apply to Books I and III of
ACC also apply to their translation in the Pratike. Special mention must be made of the rest of the treatise. What I have
labeled Book V of the Pratike contains the geometric additions
of the Pratike. A small part deals with areas.[50] However, the
majority of the propositions concern themselves with measuring
the volume of a barrel.[51] One proposition deals with length,
computing the diagonal of a known square.[52] The geometric part

---

[50]Pratike, V, 1 treats squares and triangles circumscribed
about and inscribed in circles (the triangle having as its side
the side of the square), and gives the areas of each part of
each figure. Propositions V, 9 and 10 seem to deal with areas.

[51]Pratike, V, 2-8. In all of these propositions, the volume
of one particular barrel is computed in a variety of ways.

[52]Pratike, V, 11.

of the Pratike seems a miscellany. It does function as a
supplement to ACC; its materials are an extension of the concerns of the practical geometry in ACC.

The extension, however, is an odd one. The author does not
enlarge the scope of the treatise to cover new figures or more
general techniques of calculation. Instead, his additions
provide greater specificity. That is, he offers particular
techniques for computing volumes and areas; his additions
deal with individual objects. For example a variety of computations are made of the volume of a certain barrel, but no
clear general technique is enunciated. Furthermore no proof
is given. In other words, the simple proofs that are one of
the hallmarks of ACC do not occur in the French additions.
The author of these additions seems to have felt that repeat-
ing a calculation in slightly different forms is an effective
pedagogic method--perhaps more effective than offering a
proof or justification of a procedure.

The rest of the Pratike, edited as Book VI, may be called
mercantile arithmetic. By that I mean that the added material
seems to be of utility to merchants and traders. Its concerns may be divided into two main areas: arithmetic of integers and calculations of monies. The arithmetical section[53] deals only with multiplication of integers and largely
with their squares. A brief discussion of the positional
notation of Arabic numerals completes the treatment of arithmetic;[54] positional notation is explained through multiplica-
tion--multiplying by ten moves the number over one position.

The arithmetic in the Pratike is extremely elementary and
even superfluous. A reader could hardly have made his way
through the area and volume calculations of ACC or the Pratike without knowing how to multiply. It is difficult to determine the author's motivations in appending such elementary
material to a reasonably sophisticated treatise. Were the

---

[53]Pratike, VI, 7 and 8.

[54]Pratike, VI, 8.

additions found at the beginning,[55] one could claim they func-
tioned as an introduction to or review of arithmetic. But
since they appear in the middle of the section on calculating
monies, we must conclude their inclusion is largely by accident.
It seems that an odd folio of arithmetic worked its way into a
series of mercantile problems to be copied over with the mercantile additions, thereby becoming part of the Pratike.

The remainder of the Pratike, the part of Book VI that does
not deal with the arithmetic of integers, explains various calculations with monies. These are of several different types,
but they all presume coinages based on the old Roman system in
which the pound, libra or libre, is divided into 20 shillings,
solidi or sous, each of which is made up of 12 pence, denarii
or deniers. The denier is further divided into fractional parts;
one-half a denier is an obole; and half an obole is a pugoise,
poitevine, or pite.

The monetary problems deal with exchange rates, alloys or a
combination of both. Many of them take the form in which the
value of pure metal is given (in sous per marc or per once of
gold or silver) and the desired result is either the value of
a coin having a certain proportion of the pure metal or the
value of a marc of bullion with a given part of the metal. In
a sense, this book consists of a series of problems in converting units--the units being money, metal, or weights of commodities. The proportion of fine gold or silver to alloy is given
in terms of deniers (or parts of a denier) per sou; here the sou
functions as a whole of which the denier is one-twelfth. In
other words, these calculations are made in terms of fractions
having a denominator of twelve. The clearest illustration of
these fractions of twelve is the table of coinage rates in Pratike, VI, 6. That table is a list of various coinages in terms
of their proportion of silver or gold. Other fractions also
occur in the problems because the weights of coins are given as

---

[55]The Pratike is preceded in both manuscripts by an arithmetic, "Le plus ancien traité français d'algorisme," ed. Mortet,
1909.

parts of an ounce, because the _livre_ contains 20 _sous_, and
because the _marc_ may be a variety of sizes: 4 s., 5 s., 6 s.
8 d., 10 s., or 13 s. 4 d.

Since Book VI contains so much material on fractions, it
may be seen as a treatise on fractional arithmetic. As a
work on fractions, it at least partially fulfills the goal
of the third part of the _Pratike_. As specified in the introduction, the third part is supposed to teach "a trover les
minuces de gyometrie et d'astronomie covignable as ii parties
devant."[56]

Although the problems on coinages manage to treat fractions, the fractions are common fractions or "minuces de
gyometrie." Astronomical or sexagesimal fractions are omitted from the _Pratike_. Both common and sexagesimal fractions
are, however, covered in the original Latin treatise.[57] One
is left to wonder why the _Pratike_ substituted for _ACC_'s more
general treatment of fractions the specific monetary fractions of the French version. A possible reason is that the
money problems are more concrete and therefore pedagogically
more effective than the treatment in the Latin. But for the
problems to be more concrete and more easily accessible to
the reader, the presumed reader must have had a wide experi-
ence with diverse coinages, exchange rates and monetary matters, or at the very least, a desire to know how to handle
commercial matters. Furthermore, the author of the French
adaptation must have had a group of people familiar with commercial matters in mind as a possible audience for his book.
This commercial or mercantile group for whom the book was
written would presumably benefit from an extensive mathematical treatment of exchange rates and other monetary problems.

As the adaptation is written in Picard dialect, the region
of its author and his audience is reasonably well circumscribed. The region of Picard dialect, comprising parts of
what are now Northern France and Belgium, was an important

---

[56]_Pratike_, I, 0. 7-8.

[57]_ACC_ IV. Also see pp. 81-82.

trading and commercial area in the thirteenth century, when
the Pratike was written. It seems very likely that the parts
of the Pratike dealing with monetary problems were intended
for the commercial class of this region. It is also probable
that the entire Pratike was meant as a textbook in mathematics for would-be merchants. If so, the treatise is a valuable
resource for the content of mercantile education in this rather
early period.[58]

---

[58]Pirenne, 1929, establishes the existence of lay schools
for merchants in the adjacent region of Flanders at this time.
Pirenne does mention instruction in commercial correspondence
but does not deal with mathematical instruction. Fanfani, 1951,
treats primarily the mathematical education of merchants, but
he confines his discussion to the libri d'abbaco of the Italian
tradition.

III.

PROCEDURES IN THE EDITIONS, TRANSLATION, AND COMMENTARY

A few symbols have been used in the edition of <u>Artis</u> <u>cuius-</u>
<u>libet</u> <u>consummatio</u> and its French adaption, the <u>Pratike</u> <u>de</u> <u>geo-</u>
<u>metrie</u>. Square brackets [ ] contain material added by the
editor as an aid in understanding the text. The material
within angle brackets < > was probably present in the archetype but is not extant in any of the manuscripts used in the
edition. The variants are listed after a line number, referring to the text on the same page. In addition, the word or
words to which the variant applies, or the lemma, is always
indicated. Where the lemma is ambivalent, that is, if the
word appears more than once in the given line, a number in
parentheses ( ) indicates which occurrence is meant. A question mark indicates that I am unsure of the reading. The
following are abbreviations for verbs of which the subject is
that manuscript or those manuscripts listed in the sigla immediately following the abbreviation:

<u>add</u>.	addidit, addiderunt
<u>corr</u>.	correxit, correxerunt
<u>hab</u>.	habuit, habuerunt
<u>iter</u>.	iteravit, iteraverunt
<u>om</u>.	omisit, omiserunt
<u>scr</u>.	scripsit, scripserunt
<u>tr</u>.	transposuit, transposuerunt

A few other abbreviations are used:

<u>ed</u>.	editore
<u>mg</u>.	margine
<u>lac</u>.	lacuna

Editing Artis cuiuslibet consummatio

Before editing the text, I established a stemma of the known
manuscripts of ACC. Having a stemma offers distinct advantages
to the editor of a text. He need not confuse the reader with
superfluous variants when it is the original text which is of
primary concern. The stemma permits a more objective decision
about which readings to use in editing a text. My text based
on four manuscripts of ACC is probably closer to the original
than a text based on all the manuscripts would have been, since
the stemma effectively eliminates the need for consulting deri-
vative manuscripts.

In what follows, this stemma is justified:[1]

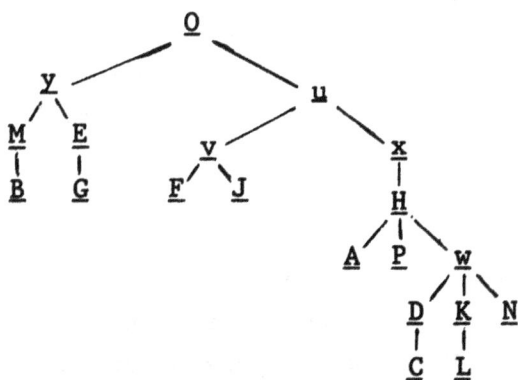

Almost all of the examples used to justify the stemma have
been taken from Part I; there is no reason to believe that the
later parts would invalidate this stemma. Whenever a whole
group is cited by its siglum, the reference is to those members
of the group containing the line in question, as indicated in
the table below. A simple 0 means that the proposition in question is missing; the annotations on the other 0's explain either
that part of a proposition or that a series of propositions is
lacking. A dash between two proposition numbers stands for a
part of the table left out in which there are no changes in
which manuscripts are missing propositions. Manuscript N contains only propositions I, 1-I, 4 and therefore is excluded
from this table.

---

[1]The manuscripts are identified by their sigla, given in
the list on pp. 105-106, below.

Table of missing sections of Part I

Propo-sitions	Manuscripts											
	M	B	E	G	F	H	A	L	K	D	C	J	P
I, 0	0^a												
--													
3	0^a												
--													
9										0	0		
--													
20			0^b			0^c	0^c	0^b	0^b	0^c	0^c	0^c	0^c
--													
29						0^d	0^d	0^d	0^d	0^d	0^d	0^d	0^d
--													
33											0		
--													
36					0	0	0	0	0	0	0	0	0

[a] I, 0 - I, 3.6 om.M   [b] I, 20.1-4 om.ELK

[c] I, 20.1-3 om.HAPDCJ   [d] I, 29.1 om.HAPLKDCJ

In the critical apparatus to the text, the variants are negative, that is, only the deviations from the adopted read-ing are noted. Here, the apparatus gives the sigla with the readings for the various traditions. Minor variants are in-dicated within parentheses. I have used x to refer to the group of manuscripts ACDHKLNP, v for FJ, y for BEGM. To shorten reference to the two groups v and x, I have adapted the siglum u and have used w to represent KLDCN.

An archetype 0 separate from any of the known manuscripts may have existed. If so all the manuscripts may be shown to derive from it, since all copies of the treatise contain the following errors with respect to my hypothetical archetype:

```
II, 32.25 transiens kh in n 0 : om.xy (II, 32 om.FLJ)
II, 35.5 altitudinem 0 : latitudinem xy (II, 35 om.FLJ)
IV, 20.21 deduplentur 0 : duplentur xy (IV, 20 om.MEGFL)
```

The two groups y and u seem to have their origin in the
archetype O, schematically

Group u has a different ordering of the four parts of the treatise from the one adopted in the edition; it puts the part on
altimetry, book II in my numbering, last, after the parts on
volume and fractional computation. Thus u may be said to have
the order I, III, IV, II. The description of the treatise takes
account of the alternate ordering.[2] These two different order-
ings provide a simple means of differentiating the two groups.
There are also differences in readings to characterize them,
as follow:

I, 0.8 fructus uberrimos ubi seminaverunt non metentes y :
      om.u
I, 0.12 scil. radicem et originem ortum y : om.u
I, 0.21 iocundum u(?J) : om.y
I, 0.27 in tercia capacitates (om.EG) corporum et (om.EG)
      crassitudines invenire docemus (docebimus EG) y :
      in tercia geometricas et (vel LKDC) astronomicas (et
      astronomicas om.J) minutias ad predicta necessarias
      docere promittimus u
I, 0.28 in quarta geometricas et astronomicas minutias
      ad predicta necessarias docere promittimus (promitibus
      G) y : in quarta altimetriam mensurare alta (alia C)
      docebimus u
I, 1.22 patet y : videbis u
I, 1.37-39 sed nota . . . be ad ba y : om.u
I, 2.55 quod est contra premissa y : quod inprobatum est
      u
I, 4.2 triangulum y : om.u
I, 4.9 sic (om.G) y : per hanc regulam u
I, 7.1 reperire y : invenire u
I, 11.12 96 . . . propositiones y : om.u
I, 12.3 reliquum y : aliud u
I, 13.1 cognoscere y : rimari u
I, 15.2 perquirere y : reperire u
I, 15.4-5 a summa excrescente subtrahatur unum latus semel
      y : et summe (summeque LK) excrescenti unum latus semel
      subtrahatur u
I, 15.7-8 quodlibet latus sit (fit M) 6 pedum y : 7 pedum
      est (sit LK) quodlibet latus u
I, 16.3 ab illa summa y : illi summe u
I, 23.2 multiplicetur y : feriatur u

---

[2]See ACC I, 0.26-30 and variants.

I, 25.13 a centro, in intersectione scil. y : ab in-
    tersecatione u
I, 36.1-6 civitātis . . . probatum est y : om.u

The group x is really a separate redaction of the treatise, based on u with all of the manuscripts deriving from H.
The author or scribe of H made a number of changes from the
original. He was a good mathematician and corrected some of
the contents of the treatise. Since his corrections remove
the text from its original form, such changes are put in the
variants but are not adopted in the reconstructed text.[3] The
remainder of group u, the unrevised part, will be known as
group v. Therefore we have

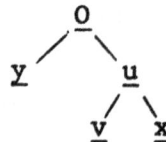

$$
\begin{array}{c}
O \\
\diagup \quad \diagdown \\
y \qquad u \\
\diagup \quad \diagdown \\
v \qquad x
\end{array}
$$

Among the revisions made in x are the following:
    I, 1.39-41 sed nota . . . longitudinem vy : om.x
    I, 3.7 post equidistantibus add. catheto et basi x :
        om.vy
    I, 14.4-5 80 . . . 60 . . . 100 vy : 8 . . . 6 . . .
        10 x
    I, 22.35-36 aream circuli invenire vy : aree (add.
        scil. LK) que circuli est invenire quantitatem x
    I, 34.6-7 iunctarum latitudinum medietas dividatur
        per 30 vy : medietas latitudinum (latitudine DC)
        per 30 x

The following examples show that F and J derive independently from v, or that their relationship is

$$
\begin{array}{c}
v \\
\diagup \quad \diagdown \\
F \qquad J
\end{array}\ .
$$

Note that at times F reads with y and x against J and at times
it is J that reads with the main traditions while F does not.
    I, 0.2 consummatio yxF : consideratio J
    I, 0.5 rerum subtilium yxF : om.J
    I, 1.4-5 respiciat . . . orthogonio yxF (respiciat
        om.x) : om.J
    I, 1.41 quadrantem yxJ : quantitatem F
    I, 2.2-3 superficialem yxJ : om.F

---

[3]Reference to the source of some of the propositions shows
that my reconstructed text is close to the original and that
the mathematical improvements are part of a later redaction.
See ACC I, 34, n. 95.

I, 2.3 invenies yxF : reperies J
I, 2.16 dividatur equales y : equales dividatur xJ
   dividatur F
I, 2.29 fit ex yxJ : om.F
I, 2.31 trianguli yxJ : om.F
I, 2.33-34 caput . . . ad yxF : om.J
I, 2.49 est propositum yxJ : erat probandum F
I, 2.63-64 et est . . . est yxF : om.J

The main group y can be divided into two traditions MB and

EG, or

$$\begin{array}{c} y \\ \diagup \diagdown \\ MB \quad EG \end{array}$$

Some of their characteristics are:

I, 0.25 instruimus, superficierum quantitates xB : et
   superficierum quantitates instruemus EG
I, 1.9-10 unus magnus et unus (alius F) parvus xvB :
   om.EG
I, 9.1 sequitur ut doceamus de quadrangulis xMB : om.EvG
I, 23.5 metiatur xMB : ducatur per EG
I, 25.4 circini xMB : om.EG
I, 25.5 super b x : super a et b vMB : ad mediatatem
   spatii b EG
I, 25.5-11 fiat quedam sectio . . . erit in intersectione
   xMB : om.EG
I, 25.11 (note) in b intersecantes . . . posito pede
   add.EG : om.MB
I, 30.5-6 et circuli . . . minoris quadrati xMB : om.EG
I, 30.7 ad minus xEG : ad ipsum MB / duplum ad ipsum EG :
   duplum ad minus MB
I, 32.1 superficiem invenire xMB : noti capacitatem
   arpennorum concludere EG
I, 35.2 comprehendere MB : concludere xv : comprehendere
   vel concludere EG
I, 35.5-6 20 medietas frontis . . . dividatur per xMB :
   om.EG
I, 35.7 1000 xMB : 100 vEG
I, 36.2-3 20 in latitudine G add.mg.E : om.MB

Within the subgroup MB, it is clear that B is inferior to

M, symbolically

$$\begin{array}{c} y \\ | \\ M \\ | \\ B \end{array}$$

Perhaps it derived from M while M was still complete. In the
edition, I have used B to substitute for the missing parts of
M. Evidence of the relationship of B to M follows:

    I, 4.17 maius MEu : om.B
    I, 11.7-8 latus se MEu : tr.B
    I, 12.7 illis 4 MEu : tr.B
    I, 14.2 minus latus se MEu : om.B
    I, 15.2 vel (om.u) equis lateribus MEu : om.B
    I, 15.6 summe MEu : om.B
    I, 17.7 3 ysopleuris MEv : tr.B
    I, 19.2 septies MEu : sexties B

In the subgroup EG, E is superior to G, or

```
          y
          |
          E
          |
          G
```

The following examples should serve to indicate their ordering:

    I, 0.2 consistit BEu : existit G
    I, 0.30 ita BEu : om.G
    I, 0.32 exordiamur BEu : ordiamur G
    I, 1.2 post planimetra add. in capite rei G : om.BEu
    I, 1.7 se habebit BEu : tr.G
    I, 1.25 vel fluvii BEu : om.G
    I, 2.1 embaldale BEu : om.G
    I, 2.28 post lateribus add. illud G : om.BEu
    I, 2.34 linea BEu : om.G
    I, 2.42 faciunt BEu : om.G

The family y may therefore be represented by

```
              y
            /   \
          M       E
          |       |
          B       G
```

    Within group x, all of the manuscripts seem to derive from
H. A and P are very close copies of H and may have been
copied directly from it, or

```
          x
          |
          H
          |
          AP
```

A is so close that determining relative priority is extremely
difficult. Perhaps the most convincing demonstration that A
derives from H and not vice versa is that the additions of H_a
and some of the revisions characteristic of group x occur in
the margins and between the lines in H. These same additions
sometimes become part of the text of A and P. These marginalia are not copyist's corrections because they are uniformly
and nearly exclusively additions to the original text of the
treatise. Examples of the inclusion of marginalia (H_a) in
the copied texts of A and P follow:

I, 28.5 <u>post</u> est (2) <u>add. super</u> inscribentis circulum
ad circulum inscriptum $\underline{H}_a\underline{P}$: <u>add.</u> inscribentis . . .
inscriptum <u>A</u>

I, 32.14 <u>post</u> arpentum <u>add. super</u> secundum nos $\underline{H}_a\underline{P}$:
<u>add.</u> secundum nos <u>A</u>

I, 34.6 <u>post</u> per <u>add.</u> in mg. 40, medietas latitudinum
per $\underline{H}_a$: <u>add.</u> 40 . . . per <u>AP</u>

I, 35.4 <u>post</u> superficiei add. <u>super</u> triangule $\underline{H}_a$: <u>add.</u>
triangule <u>P</u> : <u>post</u> scil. (1) (I, 35.5) <u>add.</u> triangule <u>A</u>

A few examples will show that <u>A</u> and <u>P</u> each derive independently from <u>H</u>, thus establishing the relation

In some cases <u>H</u> and <u>A</u> agree, with <u>P</u> in opposition; in others <u>H</u>
and <u>P</u> read together against <u>A</u>.

I, 0.17 examussim yvHA : examu (followed by blank) <u>P</u>
I, 2.4 possit <u>yvHA</u> : <u>om.P</u>
I, 2.21 <u>post</u> additis <u>add.</u> que <u>A</u> : <u>om.yvHP</u>
I, 2.43 equalia latera <u>yvHA</u> : <u>equilatera</u> <u>P</u>
I, 2.46 trigonus <u>yvHP</u> : triangulus <u>A</u>
I, 3.5 ut si basis <u>y</u> : eius basis si <u>F</u> que basis si <u>HP</u>
quem basis si <u>A</u>

In some places the reading of <u>KLDCN</u> is different from that
of <u>HAP</u>, indicating that a copy of <u>H</u>, which we shall call <u>w</u>,
was the immediate source of <u>KLDCN</u>, as in this diagram:

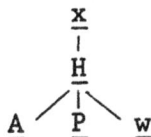

I, 1.24 et hoc <u>yvHAP</u> : quod <u>KLDCN</u>
I, 4.14 dividas (divides <u>FHAP</u>) per basim <u>yvHAP</u> : per basim
divides <u>KLDCN</u>
I, 22.15-16 circumferentie <u>yvHA</u> : <u>om.KLDC</u>
I, 22.20-23 protrahatur equedistans . . . quod sic proba-
tur <u>yvHAP</u> : <u>om.KLDC</u>
I, 26.13 differentie (differre ?<u>P</u>) quasi <u>yvHA</u> : <u>om.KLDC</u>
I, 34.11 note <u>yvHAP</u> : <u>om.KLDC</u>

This group <u>w</u> is the source of three subgroups, one compris-
ing <u>K</u> and <u>L</u>, another consisting of <u>D</u> and <u>C</u> and the third having
only the incomplete <u>N</u>. The following examples should suffice
to show that the three subgroups derive independently from <u>w</u>,
as in this schema:

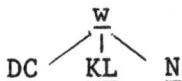

In some instances all three have separate readings. Also
note that each of the groups reads against the other two in
several examples.

    I, 0.5 moderni latini, scil. (scil. om.y) yvHDC :
    moderni, scil. N moderni KL
    I, 0.10 dulcius yvHKLN : dilectius DC
    I, 0.11 noscere yvH : agnoscere KLDC cognoscere N
    I, 0.18 excellentius yvHDCN magnificentius KL
    I, 1.19 regula enim est in geometria y : regula enim
    geometrica est v regula enim est geometrica H
    figura est geometrica DC suia (?) est geometrica N
    om.KL
    I, 1.25 aliter y : item vHDCN porro KL
    I, 1.33-34 trianguli yvHKLDC : anguli N
    I, 2.4 scil. d yvHDC : om.KLN
    I, 2.18-19 aliud latus . . . dividatur yvHKLN : om.DC
    I, 2.42 parallelogramum yvHKLN : pararellum DC
    I, 2.46 quadranguli yKL : trianguli vH trianguli vel
    quadrati DC trianguli vel quadranguli N
    I, 2.49 equilateri yHKLDC : equilatus ?F om.N
    I, 2.52 superagrigetur yvHKLDC : agrigetur N
    I, 3.2 proponit yvHDCN : docet KL
    I, 3.2 superficiem yvHDCN : aream isoscelis, i.e. KL

In the subgroup KL, it appears that L is inferior to K, or

$$\begin{array}{c} w \\ | \\ K \\ | \\ L \end{array}$$

Readings which show their positions follow:

    I, 0.3 perceptione yvHK : om.L
    I, 0.19 secreta yvHK : serta L
    I, 0.27-28 minutias vHK (K add. in mg.) : om.L
    I, 1.36 sicut yvHK : sic L
    I, 2.32 ysopleuro yvHK : Isto pleuro L

Similarly in DC we see that C belongs below D, or

$$\begin{array}{c} w \\ | \\ D \\ | \\ C, \end{array}$$

as the following examples show:

    I, 0.17 eorum yvHD : om.C
    I, 1.31 magni yvHD : magis C
    I, 2.9 erit yvHD : est C
    I, 2.14 quadrangulum EHD : triangulum C
    I, 2.40 probo yvHD : om.C
    I, 2.60 computat yvHD : capit C
    I, 3.10 basis medietas yvHD : tr.C

Schematically, the family $\underline{x}$ is shown by

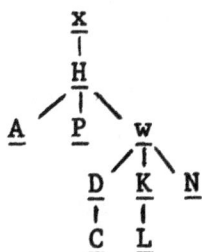

```
        x
        |
        H
      / | \
  A   P   w
  _   _  /|\
       D K N
       | | _
       C L
       _ _
```

For the purposes of the edition, I have collated manu-
scripts $\underline{M}$, $\underline{B}$, $\underline{E}$, $\underline{F}$, and $\underline{H}$. In most cases, the chosen text
belongs either to $\underline{M}$ or $\underline{E}$. $\underline{B}$ has been used only to fill in
for missing sections of $\underline{M}$. The spellings of $\underline{M}$, which are
fairly standard medieval ones, have been adopted. $\underline{M}$ sometimes
spells the same word differently in several places, e.g.,
equidistans, equedistans; I have retained this curious inconsistency. All of the abbreviations have been expanded except
those for scilicet, etcetera, and some forms of numbers. The
punctuation is entirely my own.

Manuscript $\underline{H}$ contains a fair number of marginal notes. Some
of these are interesting comments and glosses on the text. They
probably date from the thirteenth century. I have included them
in the critical apparatus, giving them the siglum $\underline{H}_a$. They provide an indication of nearly contemporary reaction to the treatise. Many of these marginal glosses were copied along with
the text in manuscript $\underline{A}$.

Editing the Pratike de geometrie

There are two manuscripts of the Pratike. Both are used in
this edition. They are

$\underline{R}$ Paris, Bibliothèque nationale, français 2021, ff. 155r-165r,
    thirteenth century. Description: Catalogue général des
    manuscrits français . . . Vol. 1 (Paris, 1868). See also
    V. Mortet, 1909: pp. 55-56.

$\underline{S}$ Paris, Bibliothèque Ste. Geneviève, 2200, ff. 151v-163v,
    a. 1276-1277. Description: Boinet, 1921: pp. 47-61;
    Kohler, 1896: pp. 283-285; and Henry, 1882: pp. 50-51.

Neither manuscript is the archetype of the Pratike. Each
manuscript seems to be copied from the original, not from the
other. The independent derivation may be shown by a few
examples as each manuscript omits material included in the
other but presumably present in the original.

    I, 3.4 l'aires R : om.S
    I, 14.1 du quadrangle R : om.S
    V, 2.13 et ensi tient li piés 12 quarrés S : om.R
    VI, 3.26 em com xxxvi libres xiii R : om.S
    VI, 3.36-37 xxvi ismes. A xxvii en descent li moitiés
        et li S : om.R
    VI, 3.41 les ii pars. A xl descent S : om.R
    VI, 10.4 xvi d. li xv isme; et de le libre le S : om.R
    VI, 10.11 de le libre, ii s d., le iiiiˣˣ S : om.R

Thus neither manuscript offers a complete reading by itself.
I have therefore used readings from both manuscripts in the
edition. The two manuscripts are similar in language; both
use the Picard dialect. There are some orthographic differences between them. S seems to have more consistent Picard
spellings, whereas the spellings of R show Francien influ-
ences. For example, S usually has ke, ki, and che where R
more frequently has que, qui, and ce. S also uses le for the
feminine singular article in the subject case, a Picard peculiarity, where R generally uses la. There are several other
reasonably consistent differences in spelling as indicated in
the following chart.

S	R
kerré	querré
k'il	qu'il
lakele	laquele
saches	saces
rachine	racine
misme	meisme
menour	menor
ou	u
hautesches	hauteces
perfondesces	perfondeces
rieule	riule

One special peculiarity in spelling deserves note. Starting
with Pratike I, 15 R writes monteplie for multeplie, where it
had multeplie a few propositions back. This curious spelling
continues through III, 8.

Since S is generally more complete and has more consistent spelling, I have used it as a base manuscript, including occasional readings from R where sense demanded it. I have put variant readings in the notes for the most part. I have not, however, noted variant forms of abbreviation. For example, I have not noted variants for un, ung, or une written as 1 or i. S spells this article out more consistently than R. I have spelled out most abbreviations, except for the final x standing for us. I have written com for ꝯ and comme for ꝯme, following S and ignoring the spelling variants in R. In problems dealing with fractions and with monies, I have kept the numeric forms and monetary abbreviations of the manuscripts.

Following standard procedures for the edition of Old French texts, the only accent I have included is the acute. I have used the acute accent wherever it seemed necessary for the sense of the text, e.g. to avoid reading pié (foot) as pie (magpie) or après (after) as âpres (bitter).

The manuscripts are irregular about distinguishing u and v; I have chosen to use whichever one makes the text more intelligible to the modern reader. An analogous situation and choice hold for c and t and i and j.

What I have called Pratike I and III are the Old French adaptations of ACC I and III respectively. I have used the same numbers for both; thus Pratike I, 1 is the French version of ACC I, 1. Since there are no extant French versions of ACC II and IV, I have no books II or IV of the Pratike. The material of the Pratike for which there is no original in ACC, I have called Pratike V and VI, V being geometrical and VI arithmetical and mercantile. The Pratike does not contain translations even of all of ACC I and III. Thus Book I of the Pratike contains no propositions 10, 18, 19, 20 and 25. Pratike III is missing versions of ACC III, 4, 5, 5bis, 6, 10, 18, 19, 20 and 21.

Since the material comprising Pratike V and VI comes from various places in the manuscripts, a description of the propositions in the order found in the two French manuscripts should be useful. Pratike I occurs in the same order as ACC I (omitting those untranslated propositions listed above), except for

the addition of <u>Pratike</u> VI, 1 after <u>Pratike</u> I, 30. Then follow <u>Pratike</u> III, 1-11, after which come <u>Pratike</u> V, 2-8 before
<u>Pratike</u> III, 12-17 resume (III, 14 preceding III, 13 in the
manuscripts). Next in order are <u>Pratike</u> V, 9-11 and then VI,
2-11.

Translating the Texts

The translations of <u>ACC</u> and the <u>Pratike</u> are an attempt to
tread the narrow line between literality and literacy. Technical texts of this sort contain usages which certainly would
not have been common and might have been awkward even at the
time they were written. I have not consistently translated
the same Latin or Old French phrases by the same English ones;
I have, however, tried to supply the English technical terms
wherever doing so did not transfigure the author's intent and
did not imply knowledge or techniques he would not have had.
A word for word translation is seldom desirable; any residual
utility is obviated in this case because the translation faces
the text.

In the translation, I have liberally added material to help
make the text more comprehensible to the modern reader; such
material has been put in square brackets, just as in the texts.

The figures accompanying <u>ACC</u> are taken from <u>H</u>, the only one
of the edited manuscripts to include figures. I have not tried
to draw the figures proportional to those of <u>H</u>, and I have not
used a scale representing the geometrical quantities given in
the examples. Where I have added a figure to clarify the text
or notes, I have put its title in square brackets. Thus, only
those without brackets belong to <u>H</u>. All the figure numbers
are my own. Any additions to the figures of <u>H</u> have been en-
closed in square brackets. I have not included the few drawings in the <u>Pratike</u> as they are decorations rather than geometric figures.

In a few cases, I have felt it desirable to tabulate some
of the material for the notes. I have put the titles of such
tables in square brackets, indicating that the tables are not
part of the text.

About the Commentary

The commentary on the text, found in the notes, is founded on some rather idiosyncratic principles. The purpose of the notes is primarily an aid to understanding the text. Comparable passages in other texts are cited so that the reader may refer to other treatments of similar material. In a few instances, I have indicated that I felt these other materials were the immediate source of our text. Where a probable source is indicated, I have cited fewer analogous texts. Thus, because the astronomical material in Book II of ACC derived from as yet undetermined sources and because it is rather complex, I have provided more references.

I have not generally referred to materials which would not have been available to our author. Thus I have cited only works written in Latin or translated into Latin before the end of the twelfth century. I do not, in any case, claim to have exhaustively investigated even all the published texts relating to medieval mathematics and astronomy; I have made no effort to use any of the superabundant unpublished materials. In referring to edited texts in the explanatory notes, I have most often cited them according to their form in the bibliography, followed by page and line number separated by a period; elsewhere the form of citation should be self-explanatory.

I have used at least reproductions (microfilm or photoprint) of all of the manuscripts as listed below. All of the manuscripts have been used to establish the stemma. All of EFHM and parts of B (where M is incomplete) were collated for the text.

A Oxford, Bodleian Library, Auct. F. 5.28, ff. 133v-144v, 13th c. Description: Madan, Craster and Denholm-Young, 1937.

B Paris, Bibliothèque nationale, lat. 16198, ff. 156r-162v, 14th c. (before 1362, v. f. 198v) Description: Thorndike, 1957b: pp. 148-149. Note that Thorndike's conjectures that ACC might be by Campanus and was addressed to the Pope are both wrong. Campanus flourished at least 50 years later. Pape cannot be the vocative of Papa, even if one had the disrespect to address the Pope in this manner.

<u>C</u> Paris, Bibliothèque nationale, lat. 7381, ff. 177r-194v,
    15th c. Description: <u>Catalogus</u> <u>codicum</u> <u>manuscriptorum</u>
    <u>Bibliothecae</u> <u>Regiae</u>, 1744.

<u>D</u> Paris, Bibliothèque nationale, lat. 7420A, ff. 147r-153v,
    14th c. Description: <u>Catalogus</u> <u>codicum</u> <u>manuscriptorum</u>
    <u>Bibliothecae</u> <u>Regiae</u>, 1744.

<u>E</u> Erfurt, Wissenschaftliche Bibliothek, O. 82, ff. 157r-
    169r, late 13th c. Description: Schum, 1887.

<u>F</u> Florence, Biblioteca Medicea Laurenziana, plut. 29, cod.
    19, ff. 51r-56r, 13th c. Description: Bandinius, 1775.

<u>G</u> Prague, Knihovna Metropolitní Kapituli, 1460, ff. 1r-13r,
    15th c. (1432, v. guard sheet). Description: Podlaha,
    1922.

<u>H</u> London, British Museum, Harley 1, ff. 41r-46v, 13th c.
    Description: Nares <u>et al</u>., 1808. Formerly in St. Aug-
    ustine's Abbey, Canterbury, see James, 1903: pp. 520,
    325-326.

<u>H</u>_a Marginalia to <u>H</u>, probably 13th c.

<u>J</u> Vatican City, Biblioteca apostolica vaticana, Ottob. lat.
    1430, ff. 36r-38v, 14-15th c. Description: Daly, 1964.

<u>K</u> Vienna, Oesterreichische Nationalbibliothek 5277, ff.
    154r-164v, 15-16th c. Description: <u>Tabulae</u> <u>codicum</u> <u>manu-</u>
    <u>scriptorum</u> <u>praeter</u> <u>graecos</u> <u>et</u> <u>orientales</u> <u>in</u> <u>Bibliotheca</u>
    <u>Palatina</u> <u>Vindobonensi</u> <u>asservatorum</u>, 1870.

<u>L</u> Vienna, Oesterreichische Nationalbibliothek, 10954, ff.
    202r-211v, 15th c. Description: <u>Tabulae</u> <u>codicum</u> <u>manu-</u>
    <u>scriptorum</u> <u>praeter</u> <u>graecos</u> <u>et</u> <u>orientales</u> <u>in</u> <u>Biblio-</u>
    <u>theca</u> <u>Palatina</u> <u>Vindobonensi</u> <u>asservatorum</u>, 1873.

<u>M</u> Paris, Bibliothèque Mazarine, 3635, ff. 113r-122v, 13-
    14th c. Description: Molinier, 1890.

<u>N</u> Kues, Handschriften-Sammlung des Hospitals, 205, ff.
    188v-189r, 14th c. (?). Description: Marx, 1905.

<u>P</u> Dublin, Trinity College, 403 (D.2.29), ff. 192r-
    14th c. (?). Description: Abbott, 1900.

ARTIS CUIUSLIBET CONSUMMATIO

EDITION, TRANSLATION AND COMMENTARY

[I, 0] [PROLOGUS.]

Artis cuiuslibet consummatio in duobus consistit,
in theorice et practice, ipsius integra perceptione.
Qui autem altero istorum membrorum mutilatur, semiartifex nuncupatur. Verum moderni latini rerum subtilium   5
fugas emulantes in quadruviali scientia virtutum sapore
condita sapientes faciente, eius theorice invigilant
practicam postponentes fructus uberrimos ubi seminaverunt
non metentes quasi florem vernum sine fructu legentes.

Pape! Quid dulcius quam numerorum cognitis vir-   10
tutibus per arismeticam eorum infinitos noscere effectus
calculatione subtili, scil. radicem et originem ortum,
omni scientie suppeditantem necessario. Quid iocundius
quam eorum nota proportione per musicam sonorum consonan-

---

4 altero istorum membrorum: alteram istarum FH / mutilatur:
    intitulatur E inutilitater ?H
5 post verum add. temporis nostri F nostri temporis H / post
    latini add. scil. FH
6-7 virtutum . . . faciente om.E
8-9 fructus . . . metentes om.FH
9 vernum: vitium ?E vacuum B
11 eorum: eius F / effectus: affectus H
12 scil. . . . ortum om.H
13 omni om.E / scientie iter.H / suppeditantem: subpeditantem
    EF
14 musicam sonorum: musicas sororum E

I, 0. PROLOGUE.

The perfection of any art, seen as a whole, depends on
two aspects: theory and practice.[1] Anyone deprived of
either of these aspects is labeled semiskilled. Truly the
modern Latins, striving to equal the flights of subtle things
in quadrivial science that makes men wise when seasoned by
the taste of achievements,[2] give attention to its theory,
neglecting the practice, failing to reap where they sowed the
richest fruits as if picking a spring flower without waiting
for its fruit.[3]

Wonderful! What is sweeter[4] when once the qualities of
numbers have been known through arithmetic than to recognize
their infinite dispositions by subtle calculation, the root,
origin, and source necessarily available for every science.
What is more pleasant when once the proportion of sounds has

---

[1]"theorice et practice" (I, 0.3). A partial history of
this distinction as it relates to geometry is found in the
Introduction.

[2]"virtutum . . . sapientes" (I, 0.6-7). The verbal play
in these words is impossible to translate but is worthy of
comment for it is only in the Prologue that the author seems
to take pains with his style. Sapere, recalling sapor and
conditus, has as a meaning "to taste of" as well as "to understand" or "to be wise." The gist of the phrase seems to be
that wisdom results from a combination of science with practical achievement, just as the perfection of any art demands
theory and practice.

[3]"Fructus . . . legentes" (I, 0.8-9). The word play continues here with the extended conceit comprising sowing, picking (legere meaning "to read" as well as "to pick"), reaping,
flower and fruit. The multiple meanings serve to give depth
to the image.

[4]"Quid dulcius . . . rimari (I, 0.10-20). The author

tias auditu discernere. Quid magnificentius quam super-   15
ficierum et corporum lateribus et angulis probatis per
geometriam, eorum quantitates scire examussim et investigare. Quid gloriosius excellentius ve stellarum motu
cognito per astronomiam quam eclipses et artis secreta
rimari. Nos igitur super geometrie practicam tractatum   20
iocundum et fructum memorem instituimus, ut quod de
magistri nostri fonte dulcius hausimus sitientibus
propinemus.

Opus autem nostrum in 4 distinguemus particulas.
In prima, planismetriam instruimus, superficierum quan-   25
titates investigando. In secunda, altimetriam, mensu-

---

17 quantitates: quantitatem H̲ / examussim et: et examussiva
   ?B̲ tr.FH̲
19 per astronomiam om.E̲ / artis: cetera FH̲
20 nos: vobis F̲ vos H̲
20-21 tractatum iocundum: iocundum om.BE̲ tr.H̲
21 et fructum memorem om.E̲ / instituimus: instruimus FH̲ /
   quod: aliud ?E̲
22 dulcius hausimus: dulcibus haustibus ?E̲
24 distinguemus particulas: distinguimus partes FH̲
25 planismetriam: plani necessarium H̲ / instruimus: et E̲
26 post̲ quantitates add.̲ instruemus E̲

been known through music than to discern their harmonies by
hearing. What is more magnificent when once the sides and
angles of surfaces and solids have been proved through geometry than to know and investigate exactly their quantities.
What is more glorious or excellent when once the motion of
the stars has been known through astronomy than to discover
the eclipses and secrets of the art. We prepare for you
therefore a pleasant treatise and delightful memoir on the
practice of geometry so that we may offer to those who are
thirsty what we have drunk from the most sweet source of our
master.[5]

We divide our work into four parts.[6] In the first, we
teach planimetry, investigating the quantities of area. In
the second, we shall teach altimetry in order to measure

---

gives an elaborate and extended description of the greater
pleasures one will derive in working in each of the branches
of the quadrivium when he has first attained a degree of
theoretical understanding of that science.

[5]"ut quod . . . propinemus" (I, 0.21-23). Our only explicit evidence of the training of the author is contained
in this poetic acknowledgment that he is continuing his
master's work.

[6]"Opus . . . promittimus" (I, 0.24-30). Traditionally
there are three parts in a practical geometry. See R. Baron,
1957: pp. 30-32 and L. Thorndike, 1957: p. 458. The author
provides us with ancillary material on fractional computation, more normally part of arithmetic. Geometrical fractions
are common fractions; astronomical fractions are sexagesimal.

rare alta, docebimus.   In tercia, capacitates corporum

et crassitudines invenire docemus.   In quarta, geometricas

et astronomicas minutias ad predicta necessarias docere

promittimus.   Ita ut principaliter geometrie, secundario   30

astronomie, hoc opus deserviat.   Theoreumatice igitur

exordiamur.

26-27 altimetriam . . . docebimus: capacitates corporum et
  crassitudines invenire docemus FH (cf. versus 27-28) /
  mensurare alta docebimus: altimetriam et profundimetriam
  erudiemus E

27 capacitates om.E

27-28 capacitates . . . docemus: geometricas et astronomicas
  minutias ad predicta necessarias (necessaries F) docere
  promittimus FH (cf. versus 28-30)

28 et om.E / docemus: docebimus E / geometricas: geometrias B /
  quarta: quod ?H

28-30 geometricas . . . promittimus: altimetriam mensurare
  alta docebimus FH (cf. versus 26-27)

30 post geometrie add. et E

heights. In the third, we teach how to find the capacities

and volumes of solids. In the fourth, we promise to teach

about the geometrical and astronomical fractions necessary

to the aforesaid. Thus this work serves principally geo-

metry, secondarily astronomy.[7] Therefore we begin theoreti-

cally.[8]

---

[7]"secundario astronomie" (I, 0.30-31). The use of astronomical instruments as an aid to terrestrial measurement is
common in practical geometries. It is supplemented in this
treatise with instruction on using these instruments in
simple astronomy. The section on fractions is an aid to
astronomical computation as well.

[8]"Theoreumatice . . . exordiamur" (I, 0.31-32). We have
been shown the importance of a theoretical foundation above
(see n. 4).

[I, 1] LINEE RECTE QUANTITATEM PODISMARI.

Esto igitur linea mensuranda a. Stetque planimetra
in capite eius, tenensque astrolabium. Per ambo foramina
mediclivii respiciat terminum a. Videatque quantum
capiat allidada de digitis catheti in orthogonio, et   5
qua proportione se habebunt ad suam basim, scil. ad 12.
Eadem proportione se habebit statura eius ab oculis
infra ad a.

Quod sic probatur. Ibi sunt duo trianguli, unus
magnus et unus parvus. Magnus cuius basis est a; cathe-   10
tus, planimetra; ypothenusa, radius visualis transiens
per allidade foramina. Et parvius triangulus est in
orthogonio quadrato walzagore, cuius basis est latus
12 digitorum; cathetus, tot digitorum quot capit alli-

---

1   recte om.B
2   esto: est BH / a om.F / post planimetra add. metra H
3   tenensque: tollensque E / post astrolabium add. respiciat H
4   mediclivii super scr. vel allidada H_a / respiciat om.H
6   ad (2) om.E / 12: 1 H
7   proportione om.FH
8   infra: in foramine E
9   ibi: isti ?H
9-10 unus . . . parvus om.E
10 et unus: alius F / magnus cuius: magni FH
11 transiens tr. post allidade FH
12 et om.FH / parvius: parvus EF / est om.EH
13 orthogonio quadrato tr.FH / cuius: eius E
14 post cathetus add. scil. F / digitorum: digiti FH

I, 1. TO MEASURE THE QUANTITY OF A STRAIGHT LINE.[9]

Let the line to be measured, therefore, be a (fig. 1).
Let the planimeter stand at one of its ends, holding the
astrolabe.[10] Let him look at the end of a through both
holes of the "mediclivium" [or alidad]. And let him see
how many digits the alidad reaches in the perpendicular, and
what ratio they have to its base, that is to 12. His height,
from his eyes down, has the same ratio to a.

This is proved as follows. There are two triangles, one
large and one small. The base of the large one is a; the altitude, the planimeter; the hypotenuse, the line of sight
across the holes through the alidad. The smaller triangle
is in the square of the walzagora, the base of which is the
side having 12 digits; the altitude, as many digits of the
side [which is also] divided into 12 parts as are marked off

Fig. 1   [Fig. 2]

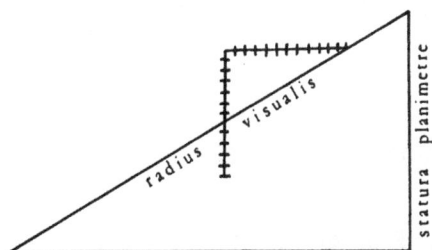
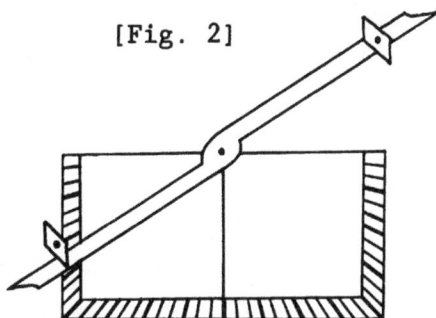

---

[9]"Linee . . . infra ad a." (I, 1.1-8). Cf. Geom. incert.,
ed. Bubnov, 1899: 320.15-2Ī and Hugh of St. Victor, ed. Baron,
1966: 47.9-24.

[10]"astrolabium" (I, 1.3). The shadow quadrant on the back
of an astrolabe (fig. 2) is simply a rectangle which is twice
as wide as high, made up of two squares each having two of
its edges, the lower and outside ones, divided conveniently.
For most western astrolabes the divisions are into twelfths.

dada de latere secto in 12 partes; ypothenusa est   15
allidada sive pars radii visualis. Inde sic, isti duo
trianguli orthogonii super eandem ypothenusam sunt constituti, ergo eandem habent proportionem latera inter se.
Regula enim est in geometria. Si igitur cathetus parvi
est subtriplus ad suam basim, ut pote 4 digitorum, sta-   20
tura planimetre erit subtripla ad a quod oculata fide
patet in astrolabio. Cum ergo sciat planimetra quot
pedum sit sua statura, sciet necessario quot pedum erit
a. Et hoc est propositum.

   Aliter. Lineam quamvis vel fluvii latitudinem per   25
cannas mensurabis sic. Esto linea mensuranda be,
erigasque cannam in capite eius, scil. ab, ex transverso

---

15 partes: digitos E
16 radii om.H / inde sic: om.E et si sic F / isti: illi F
17-18 sunt constituti tr.F
18 habent proportionem: est proportionem habent F proportionem
      habent H
19 enim om.FH / est in geometria: geometrica est F est geo-
      metrica H
21 oculata: ocultata ?E
22 patet: videbis FH / quot: quod F
23 sua statura tr.FH / sciet: sciat F / necessario om.FH
24 et . . . propositum om.B
25 aliter: item FH / fluvii: fluvium F
26 be: abe F (corr. ex abe H)
27 post capite add. c FH / ex om.H

by the alidad; the hypotenuse is the alidad or the part of
the line of sight. Then these two right triangles are set
up on the same hypotenuse; therefore, the sides have the
same ratio between them, according to the rule in geometry.[11]
If therefore the first altitude is one-third of its base,
as 4 digits, the height of the planimeter will be one-third
of a, which is established by looking at the astrolabe.
Since the planimeter knows his height in feet, he would necessarily know how many feet a will be. And this is what was
proposed.

You will measure any line or the width of a river in
another way by means of sticks as follows.[12] Let there be
a line to be measured, be (fig. 3). Then you erect a stick,
ab, at its end, across which you set up another stick, namely

Fig. 3

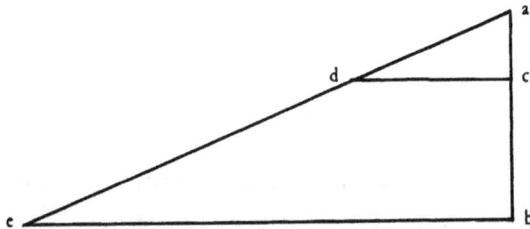

Cf. II, 36 for the construction of the shadow quadrant. The
pointer or alidad is attached to the center of the back of
the astrolabe by a pin around which it swivels. Sighting is
accomplished through 2 holes on small plates attached to the
alidad, perpendicular to its plane. Walzagora is a name
for astrolabe found only in Latin versions of Hispano-Arabic
texts (Hartner, 1960). For a more complete description of
the astrolabe see Hartner, 1960. Hugh of St. Victor gives a
description of the shadow quadrant and its use in measuring
in his practical geometry, ed. Baron, 1966: 25.55-26.80.

[11]"Regula . . . geometria" (I, 1.19). The preceding proposition is an implication of Euclid VI, 2. Hugh gives the
same proposition, ed. Baron, 1966: 22.175-179.

[12]Aliter . . . longitudinem" (I, 1.25-41). Cf. Geom.

que   erigas aliam cannam faciens rectum angulum cum ab,
scil. cd.   Ita quod radius visualis per ad videat e
terminum, scil. linee mensurande.   Eritque ade ypothenusa   30
magni trianguli, cuius cathetus est ab, basis be.   Est
etiam ibidem parvus triangulus cuius basis est cd,
cathetus ac, ypothenusa ad.   Inde sic, illi duo trianguli orthogonii constituuntur super eandem ypothenusam,
ergo eandem habent proportionem latera inter se per pri-   35
mam.   Ergo sicut se habet cd ad ca, ita se habebit be
ad ba.   Sed nota erat tibi quantitas cd.   Videas ergo
quot pedum sit; ergo quantitas ca nota erit.   Ergo
scies proportionem be ad ba.   Sed nota erit quantitas

---

28 faciens: facientem FH / rectum angulum tr.BF
29 ad: nd ?F
30 scil.: videlicet B / ade: ad e B
31 magni trianguli: trianguli mensurandi magni B / est om.FH /
   post basis add. vero F
32 etiam om.E
33 post cathetus add. est FH / post ypothenusa add. est F /
   illi om.B
35 eandem . . . latera: habent eandem proportionem laterum E /
   habent proportionem tr.FH / inter se om.E
36 se habebit: tr.H se habet F
37 erat: erit E / ergo: eius ?E
37-39 sed nota . . . be ad ba om.FH
39 post erit add. quod F

cd, making a right angle with ab. Thus the line of sight
passes through ad to e, the end of the line to be measured.
Then ade will be the hypotenuse of the large triangle, ab
its altitude, be its base. In the same place there is also
a small triangle whose base is cd, altitude ac, hypotenuse
ad. Then those two right triangles are set up on the same
hypotenuse; therefore, the sides have the same ratio between
them by the first part.[13] Thus, cd is to ca as be is·to ba.
But cd was known to you. Therefore, you may see how many
feet it is, and the quantity ca will be known. Therefore
you will know the ratio of be to ba. But ba will be known;

---

incert. ed. Bubnov, 1899: 333.15-334.3, Hugh,ed. Baron, 1966:
47.25-43 and Trac. quadrantis, ed. Britt, 1972: 186-188.

[13]"per primam" (I, 1.35-36). See above I, 1.16-19.

ab. Videas quanta sit. Ergo nota erit quantitas be.   40

Ita mensurabis omnem longitudinem. Idem per quadrantem

invenies sicut per astrolabium.

39-41 sed nota . . . longitudinem om.H

41 mensurabis: mensurabilis ?E / per quadrantem tr.post

    invenies B / quadrantem: quantitatem F

42 sicut: aut E ut FH

you may see how much it is. Therefore be will be known.

In this way you will measure every length. You will find

it by the quadrant just as by the astrolabe.[14]

[fig. 4]

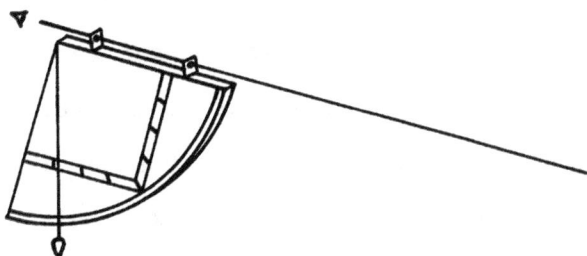

---

[14]"Idem . . . astrolabium" (I, 1.41-42). Strictly speak-
ing, the astrolabe and quadrant are used differently in finding distances. In a quadrant (fig. 4) one sights along the
edge of the quadrant as it pivots around its corner; the
reading is taken where a plumb line, attached to that corner,
crosses the edge of the shadow quadrant. In an astrolabe,
the shadow quadrant remains fixed while the alidad moves.
Once the number of digits has been read, the computation is
the same. Cf. Trac. quadrantis, ed. Britt, 1972: pp. 184-185.
Cf. II, 37 for the construction of the shadow quadrant.

[I, 2] ISOPLEURI SPATIUM REPERIRE EMBALDALE.

Esto triangulus equilaterus abc. Quantitatem superficialem sic invenies. Dividatur triangulus per cathetum,
scil. d, in duos equales triangulos. Quod ita fieri
possit patet per laterum equalitatem. Sed in equilatero   5
latus superat suum cathetum septima sui parte, ut probat
Girbertus in sua Geometria et per quam potest probari,
et per primam Euclidis. Ergo si latus quodlibet equilateri sit 7 pedum, cathetus erit sex. Multiplicetur
ergo unum latus per medietatem catheti, scil. septenarius,   10
per ternarium; vel cathetus per medietatem lateris.
Productum dabit quanta sit superficies ipsius, scil. 21
pedum si latus 7 fuerit.

2-3 superficialem: superficiales B om.F
4 equales triangulos tr.H / fieri om.H
4-5 fieri possit tr.F
7 Girbertus: Gilbertus BE / quam: primam EF
8 et per primam Euclidis om.FH
8-9 latus quodlibet equilateri: quodlibet latus trianguli F
    quodlibet latus equilaterum H
9 post cathetus add. ergo H / multiplicetur: multiplicet E
10 per om.E / per . . . catheti: septenarium medietate
    catheti F medietate catheti H / scil. septenarius
    om.F tr.H.
11 per ternarium: scil. ternario F / per medietatem:
    medietatem EF medietate H
12 quanta sit superficies: sit superficies quanta E /
    ipsius om.B
13 7 fuerit tr.FH

I, 2. TO ASCERTAIN THE EXTENT OF AN EQUILATERAL TRIANGLE
IN SQUARE MEASURE.

Let there be an equilateral triangle abc (fig. 5). You
find the surface area as follows. Let the triangle be divided
by the altitude, d, into two equal triangles. That this can
be done is clear by the equality of the sides. But for an
equilateral, the side is one and one-sixth times its altitude,
as Gerbert proves in his Geometry, and it can be proven using
it and also using the first proposition of Euclid.[15] Therefore if any side of an equilateral triangle is 7 feet, then
the altitude will be 6.[16] Therefore, let one side be multiplied by half of the altitude, that is, seven by three, or the
altitude by half the side. The product gives its surface, that
is 21 feet, if the side was 7.[17]

This is proved thus.[18] Let there be made a quadrangle
(fig. 6) of which one side is three feet, equal to the semi-

Fig. 5

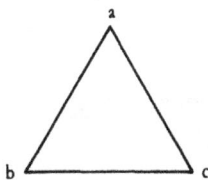

---

[15]"ut probat . . . Euclides" (I, 2.6-8). Gerbert gives
this rule not in his Geometry but in his letter to Adelbold
on the equilateral triangle, ed. Bubnov, 1899: 44.8-10. Even
though Euclid I, 1 deals with constructing an equilateral triangle, nothing close to the desired result can be derived from
it.

[16]"Ergo . . . sex" (I, 2.8-9). For an equilateral tri-
angle of side 7, the altitude will be $7 \times \frac{\sqrt{3}}{2}$, which is very
close to 6.

[17]"Ergo . . . fuerit" (I, 2.8-13). Cf. Gerbert's letter,
ed. Bubnov, 1899: 44.11-45.3.

[18]"Quod . . . 21 pedum" (I, 2.14-31). This curious proof
involves constructing a rectangle to represent geometri-

Quod sic probatur. Fiat quadrangulum cuius unum
latus sit trium pedum equale semicatheto <u>a</u>, et in tres   15
partes dividatur equales. Aliud latus sit trium pedum,
equale semicatheto <u>a</u>, et in 3 partes dividatur equales.
Aliud latus 7 pedum <u>b</u>, scil. equale lateri trianguli,
et dividatur in 7 partes equales. Protrahanturque
linee ad sectiones singulas facto quadrangulo, addi-   20
tis lateribus illis equalibus. Inde sic. Hoc quadrangulum continet 21 pedes, in quo linea <u>a</u>, trium pedum,
currit super lineam <u>b</u>, 7 pedum, quia illam metitur.
Quod probatur per equalitatem laterum angulorum singulorum, et pedum planorum inter se. Ergo illa super-   25
ficies est equalis illi superficiei que fit ex mensura

14 quadrangulum: quadrangulus <u>B</u> quadrans <u>F</u>
14-15 cuius unum latus: unum cuius latus <u>F</u> cuius latus
  unum <u>H</u>
16 dividatur equales <u>tr</u>.H / equales <u>om</u>.F
16-18 sit trium . . . aliud latus <u>om</u>.EFH
18 scil. <u>om</u>.E
19 et <u>om</u>.FH / dividatur: dicantur <u>F</u> / protrahanturque:
  protrahenturque <u>H</u>
20 sectiones singulas <u>tr</u>.FH
22 pedes: pedem <u>FH</u>
23 currit: erit <u>E</u> / <u>b</u> <u>om</u>.?E / <u>ante</u> 7 pedum <u>add</u>. similiter <u>E</u> /
  quia: <u>b</u> <u>E</u> 12 ?<u>F</u> / illam: scil. ?<u>E</u>
24 quod: qui ?<u>E</u> / <u>post</u> laterum <u>add</u>. et <u>H</u> / angulorum:
  triangulorum <u>B</u>
25 et <u>om</u>.E / <u>post</u> et <u>add</u>. sic <u>H</u> / planorum <u>om</u>.EF / <u>post</u> ergo
  <u>add</u>. quia <u>H</u>
25-26 superficies <u>mg. hab</u>. que superficies ut patet ex primo
  Euclidis equalis est superficiei ysopleuri propositi
  [?] <u>H</u>_{<u>a</u>}
26 illi <u>om</u>.EFH / <u>post</u> fit <u>add</u>. tibi ?<u>B</u>

altitude a, and let it be divided into three equal parts.
Let the other side of 7 feet, namely b, equal to the side
of the triangle, be divided into 7 equal parts. And then
let lines be extended to the individual sections in the
quadrangle made by adding lines equal to its sides. Then
this quadrangle contains 21 feet, in which line a, three
feet is multiplied by line b, 7 feet, since it multiplies
that one. This is proved by the equality of sides and individual angles, and of the square feet between them. Therefore that surface is equal to a surface which is made by
the multiplication of half the altitude by a side of 7 feet.
For if the sides are equal to the sides, that which arises
from multiplying one line by the other in one plane, will be
equal to the one arising from the multiplying of the latter
by the former, that is by the straight lines in the other

Fig. 6

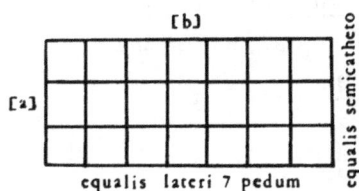

cally the product of the two quantities, base and semi-
altitude. The rectangle has the same area as the triangle,
since they are both the product of the same two lengths,
even though in different surfaces. If the demonstration
is considered a graphic representation rather than a formal proof, it is quite satisfying. Each of the 21 square
units making up the area of the triangle can be counted
in the rectangle.

semicatheti in latus septem pedum. Si enim latera sint
equalia lateribus, quod fit ex ductu unius in alterum in
una superficie, erit equale ei quod fit ex ductu alterius
in reliquum, in alia, scil. in rectis lineis. Ergo   30
superficies predicti trianguli patet esse 21 pedum.

   Istud aliter probatur. In prefato ysopleuro ducatur linea super caput catheti f equalis basi b. Et a
capite eius ducatur linea ad caput basis equalis catheto
d, et vocetur g. Similiter ducatur alia ei equalis ex   35
opposita parte, et vocetur h. Inde sic. f, septem pedum, metitur g, sex pedum. Ergo spatium superficiale
quadranguli, scil. area, continet 42 pedes per precedentem. Patet ergo quod area equilateri trianguli que

---

27 sint: sunt FH
29 fit ex om.F
30 scil. om.E / rectis lineis tr.E
31 superficies: superficiem H / trianguli om.F
32 istud: item E   illud FH
33 f: r ?F / post f add. scil. FH
34 eius: illius FH / equalis: equale B
34-35 catheto d tr.FH
35 alia ei tr.H
36 f: r ?F
37 superficiale: superficias ?F
38 continet om.F / 42 pedes: 34 pedes B   ex pedes F
39 quod om.F / area: aera E / que: qui E

plane. Therefore the surface of the aforementioned triangle
is seen to be 21 feet.

This is proved in another way.[19] In the aforementioned
equilateral triangle let a line, f, be drawn through the end
of the altitude equal to the base b (fig. 7). And from its
end let a line be drawn to the end of the base, equal to the
altitude d, and let it be called g. Similarly let another
be drawn equal to it from the other end, and let it be
called h. Then f, 7 feet, multiplies g, 6 feet. Therefore
the surface of the quadrangle, that is the area, contains 42
feet by the preceding argument. It is clear, therefore that
the area of the equilateral triangle which is its half contains 21 feet. I prove that it is half; d and half of f,
along with g and half of b, make up a parallelogram divided
diagonally. Therefore the triangles are equal, since they
have equal sides and a common hypotenuse. Similarly it will
be proved from the other side. Therefore the exterior triangles

Fig. 7

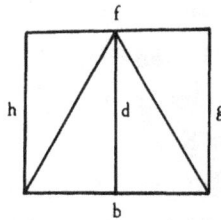

19"Istud . . . propositum" (I, 2.32-49). This proof is
an elaboration of the one suggested in Gerbert's letter to
Adelbold, ed. Bubnov, 1899: 45.20-25. Gerbert's drawing
is the same as the one in our text. Note that the proof is
general for all triangles, not only for the equilateral.

est eius medietas, continet 21 pedes. Probo quod sit   40

eius medietas. d̲ et medietas f̲, cum g̲ et medietate b̲

faciunt parallelogramum diagonaliter divisum. Ergo

trianguli sunt equales, cum habeant equalia latera et

communem ypothenusam. Similiter probabitur ex alia

parte. Ergo exteriores trianguli sunt medietas totius   45

quadranguli. Ergo trigonus ysopleurus prefatus continet

21 pedes. Et similiter, reliqua, scil. medietas. Ergo

notum est embaldale spatium, i.e. area, trianguli

equilateri. Et hoc est propositum.

   Obicitur contra de regula Boetii in sua geometria   50

que talis est. Ysopleuri quantitatem invenire sic.

Metiatur unum latus se ipsum, et producto superagrigetur

---

40 eius o̲m̲.BH / medietas: medias F̲ / 21 pedes: 22 pedes B̲
      ei pedem F̲ 21 pedem H̲
41 eius o̲m̲.H / medietate: medietatem E̲
42 parallelogramum: ?F̲
44 probabitur: probabis EFH̲
45 medietas: medias F̲
46 quadranguli: trianguli FH̲
47 21 pedes: ei pedem F̲ 21 pedem H̲ / et similiter o̲m̲.FH /
      scil. o̲m̲.E / medietas: medias F̲
48 notum: totum BF̲ tantum H̲
49 equilateri: equilatus F̲ / est propositum: erat probandum F̲
50 contra o̲m̲.H / in sua geometria o̲m̲.E
51 quantitatem invenire sic: latus E̲
52 a̲n̲t̲e̲ metiatur a̲d̲d̲. si E̲ / unum latus o̲m̲.E / p̲o̲s̲t̲ latus
      a̲d̲d̲. sic H̲

are half of the whole quadrangle. Therefore the aforementioned equilateral triangle contains 21 feet; and similarly
the other triangles, of course, [contain] half. Thus is known
the space in square measure, that is the area, of an equilateral triangle. And that is what was proposed.

We object on the contrary to the rule of Boethius in his
geometry, which is as follows. To find the quantity of an
equilateral triangle.[20] Let one side be multiplied by itself,
and to the product add [the quantity of] one side. Half of
the sum gives the area of the triangle. According to this,
if a side is 7 feet, its area will be 28, which is contrary
to what went before.

---

[20]"Obicitur . . . premissa" (I, 2.50-55). The example and
method used here are found in Gerbert's letter to Adelbold,
ed. Bubnov, 1899: 45.4-8. Gerbert does not, however, attribute
the erroneous solution to Boethius. This same method, with
an equilateral triangle of side 28 as an example, is found in
the pseudo-Boethian geometry, ed. Folkerts, 1970: 149.591-597.
as well as in Epaphroditus and Vitruvius Rufus, ed. Bubnov,
1899: 532.5-11. Cf. Geom. Incert., ed. Bubnov, 1899: 345.17-19.
It is likely that our author used both the Gerbert version,
with its explanation of the error (see below) and the "Boethian" geometry. The origin of the confusion seems to be the
problem of finding the nth triangular number, which Boethius
discusses in his De Institutione Arithmetica, ed. Friedlein,
1867: 92.11-95.6. Boethius describes the process of summing
the successive digits up to n in order to find the nth triangular number (which Boethius calls the (n-1)st triangular
number since he does not count 1 as a triangular number). He
gives no simple formula for the sum, but the procedure given
is equivalent to the formula $A_3 = (n^2 + n)/2$. This formula,
although clearly wrong for the area of an equilateral triangle,
is correct for the nth triangular number. It is the formula
for the sum of the first n natural numbers. Thus the arithmetic rule for the triangular number of order n was somehow
taken as the geometric formula for the area of an equilateral
triangle of side n.

unum latus. Medietas summe dabit aream trianguli.

Secundum hoc, si latus est 7 pedum, area ipsius erit 28,

quod est contra premissa.   55

   Solutio. Boetius in arismetica computat pedes

semiplenos pro integris per synodochen. Et ita, 28

pedes, tum integraliter, tum partialiter, continentur

in illo trigono, ut patet in figura. In geometria

autem computat pedes integros ex recissuris integrum   60

pedem constituens nil synodoche faciens.

   Et nota est quod pes longus 4 palmorum, ad lineas

mensurandas; pes quadratus ad superficies. Et est cura

geometrie tantum in mensura superficierum. Est etiam

pes crassus sive cubicus, ad mensurandum crassitudinem.   65

---

53 unum: ?F / medietas: medias F
54 ipsius om.E / 28: 38 B ?F
55 est contra premissa: improbatum est FH
56 in: et FH / post arismetica add. ibi ?H / computat:
     computant FH
57 pro integris: per integrum B / 28: ?F
58 partialiter: particulariter E
59 ut: non F   quod H / in om.FH
60 autem: vero FH / post computat add. tantum E
60-61 integrum pedem tr.FH
62 4: quod H
63 cura: cum B
64 geometrie: geometrice F / tantum: ?F / etiam: ?E om.FH
65 mensurandum: mensuram E

The explanation[21] [is that] Boethius in the Arithmetic
reckons partly complete feet for whole ones by synecdoche
[i.e. taking the part for the whole], and thus 28 feet are
contained, some wholly, some partially, in that triangle,
which is clear in the figure (fig. 8). In the geometry,
however, he reckons whole feet, making up a whole foot from
pieces cut off, committing no synecdoche.

It is to be noted[22] that a linear foot of 4 palms is for
measuring lines; a square foot, for surfaces, and as such it
is the concern of geometry in the measurement of surfaces.
A solid or cubic foot is for measuring volume.

Fig. 8

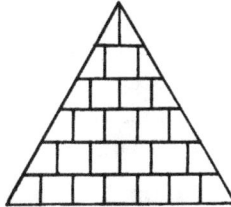

---

[21]"Solutio . . . faciens" (I, 2.56-61). In the Arithmetic,
ed. Friedlein, 1867: 92.11-95.6, Boethius is not dealing with
measure and thereby says nothing about either whole or partly
complete feet. Gerbert's letter, ed. Bubnov, 1899: 45.9-20,
talks about an "arithmetical rule" as being the source of the
erroneous reckoning with part feet as if they were whole ones.
It seems that our author attributed to Boethius what Gerbert
had only called an "arithmetical rule." The term synecdoche,
normally used in grammar and rhetoric, is our author's erudite
elaboration of Gerbert's straightforward description of computing with whole feet where only parts were needed.

[22]"Et nota . . . crassitudinem" (I, 2.62-65). Cf. Gerbert,
Geom., ed.Bubnov, 1899: 56.14-57.21 and "Boethius" Geom. II,
ed. Folkerts, 1970: 146.540 and .547-550.

[I, 3]   ISOCHELES AREAM INVESTIGARE.

Hec proponit mensurare superficiem trianguli habentis tantum duo latera equalia sic. Medietas basis b
currat super cathetum c, 24 pedum, productum dabit aream.
Ut si basis est 14 pedum, cathetus 24, area erit 168   5
pedum. Quod probatur per precedentem ductis lineis equedistantibus, c catheto f, g, et e equedistante h basi
super cathetum. Quadranguli enim magni medietas erit
ysocheles, quod precedens probat.

Cathetum sic invenies. Si basis medietas in se   10
ducatur, scil. 7, et productum tollatur a producto
lateris se ipsum metientis, residui radix dabit cathetum,
c scil. Ut si basis est 14 pedum, latus utrumque 25,
cathetus erit 24 pedum, secundum hanc regulam.

---

3 tantum tr.post equalia FH / latera equalia tr.F /
   sic: si E / medietas: medias F / b om.FH
4 currat: currit H mg. hab. scil. ducitur in cathetum H$_a$ /
   24: ?F   34 H om.E / pedum om.E
5 ut si basis: eius basis si F   que basis si H / 14: ?F /
   24: est etiam quatuordecim B   ?F / 168: 18.24 ?B   ?F
6 pedum om.FH / per hic coepit M / ductis: duobus F / lineis
   om.E
7 post equedistantibus add. catheto et basi H / c om.EFH /
   e om.EFH / equedistante: equidistantem H / h basi tr.FH
8 enim: vero F / magni mg. hab. qui est 336 pedum planorum,
   ex catheto qui est 24 in basim 14 productus cuius medietas
   168 pedum H$_a$ / medietas erit tr.FH
10 invenies: inveniens M / medietas: medias F
12 metientis mg. hab. i.e., in se ipsum ducti H$_a$ / residui
   radix: residuum radix M   radix residui H
13 c scil. om.ME / pedum om.F / 25: in 5 M
14 24: 34 M / pedum om.E

I, 3. TO INVESTIGATE THE AREA OF AN ISOSCELES.[23]

This proposes to measure the surface of a triangle having only two equal sides as follows: Half the base b is multiplied by the altitude, c, 24 feet, the product gives the area. So that if the base is 14 feet, the altitude 24, the area will be 168, which is proved by the preceding,[24] by means of lines drawn parallel to the altitude and base (fig. 9); to the altitude c [they are] f, g; and the parallel to the base h [is] e upon the altitude. Half of the large quadrangle, therefore, will be the isosceles, which the preceding proves.

You will find the altitude in this way. If half the base, namely, 7, be multiplied by itself, and the product taken from the product of a side multiplied by itself, the root of the remainder gives the altitude, namely c, so that if the base is 14 feet, either of the sides 25, the altitude will be 24 feet, according to this rule.

Fig. 9

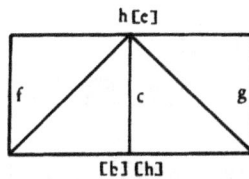

---

[23]"Isocheles . . . regulam" (I, 3.1-14). This proposition is similar to the ones in "Boethius" Geom. II, ed. Folkerts, 1970: 150.609-618 and in Geom. incert., ed. Bubnov, 1899: 343.19-24. Finding the altitude is ancillary to determining the area in this proposition; the analogous propositions in the "Boethian" geometry and the Geom. incert. calculate the altitude first.

[24]"quod . . . probat" (I, 3.6-9). The proof in the preceding proposition (I, 2.32-49) is applicable here too because it refers only to properties common to all triangles. It is not found in the similar propositions of the "Boethian" geometry and the Geom. incert.

[I, 4] SCALENONIS EMBALDUM RIMARI.

Euclides hunc triangulum trium laterum inequalium
tertium ponit, cuius si cathetus $\underline{d}$ metiatur basim $\underline{c}$,
medietas producti dabit aream. Ut si basis $\underline{c}$ sit 25
pedum, cathetus <$\underline{d}$> 12,maius latus $\underline{e}$ 20, minus latus $\underline{b}$ 5
15, area erit 150. Quod probatur per secundam ductis
equedistantibus $\underline{g}$, $\underline{h}$ catheto $\underline{d}$ et basi $\underline{c}$ equedistante
$\underline{n}$, ut patet.

Cathetum eius sic invenies: minus latus metiatur se,
basis similiter se, et summe ille agregentur. Maius   10
latus se similiter metiatur. Sumaturque excessus
maioris summe, scil. basis et lateris minoris, ad
productum a maioris latere. Deinde sumatur medietas
excessus, quam medietatem dividas per basim. Et

---

1 scalenonis: scalenois $\underline{H}$
2 triangulum $\underline{om.FH}$ / laterum inequalium $\underline{tr.F}$
3 cathetus $\underline{d}$ $\underline{tr.H}$ / $\underline{d}$ $\underline{om.EF}$ / $\underline{c}$ $\underline{om.EH}$ / metiatur basim $\underline{super}$
   $\underline{scr}$. ducatur in basim $\underline{H}_{\underline{a}}$
4 basis $\underline{c}$ $\underline{tr.FH}$ / sit: fit $\underline{M}$
5 pedum $\underline{om.FH}$ / $\underline{d}$ ($\underline{ed}$): $\underline{a}$ $\underline{M}$ $\underline{om.EFH}$ / 12: 13 $\underline{M}$ / latus (2)
   $\underline{om.E}$
7 $\underline{post}$ equedistantibus $\underline{add}$. scil. $\underline{E}$ / $\underline{d}$ $\underline{om.EF}$ / $\underline{c}$: $\underline{d}$ $\underline{E}$ $\underline{om.F}$ /
   equedistante $\underline{om.EF}$
8 $\underline{n}$: $\underline{h}$?$\underline{H}$
9 sic: per hanc regulam $\underline{FH}$ / se $\underline{om.M}$
10 ille $\underline{om.FH}$
11 similiter metiatur $\underline{om.FH}$ / metiatur: mentiatur $\underline{E}$
12 scil. $\underline{om.H}$ $\underline{tr.post}$ basis $\underline{F}$
12-13 ad . . . deinde: aliud productum a maiori deinde $\underline{M}$ ad
   hanc $\underline{FH}$
13 $\underline{post}$ sumatur $\underline{add}$. que $\underline{FH}$
13-14 medietas excessus $\underline{tr.FH}$
14 dividas: divides $\underline{FH}$ / et $\underline{om.FH}$

I, 4. TO DISCOVER THE SQUARE MEASURE OF A SCALENE.[25]

Euclid puts this triangle of three unequal sides in the
third place. Let its altitude d be multiplied by the base c;
half the product gives the area. So that if the base c be
25 feet, the altitude d 12, the larger side e 20, the smaller
side b 15 (fig. 10); the area will be 150. This is proved by
the second proposition by means of the parallels, g and h,
drawn to the altitude d and the parallel n to the base c as
is shown.

You will find its altitude as follows.[26] Let the smaller
side be multiplied by itself, the base similarly by itself.
And let the results be combined. Let the larger side similarly be multiplied by itself. Then subtract the product of
the larger side [by itself] from the larger sum, namely that

Fig. 10   [Fig. 11]

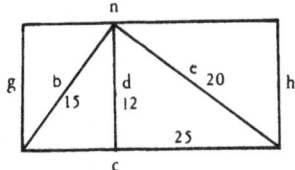
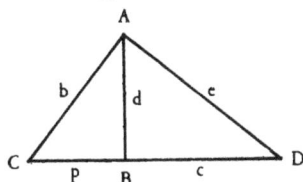

[25]"Scalenonis . . . cathetus d" (I, 4.1-20). The same problem is found in "Boethius," ed. Folkerts, 1970: 150.619-151.639,
and in Ep. et Vit. Ruf., ed. Bubnov, 1899: 530.9-531.10. Our
author's source seems to be the pseudo-Boethian geometry, for
Ep. et Vit. Ruf. does not mention Euclid in the context of this
proposition. Cf. Geom. incert., ed. Bubnov, 1899: 344.1-12
which, also failing to mention Euclid, finds the "precisura"
and altitude but does not go on to find the area. "Boethius,"
Geom. II, ed. Folkerts, 1970: 115.41-116.48 is a translation
of Euclid I, defs. 20 and 21, which gives the six sorts of triangles with their names as found in our text. Cf. ibid.:
148.576-578 and Gerbert, Geom., ed. Bubnov, 1899: 72.9-10 and
73.4-5.

[26]"Cathetum . . . cathetus d" (I, 4.9-20). The altitude
was taken as known in the first part of this proposition. The

denominatio dabit precisuram, i.e. minorem partem   15

basis. Ut si 15 est <u>b</u> minus latus, scil. basis <u>c</u>

25, maius latus <u>e</u> 20, precisura erit 9 pedum. Que

si se ipsam metiatur, et productum auferatur a duc-

tu minoris lateris in se, residui radix erit cathetus.

Secundum hanc regulam, 12 pedum erit cathetus <u>d</u>.   20

---

16 <u>b</u> <u>tr.post</u> minus latus <u>E</u> / scil. <u>om.EFH</u>
16-17 basis <u>c</u> 25: basis <u>c</u> 35 <u>M</u> <u>c</u> basis 25 <u>FH</u>
17 <u>e</u> <u>om.F</u> / 20: 9 <u>M</u> / precisura erit 9 pedum: 9 pedum erit
     precisura <u>FH</u>
18 se <u>om.M</u>
18-19 ductu minoris: ducti <u>M</u>
20 12: 11 <u>?H</u> / <u>d</u>: <u>a</u> <u>M</u> <u>om.E</u>

of [the squares of] the base and smaller side. Take half the excess, which half you divide by the base. And the result gives the precisura, that is the smaller part of the base. So that if the smaller side b is 15, the base c 25, the larger side e 20, the precisura will be 9 feet. If the precisura be multiplied by itself and the product subtracted from the product of the smaller side by itself, the root of the remainder will be the altitude. According to this rule the altitude d will be 12 feet.

procedure for finding the "precisura" is equivalent to the formula

$$p = \frac{(b^2 + c^2) - e^2}{2c}$$

The formula is easily derived as follows   (fig. 11):

In right triangle ABC, $d^2 = b^2 - p^2$.   (1)

In right triangle ABD, $d^2 = e^2 - (c-p)^2$.   (2)

From (1) and (2), $b^2 - p^2 = e^2 - c^2 + 2cp - p^2$

$$b^2 - e^2 + c^2 = 2cp$$

$$p = \frac{(b^2 + c^2) - e^2}{2c}$$

Once one has the "precisura," finding the altitude is a simple matter of applying the Pythagorean theorem.

[I, 5] ORTHOGONII PODISMUM CONTEMPLARI.

Currat cathetus super basim; medietas producti dabit
podismum, i.e. aream eius.   Quod probatur per secundam
propositionem ductis equidistantibus catheto et basi.

Hac regula invenies latera per cathetum. Medietas   5
catheti metiatur se, si autem a producto dematur unitas,
habes basim. Si addatur 1, habes ypothenusam. Ut si
8 est cathetus, 15 erit basis, 17 ypothenusa, quod probatur
per secundam vel per primam.

---

5 latera per cathetum: per cathetum latera H
6 autem om.FH
7 1 om.FH
8 erit om.FH
9 per secundam vel om.FH

I, 5. TO CONSIDER THE MEASURE OF A RIGHT TRIANGLE

Let the altitude be multiplied by the base. Half of the

product gives the extent, i.e. its area. This is proved by

the second proposition[27] once lines have been drawn parallel

to the altitude and to the base (fig. 12).

By this rule you will find the sides by the altitude.[28]

Let half the altitude be multiplied by itself; if one is

subtracted from the product, you have the base. If one is

added, you have the hypotenuse. So that if 8 is the alti-

tude, 15 will be the base, 17 the hypotenuse, which is

proved by the second proposition or by the first.[29]

Fig. 12

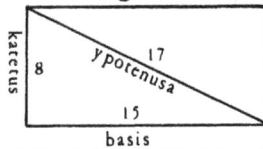

_____

[27]"Quod . . . basi" (I, 5.3-4). Cf., once again, I,
2.32-49.

[28]"Hac regula . . . ypothenusa" (I, 5.5-8). The rule for
finding the hypotenuse and side of a right triangle, given
one of the sides , a, may be derived from the identity

$a^2 + (\frac{a^2}{4} - 1)^2 = (\frac{a^2}{4} + 1)^2$. The rule is given in "Boethius"
Geom. II, ed. Folkerts, 1970: 152.643-652, Geom. incert. and
Ep. et Vit. Ruf., ed. Bubnov, 1899: 357.8-12 and 531.12-532.4,
in all of which it is said to apply only when the altitude
is an even number. When the number is even, the three sides
of the right triangle, a, $(a^2/4) - 1$, $(a^2/4) + 1$, will all
be integers. Nonetheless, the same formulas will produce a
Pythagorean triple no matter what the value of a. Cf. Geom.
incert., ed. Bubnov, 1899: 339.22-340.2 and "Boethius" Geom.
II, ed. Folkerts, 1970: 152.653-156.772.

[29]"quod probatur . . . primam" (I, 5.8-9). Neither of
Propositions 1 or 2 has anything to do with finding the sides
of a right triangle.

[I, 6]   DYAMETRUM CIRCULI IN ORTHOGONIO INSCRIPTI INVENIRE.

Hoc invenit Archites.   Inscribatur circulus in orthogonio, omnia tangens latera.   Iungatur cathetus basi,
a summa subtrahatur corausta, i.e. ypothenusa, residuum
est dyameter circuli.   Ut si 12 est cathetus, 8 basis,   5
15 corausta, 5 erit dyameter, quod probatur per primam.

---

1 invenire: reperire H
4 subtrahatur: subtrahantur M
6 post corausta add. et FH

I, 6.   TO FIND THE DIAMETER OF A CIRCLE INSCRIBED IN A
RIGHT TRIANGLE.[30]

Archytas discovered this.[31] Let a circle be inscribed in

a right triangle, touching every side (fig. 13). Let the

altitude be added to the base. From the sum, let the upper

line, i.e. the hypotenuse, be subtracted; the remainder is

the diameter of the circle. So that if the altitude is 12,

the base 8, and the hypotenuse 15, the diameter will be 5.

This is proved by the first proposition.[32]

Fig. 13   [Fig. 14]

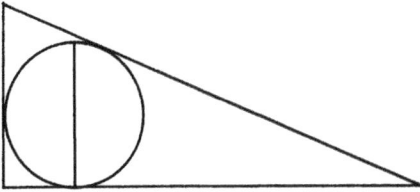 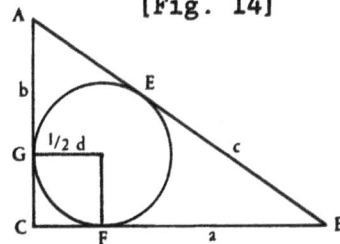

---

[30]"Dyametrum . . . dyameter" (I, 6.1-6). The procedure
for finding the diameter of an inscribed circle is equiva-
lent to the formula d = a+b-c, where $\underline{d}$ is the diameter, $\underline{a}$
and $\underline{b}$ the sides of a right triangle and $\underline{c}$ its hypotenuse.
     Let the inscribed circle be tangent to the right tri-
angle ABC at points E, F and G, as in fig. 14. Then AE=AG,
BF=BE, CF=CG=1/2d.

         c = AB=AE + BE
           = AG + BF, but AG=b-1/2d, BF=a-1/2d
           = a + b - d
       ∴d = a + b - c
The proposition seems to derive from "Boethius" Geom. II, ed.
Folkerts, 1970: 156.723-157.733, where the same numerical
values are found. Although the computation for $\underline{d}$ is formally
correct, the given values comprise an oblique, not a right,
triangle. Cf. Geom. incert. and Ep. et Vit. Ruf., ed. Bubnov, 1899: 347.13-15 and 547.10-548.2 where no numerical
values are given.

[31]"Hoc . . . archites" (I, 6.2). How this proposition
came to be erroneously attributed to Archytas is discussed
in Bubnov, 1899: p. 155, n. 3, and in Folkerts, 1970: p. 104.

[32]"quod . . . primam" (I, 6.6). It is not clear how the
author intended to use the first proposition in proving this.

[I, 7]   AMBLIGONI SUPERFICIEM REPERIRE.

Currat cathetus super basim; medietas summe reddit aream. Ductis enim equedistantibus catheto et basi, probatur hoc per secundam.

---

1 reperire: invenire FH
2 reddit: reddet E dabit F
3 aream: raream H
4 hoc om.E

I, 7. TO ASCERTAIN THE SURFACE OF AN OBTUSE TRIANGLE[33]

Let the altitude be multiplied by the base; half the sum gives the area. This is proved by the second proposition[34] once lines are drawn parallel to the altitude and base (fig. 15).

Fig. 15

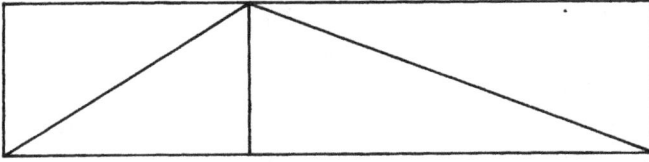

---

[33]"Ambligoni . . . area" (I, 7.1-3). Cf. "Boethius" Geom. II, ed. Folkerts, 1970: 157.734-158.747 and Geom. incert., ed. Bubnov, 1899: 338.19-22.

[34]"Ductis . . . per secundam" (I, 7.3-4). The proof technique, once again, is that of I, 2.32-49.

[I, 8]   OXIGONII PODISMUM PERQUIRERE.

Medietas basis metiatur cathetum. Productum dabit
aream. Quod probatur ductis equidistantibus catheto
et basi per secundam propositionem.

Et sic superficies omnium generum triangulorum   5
secundum ordinem Euclidis docuimus.

Sequitur ut doceamus de quadrangulis.

---

1 oxigonii: exigonii M
5 omnium generum triangulorum: omnium triangulorum generum
   F triangulorum omnium generum H
7 sequitur . . . quadrangulis om.EF

I, 8. TO INQUIRE INTO THE MEASURE OF AN ACUTE TRIANGLE[35]

Let half the base be multiplied by the altitude. The product will give the area. This is proved by the second proposition[36] once lines are drawn parallel to the altitude and
base (fig. 16).

And thus following the order of Euclid,[37] we have taught
about the surfaces of all the types of triangles.

There follows material so that we may teach about quadrangles.

Fig. 16

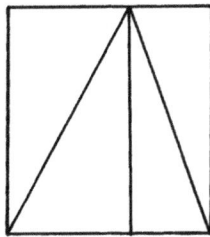

---

[35]"Oxigonii . . . aream" (I, 8.1-3). Cf. "Boethius" Geom.
II, ed. Folkerts, 1970: 159.767-771, Geom. incert. and Ep. et
Vit. Ruf., ed. Bubnov, 1899: 352.8-10 and 534.6-9.

[36]"Quod probatur . . . propositionem" (I, 8.3-4). Cf.
I, 2.32-49.

[37]"secundum ordinem Euclidis" (I, 8.6). The names of the
six types of triangles are found in "Boethius" Geom. II, ed.
Folkerts, 1970: 115.41-116.48. See above, n. 25.

[I, 9]   QUADRANGULI MENSURAM QUERERE.

Currat minus latus super maius latus, productum
dabit aream.   Quod probatur per secundam.

1-3 In codice E haec propositio illi I, 10 successit.

2-3 Pro his versibus codices M et F habuerunt sequentis
    propositionis versus 2-4.

2 minus om.F / latus (2) om.E

I, 9. TO FIND THE MEASURE OF A QUADRANGLE [I.E. RECTANGLE].[38]

Let the smaller side be multiplied by the greater side
(fig. 17); the product gives the area. This is proved by
the second proposition.[39]

Fig. 17

---

[38]"Quadranguli . . . aream" (I, 9.1-3). Cf. "Boethius"
Geom. II, ed. Folkerts, 1970: 160.787-793, and Geom. incert.,
ed. Bubnov, 1899: 344.20-22.

[39]"Quod . . . secundam" (I, 9.3). The area of a rectangle is not given, but is rather assumed, in the proof
of the area of a triangle given in I, 2.

[I, 10] QUADRATI AREAM INVENIRE.

Hoc docet et probat probatio secunde propositionis,
scil. si unum latus metiatur se, habebis aream
quadranguli.

1-4 In codice E haec propositio illam I, 9 antecessit.
2-4 Pro his versibus codices M et F habuerunt antecedentis
     propositionis versus 2-3.
2 probatio: ratio E
3 scil. om.E / metiatur: mentiatur E / habebis aream:
     habetur area E habebitur area FH
4 quadranguli: quadrangulus H

I, 10. TO FIND THE AREA OF A SQUARE.[40]

The proof of the second proposition teaches about and
demonstrates this;[41] that is, if one side is multiplied by
itself (fig. 18) you will have the area of the quadrangle.

Fig. 18

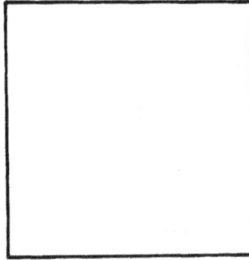

---

[40]"Quadrati . . . quadranguli" (I, 10.1-4). Cf. Geom.
incert. and Ep. et Vit. Ruf., ed. Bubnov, 1899: 340.20-22,
345.19-346.1 and 523.1-3, and "Boethius" Geom. II, ed. Folkerts, 1970: 159.777-782.

[41]"Hoc . . . propositionis" (I, 10.2). See n. 39 above.

[I, 11]   RUMBI EQUILATERI SUPERFICIEM QUERERE.

Metiatur cathetus a, scil. medietas alterius
dyagonalis, dyagonalem lineam b, productum dabit aream.
Quod probatur per secundam ductis equedistantibus
catheto et basi utrobique per secundam.   5

Cathetum rumbi sic invenies, cathetum, scil. ab
angulo ad dyagonalem lineam, protensum. Metiatur unum
latus se. Metiaturque medietas dyagonalis se. Productum
dematur a primo producto lateris. Radix residui erit
cathetus. Ut si rumbi quodlibet latus sit 10 pedum,   10
dyagonalis 12, cathetus erit 8, quod probatur per primam.

96 erit area quod probatur per dictas propositiones.

---

1 rumbi: tumbi H
2 a om.F / scil. om.MFH
2-3 medietas alterius dyagonalis om.FH
3 dyagonalem om.M
4 per secundam om.FH / ductis om.F
5 per secundam om.E
6 sic invenies tr.post protensum FH / cathetum scil. om.FH
7 lineam om.FH / metiatur: mensiatur E
8 metiaturque: menciatur E
9 residui: residua H
10 10: 20 M 12 F
11 primam: prima M
12 96 . . . propositiones om.FH

I, 11. TO OBTAIN THE SURFACE OF AN EQUILATERAL RHOMBUS.[42]

Let the altitude a̲, that is half of one diagonal, be multiplied by the other diagonal line b̲ (fig. 19); the product
will give the area. This is proved by the second proposition[43] once the lines are drawn parallel to the altitude and
the base on both sides, as in the second proposition.

You will find the altitude of the rhombus, namely the
altitude reaching from the corner to the diagonal line as
follows. Let one side be multiplied by itself. Then let
half of the diagonal be multiplied by itself. Let the product be subtracted from the first product of the sides. The
root of the remainder will be the altitude. So that if any
side of a rhombus be 10 feet, and the diagonal 12, the altitude will be 8, which is proved by the first proposition.

The area will be 96, which is proved by the specified
propositions.

Fig. 19

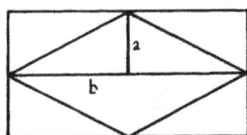

---

[42]"Rumbi . . . propositiones" (I, 11.1-12). Cf. "Boethius"
Geom. II, ed. Folkerts, 1970: 160.794-161.804, Geom. incert.
and Ep. et Vit. Ruf., ed. Bubnov, 1899: 345.10-16, and 525.14-
526.9.

[43]"Quod . . . secundam" (I, 11.4-5). Cf. I, 2.32-49,
which must be applied twice in this case.

[I, 12]   RUMBI NON EQUILATERI EMBALDUM SCIRE.

Equentur latera opposita, et currat unum equatum
super reliquum equatum. Productum dabit aream. Ut si
8 pedum est <u>c</u>, maius latus; 6 oppositum, <u>b</u>; 4, tertium;
2, illi oppositum. Equata latera sunt unum 3, reliquum   5
7. Area 21 quod probatur per secundam quia ista duo
latera equata sunt equalia illis 4 inequalibus per
equipollentiam, scil. productum ex hiis equabitur
producto ex illis.

---

1 rumbi: <u>?E</u> / embaldum: embaudum <u>H</u>
2 <u>post</u> currat <u>add</u>. et <u>F</u> / currat <u>mg. hab.</u> scil. ut si latus
   <u>c</u> equatum suo opposito, ducatur in <u>b</u> equatum suo opposito.
   Et hoc tamen non intelligo quomodo fuerit nisi per scientiam de arcu et circumferentia, nota quantitate angulorum, aut nisi ponantur duo latera equidistantia;
   reliqua duo quomodocumque, quorum sumtorum non sumatur
   nec medietas, sed sumatur vera latitudo rumbi. <u>H</u>_a
3 reliquum: aliud <u>FH</u>
5 equata: equa <u>MH</u>
5-6 latera . . . duo <u>om.H</u>
6 ista <u>om.M</u>
7 sunt: sint <u>F</u> / inequalibus <u>om.FH</u>
8 scil.: sed <u>?H</u> <u>om.E</u>
9 ex illis: illius <u>M</u>

I, 12. TO KNOW THE AREA OF A NON-EQUILATERAL RHOMBUS.[44]

Let the opposite sides be averaged, and let one averaged side be multiplied by the other averaged side. The product gives the area. So that if the greatest side c̲ is 8 feet, the opposite side b̲, 6 feet, the third, 4 feet, the side opposite it 2 feet (fig. 20). The averaged sides are 3 for the one, 7 the other. The area is 21 which is proved[45] by the second proposition because the 2 averaged sides are equal to the 4 unequal sides by equipollence, that is the product of the former will be averaged [to make] the product of the latter.

Fig. 20

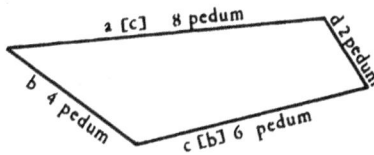

a [c]   8 pedum
d 2 pedum
b 4 pedum
c [b] 6 pedum

---

[44]"Rumbi . . . area 21" (I, 12.1-6). This proposition with the same numerical example is found in "Boethius" Geom. II, ed. Folkerts, 1970: 161.807-813, but not in Geom. incert. or Ep. et Vit. Ruf. The source of the "Boethian" version, according to Folkerts, 1970: p. 103, is "De iugeribus meti-undis," Lachmann, 1848: 355.1-7, which has different numeri-cal values. The procedure for finding the area, equivalent to the formula $A = \frac{a+c}{2} \cdot \frac{b+d}{2}$, where a̲, c̲ and b̲, d̲ are pairs of opposite sides, is not generally correct. How good an approximation it might be depends on individual cases because the problem as given is indeterminate.

[45]"quod probatur . . . illis" (I, 12.6-9). I, 2 is not applicable here. The author makes a futile attempt to prove this incorrect procedure. It is difficult to determine exactly what he is trying to do, but he seems to be begging the question.

[I, 13] TRAPEZITI ORTHOGONII EMBALDUM COGNOSCERE.

Iungatur basis p̲ vertici q̲. Medietas summe metiatur
cathetum o̲. Productum dabit aream. Ut si basis est
45, vertex 15, cathetus 20; 600 erit area. Quod
probatur ducta equidistanti c̲ catheto o̲ a termino   5
verticis ad basim, et ducta equidistanti h̲ relique
parti basis et c̲, linee ducte prius per triangulos
equales.

---

1 cognoscere: rimari F̲H̲
2 basis p̲ tr.E̲
4 cathetus 20: om.F̲ cathetus 30, vertex 20 M̲ cathetus
    30 E̲ / 600: om.M̲ 900 E̲ 500 ?H̲
5 ducta tr.post catheto o̲ F̲H̲ / catheto o̲: catheto e̲ M̲F̲
7 c̲ om.E̲

I, 13. TO LEARN THE SQUARE MEASURE OF A RIGHT TRAPEZOID.[46]

Let the base $\underline{p}$ be added to the upper base $\underline{q}$ (fig. 21).

Let half the sum be multiplied by the altitude $\underline{o}$. The pro-

duct will give the area. So that if the base is 45, the

upper base 15, and the altitude 20, the area will be 600.

This is proved by means of equal triangles[47] once a line $\underline{c}$

has been drawn parallel to the altitude $\underline{o}$ from the end of

the upper base [extended] to the base, and a line $\underline{h}$ has been

drawn parallel to the remaining part of the base, as far as

$\underline{c}$, the line previously drawn.

Fig. 21

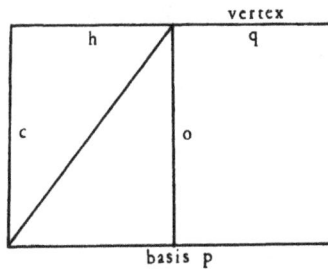

---

[46]"Trapeziti . . . area" (I, 13.1-4). The method for finding the area amounts to applying the formula $A = \frac{p+q}{2} \cdot o$.

The same procedure using the same numbers, except for an altitude of 30, is found in "Boethius" Geom. II, ed. Folkerts,
1970: 162.814-821 and in Ep. et Vit. Ruf., ed. Bubnov, 1899:
525.7-13. Cf. Geom. incert., ed. Bubnov, 1899: 342.7-12,
which has a different numerical example and Ep. et Vit. Ruf.,
ed. Bubnov, 1899: 520.15-521.2, which has different numbers
and procedure.

[47]"Quod probatur . . . equales" (I, 13.4-8). The author
seems to be appealing once again to his favorite proof technique, that of I, 2.32-49. Since a trapezoid is made up of
a triangle, having a base $\underline{p}$ - $\underline{q}$ and an altitude $\underline{o}$, and a
rectangle, having a base $\underline{q}$ and an altitude $\underline{o}$, one can derive
its area by adding the expressions for the areas of simpler
figures.

[I, 14] PARALELLOGRAMI ORTHOGONII DYAGONALEM REPERIRE.

Maius latus metiatur se, minus latus se. Iungantur
ille summe. Tetragonale latus totius summe, scil.
radix dabit diagonalem a lineam. Ut si 80 est maius latus,
60 minus, 100 erit diagonalis linea. Quod probatur per   5
primam.

Sequitur de pentagono.

---

1 orthogonii om.F
2 iungantur: iungatur F
3 ille: illis F
4 a om.E / lineam om.FH / ut om.M / 80: 8 H / post est add.
  primus F / maius latus tr.H
5 60: 80 M 6 H / ante minus add. vero latus H / 100: 10 H
7 sequitur de pentagono om.EF

I, 14.   TO FIND THE DIAGONAL OF A RIGHT PARALLELOGRAM[48]

Let the greater side be multiplied by itself, and the

smaller by itself. Let these results be added. The tetra-

gonal side of the whole sum, that is, its root, will give

the diagonal line a. So that if the larger side is 80, the

smaller 60; the diagonal line will be 100. This is proved

by the first proposition.[49]

There follows material concerning the pentagon.

Fig. 22

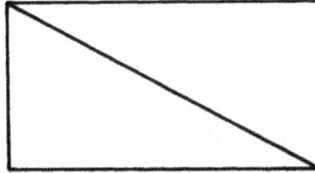

---

[48]"Paralellogrami . . . linea" (I, 14.1-5). The same
proposition, using the same numerical example, is found in
"Boethius" Geom. II, ed. Folkerts, 1970: 162.825-163.830 and
Ep. et Vit. Ruf., ed. Bubnov, 1899: 530.1-8. Cf. also Geom.
Incert., ed. Bubnov, 1899: 344.13-17 and 361.21-24.

[49]"Quod probatur . . . primam" (I, 14.5-6). Nothing in
the first proposition seems to apply in proving this one.

[I, 15] PENTAGONI PAVIMENTUM PERQUIRERE.

Esto pentagonus equilaterus vel equis lateribus
distinctus. Si unum latus se ipsum metiatur, et
productum ternario multiplicetur. Et a summa excrescente subtrahatur unum latus semel. Medietas   5
summe dabit aream. Ut si quodlibet latus sit 6
pedum, 51 erit area, quod probatur per secundam.
Ducta equedistanti lateri in summitatibus 2 laterum
faciendo quadratum, et ducta equedistanti illi ducte
in termino laterum et equedistantibus ad terminos   10
extremos laterum superiorium, probetur quantitas

---

1 pentagoni: pantagoni FH mg.hab. Per hanc demonstrationem
   de pentagono et per 4 sequentes, non investigatur nisi
   quantitas figurarum multilaterarum, quarum pentagonus
   creatur ex quadrato sui lateris et trigono equilatero
   eiusdem lateris addito. Exagonus vero ex quadrato unius
   lateris et duobus trigonis additis. Et sic consequenter
   in reliquis que omnis figure sunt non equiangule. Sed
   ut patet in subiectis cematibus, figurarum autem multilaterarum quantitas investigatur per scientiam de corda
   et arcu et noticiam angulorum--figurarum dico
   multila[tera]rum et equiangularum. H_a / perquirere:
   reperire FH
2 pentagonus: pantagonus H / equilaterus vel om.FH
3 ante si add. ut H
4-5 et a summa . . . semel: et summe excrescenti unum latus
   semel subtrahatur FH
6-7 quodlibet . . . 6 pedum: 7 pedum est quodlibet latus FH
6 sit: fit M
7 51: 70 MH ac ?F / secundam mg. hab. ut mihi videtur mel-
   ius per 3^{am} H_a
8 ducta om.F / lateri iter.M / laterum om.FH

I, 15. TO INQUIRE INTO THE AREA OF A PENTAGON.[50]

Let there be an equilateral pentagon or one distinguished
by equal sides (fig. 23).[51] If one side be multiplied by
itself and the product multiplied by three, and if, from
the result, one side is subtracted once, then half of that
result gives the area. So that if any side is 6 feet, the
area will be 51, which is proved by the second proposition.[52]
The size of the triangles may be proved once a [line] is
drawn parallel to the side at the [upper] ends of the two
sides thereby making a square, and a parallel drawn to that
[line] at the end of the sides [i.e. the vertex], and

Fig. 23

---

[50]"Pentagoni . . . area" (I, 15.1-7). The description of
the procedure is equivalent to the formula $A_5 = \frac{3n^2 - n}{2}$.
In "Boethius" Geom. II, ed. Folkerts, 1970: 163.832-838 (Cf.
Ep. et Vit. Ruf. and Geom. incert., ed. Bubnov, 1899: 534.10-
535.5 and 346.1-4), however, the numerical example is wrongly
calculated (see Folkerts, 1970: p. 102); the calculation is
corrected in certain manuscripts. This procedure gives only
a rough approximation of the area of a pentagon (see n. 52).

[51]"Esto pentagonus . . . distinctus" (I, 15.2-3). This is
not a regular pentagon, which would have equal angles as well
as equal sides.

[52]"quod probatur . . . pentagonum" (I, 15.7-15). The
proof technique of I, 2.32-49 would give a formula for the
area of this pentagon different from the one described. The
author seems to be applying a correct formula for finding

triangulorum, et habebitur propositum.   Probatur

enim hoc per probationes secunde propositionis,

scil. trianguli equilateri, et quadrati, quia ex

illis constat pentagonum.   15

---

12 habebitur: habetur M / probatur: probabitur FH
14 scil. trianguli tr.FH
15 pentagonum: pentagonium MF pantagonium H

parallels [drawn] to the outer ends of the upper sides, and
what was proposed will be had. For this will be proved by
the proofs of the second proposition, that is of the equilateral triangle, and of the square, since the pentagon is
made up from these.

---

the nth pentagonal number to the unrelated problem of finding the area of a pentagon of side n. Thereby he is adopt-
ing a technique analogous to the one he criticized at I,
2.50-61.

[I, 16] EXAGONI AREAM REPERIRE.

Esto exagonus 6 laterum equalium. Metiatur unum
latus se; productum multiplicetur per 4. Ab illa summa
quantitas unius lateris bis ducta subtrahatur. Residui
medietas summe dabit aream. Quod probatur per precedentem. 5

---

1 reperire: invenire F

2 exagonus: eaxagonum H / laterum equalium tr.FH

3 ab illa summa: illi summe FH

4 ducta om.H / residui om.FH

5 summe om.E / quod: hoc FH / per precedentem: per precedentem propositionem F per propositionem precedentem H

I, 16. TO ASCERTAIN THE AREA OF A HEXAGON.[53]

Let there be a hexagon of six equal sides (fig. 24). Let
one side be multiplied by itself; let the product be multiplied by four. From that result subtract twice the quantity
of one side. Half of the remaining result will give the
area. This is proved by the preceding proposition.[54]

Fig. 24

---

[53]"Exagoni . . . aream" (I, 16.1-5). The procedure is
like that of "Boethius" Geom. II, ed. Folkerts, 1970: 163.839-
164.844, but in the pseudo-Boethian geometry twice the side
is added instead of subtracted. This error is corrected in
some of the manuscripts. Cf. also Ep. et Vit. Ruf. and Geom.
incert., ed. Bubnov, 1899: 536.13-537.2 and 346.4-6. The
formula would be $A_6 = \dfrac{4n^2 - 2n}{2}$, which is the way of finding
the nth hexagonal number, not the required area.

[54]"Quod . . . precedentem" (I, 16.5). It is not clear
how the author intends to prove this.

[I, 17]   EPTAGONI PAVIMENTUM PERSCRUTARI.

Metiatur unum latus se; productum quinques ducatur.

A summa unius lateris quantitas ter subtrahatur.

Medietas residui pavimentum reddet. Ut si latus

quodlibet est 7 pedum, 112 erit pavimentum eius.   5

Hoc probatur per premissarum propositionem quia

constat ex quadrato et 3 ysopleuris, quorum quantitas

probata, per secundam istam probat propositionem.

---

3 ter: et M̲ cum? F̲
5 112: 114 M̲E̲ 119 H̲ / pavimentum eius t̲r̲.H
6 premissarum propositionem t̲r̲.FH
7 tribus: et F̲ / ysopleuris: exspoletus H̲ / p̲o̲s̲t̲ exspoletus
    a̲d̲d̲. ysopleuris scil. H̲ₐ
8 istam probat propositionem: propositionem probat istam E̲
    probat istam F̲H̲

I, 17. TO EXAMINE THE AREA OF A HEPTAGON.[55]

Let one side be multiplied by itself (fig. 25). Let the
product be taken five times. From the result let the quantity of one side be subtracted three times. Half of the
remainder gives the area. So that if any side is 7 feet,
112 will be its area. This is proved by the proposition
made up of the preceding ones since [the heptagon] is made
up from a square and three equilateral triangles, of which
the quantity, demonstrated by the second proposition, proves
this proposition.[56]

Fig. 25

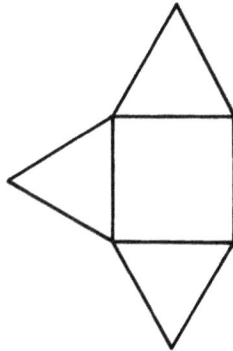

---

[55]"Eptagoni . . . eius" (I, 17.1-5). "Boethius" Geom. II,
ed. Folkerts, 1970: 164.847-852 and Ep. et Vit. Ruf., ed.
Bubnov, 1899: 538.5-10 both give a procedure and a correct
calculation. Cf. Geom. incert., ed. Bubnov, 1899: 346.6-7,
where no example is computed. The method is equivalent to
the formula for the nth heptagonal number, that is

$$A_7 = \frac{5n^2 - 3n}{2}$$

[56]"Hoc probatur . . . propositionem" (I, 17.5-8). Again
the author is assuming his method derives from the addition
of areas, which it does not.

[I, 18]   OCTOGONII AREAM PERQUIRERE.

Metiatur unum latus se; productum sexcies ducatur.
A summa quantitas lateris quater subtrahatur. Medietas
residui aream concludet, quod probatur per premissas
probationes.   5

---

2 sexcies ducatur tr.H
3 quater: quantum M
4 concludet: concludit H /quod: hec FH / post premissas add.
   earum E ?F

I, 18. TO INQUIRE INTO THE AREA OF AN OCTAGON.[57]

Let one side be multiplied by itself (fig. 26). Let the
product be taken six times. From the result let the quantity
of a side be subtracted four times. Half of the remainder
will comprise the area, which is proved by the preceding
proofs.[58]

Fig. 26

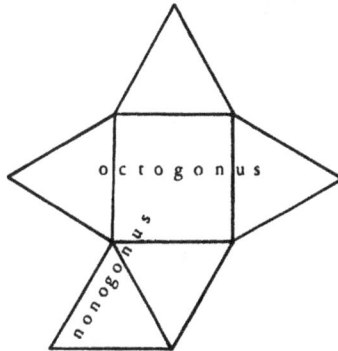

---

[57]"Octogonii . . . concludet" (I, 18.1-4). The procedure
is that for the nth octagonal number, not for the area of
an octagon. It can be expressed as $A_8 = \frac{6n^2 - 4n}{2}$.
Cf. "Boethius" Geom. II, ed. Folkerts, 1970: 164.854-859,
Ep. et Vit. Ruf. and Geom. incert., ed. Bubnov, 1899: 539.9-
14, and 346.6-7.

[58]"quod probatur . . . probationes" (I, 18.4-5). As we
have seen above, the proofs do not show what the author would
like them to.

[I, 19] NONAGONI PODISMUM CONCLUDERE.

Metiatur unum latus se; productum septies ducatur.
A summa, latus quinquies subtrahatur. Medietas residui
aream podismatur. Hoc probatur per premissas proximas.

---

1 podismum concludere: embaldum rimari E
2 unum om.E / septies ducatur tr.FH
3 latus: quantitas lateris F
4 aream podismatur: embaldum reddet E / premissas: probationes E precedentes FH

I, 19. TO DEDUCE THE MEASURE OF A NONAGON.[59]

Let one side be multiplied by itself (fig. 26). Let the
product be taken seven times. Let the side be taken five
times from the result. Half the remainder measures the area.
This is proved by the previous preceding propositions.[60]

---

[59]"Nonagoni . . . podismatur" (I, 19.1-4). Written as
$A_9 = \frac{7n^2 - 5n}{2}$, this algorithm will compute the nth nonagonal
number but not the area of a general nonagon. Cf. "Boethius"
Geom. II, ed. Folkerts, 1970: 164.860-165.865, Ep. et Vit.
Ruf. and Geom. incert., ed. Bubnov, 1899: 540.15-20 and
346.7-8.

[60]"Hoc . . . proximas" (I, 19.4). The preceding proofs
do not fulfill the author's requirements.

[I, 20]   DECAGONI EMBALDUM RIMARI.

Metiatur latus se. Productum ducatur 8^{es}. A summa,
quantitas lateris 6^{es} subtrahatur. Medietas residui
embaldum reddet. Hoc probatur per probationes proximas.

Hiis regulis omnium angularium figurarum   5
superficies invenies. Hoc pacto primum servato ut
naturali ordine procedas in multiplicatione producti
ex latere in pentagono scil. per 3, in exagono per 4
et deinceps suo ordine. Et hoc pacto servato ut in
lateris subtractione ordinem naturalem serves, scil. in   10
pentagono semel, in exagono bis, et sic deinceps. In
crescentibus figuris que sunt equalium laterum, numeri
ordinem serves. Et aream invenies, et probationes.

Et sic de angularibus figuris sufficienter dictum est.

Sequitur de circularibus.   15

1-4 decagoni . . . proximas om.E
1-3 decagoni . . . subtrahatur om.H
2 post metiatur add. unum F / ducatur 8^{es} tr.F
4 reddet: reddit F
5 ante hiis add. et ita FH / regulis: figuris F
6 post superficies add. si subtiliamus ?F / primum: primo FH
7 naturali: simili F
8 scil.: si H / pentagono: pantagonio H / exagono: oxigonio
   M oxigono E
10 ordinem naturalem tr.FH / serves: observes FH / scil.
   om.FH
11 pentagono: pantagonio H
12 numeri: unum MH
13 et (1) om.F / et (2): per F / probationes mg. hab. probatio enim in singulis plana est per 3^{am} per quam scitur
   quantitas ysochelis et sui catheti H$_a$
14 sufficienter: sufficit F / dictum est: diximus FH
15 sequitur de circularibus om.F

I, 20. TO DISCOVER THE SQUARE MEASURE OF A DECAGON.[61]

Let one side be multiplied by itself. Let the product be
taken eight times. From the result, subtract six times the
quantity of the side. Half of the remainder will give the
square measure. This is proved by the previous proofs.[62]

By these rules you will find the surfaces of all angular
figures.[63] When this rule is observed at the start you may
proceed as in the natural order: multiplying the product of
the side [with itself] by 3 in the pentagon, by 4 in the hexagon, and so on in order. And when this rule is observed
you will follow the natural order in subtracting sides, that
is in the pentagon once, in the hexagon twice, and so on.
In increasing figures, which are of equal sides, you pre-
serve the numerical order. And you will find the area and
the proofs.

And thus enough has been said about angular figures.

There follows material concerning circles.

---

[61]"Decagoni . . . reddet" (I, 20.1-4). The procedure,
equivalent to $A_{10} = \frac{8n^2 - 6n}{2}$, applies not to the area of a
decagon but to the nth decagonal number. Cf. "Boethius"
Geom. II, ed. Folkerts, 1970: 165.866-872, and Ep. et Vit. Ruf.,
ed. Bubnov, 1899: 541.16-542.4.

[62]"Hoc . . . proximas" (I, 20.4). The author appeals again
to a proof technique which does not apply here.

[63]"Hiis . . . invenies" (I, 20.5-13). The general formula
for the nth a-gonal number, which the author takes as the
area of an a-gon of side n is $A_a(n) = \frac{(a-2)n^2 - (a-4)n}{2}$
if a ≥ 3, n ≥ 1. Cf. "Boethius" Geom. II, ed. Folkerts,
1970: 165.873-166.880.

[I, 21]   CIRCULI PERIFERIAM INVENIRE.

Hec linee curve quantitatem podismatur. Si dyameter
triplicetur et ei addatur 7^a pars eius, et habetur
periferia circuli. Quod probatur per rectam lineam
equalem curve que si illi superponeretur et   5
recta curvaretur, nec excederet nec excederetur.

---

1 periferiam: pariferiam M̲
2 hec: hee H̲ / si: scil. F̲H̲
3 ei o̲m̲.F̲H̲ / pars eius t̲r̲.H̲ / Et o̲m̲.E̲ / habetur: habebitur F̲
   habebis H̲
4 periferia: periferiam H̲ / rectam lineam t̲r̲.F̲H̲
5 superponeretur: supponeretur M̲ superponentur ?̲H̲
6 excederet nec excederetur: excederetur nec excederet F̲

I, 21.   TO FIND THE CIRCUMFERENCE OF A CIRCLE[64]

This proposition measures the quantity of a curved line.

If the diameter be tripled and its seventh part added to it,

the circumference of the circle will be had (fig. 27).   This

is proved using a straight line equal to a curve.[65]   If a

straight line could be curved and if it could be superposed

to [the curve], it would neither exceed [it] nor be exceeded

Fig. 27

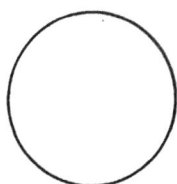

linea tripla sesquiseptima ad diametrum et equalis circumferentie

---

[64]"Circuli periferiam . . . circuli" (I, 21.1-4).   The
source of this proposition is by no means obvious, even
having searched M. Clagett's Archimedes in the Middle Ages,
v. 1:   The Arabo-Latin Tradition (Madison, 1964).   Comparable propositions may be found in the "Epistola Adelboldi
ad Silvestrum II papam," and Geom. incert., ed. Bubnov, 1899:
304.17-19 and 356.13-15; however, neither of these propositions uses the term "periferia" for the circumference of the
circle.   The proposition assumes 3 1/7 as the value of π
in the calculation which is equivalent to c = 3 1/7 d.

[65]"Quod probatur . . . habemus" (I, 21.4-7).   The assumption that a curve may be made equal to a straight line, or
rectified, is a necessary precondition to any rigorous measurement of a circle.   Its inclusion here indicates considerable mathematical sophistication on the part of the author
or his source.   The postulate the author refers to could be
Euclid, Axiom 4 [7], "Things which coincide with one another
are equal to one another."   A more directly applicable postulate may be found in several Latin emendations of the works
of Archimedes, e.g. "Cuilibet recte linee aliquam curvam esse

Hoc etiam ex petitionem habemus.   Et nota est quantitas

recte per primam, ergo curve ei equalis ex ypothesi.

---

7 <u>post</u> quantitas <u>add.</u> et <u>F</u>

8 curve <u>om.FH</u> / ei: ex <u>F</u>

[by it]. This we have by postulate. And the quantity of

the straight line is known by the first proposition,[66]

therefore the quantity of the curve, equal to it by hypoth-

esis, is also known

---

equalem et cuilibet curve aliquam rectam," Clagett, 1964:
68.3. Cf. also ibid. 171.8 and 452.15-17.

   [66]"Et nota . . . per primam" (I, 21.7-8). Proposition
I, 1 might be applied here only if it is considered a general
proof that a line is capable of being measured.

[I, 22]   CIRCULI AREAM CONCLUDERE.

Medietas dyametri ducatur in medietatem circumferentie. Productum dabit aream. Ut si dyameter
est 7 pedum, area erit 38 et dimidii pedum. Item
aliter. Ducatur dyameter in se. Productum multi-   5
plicatur per 11; summa dividatur per 14. Denominatio dabit aream circuli. Probatio huius propositionis habetur ex Archimenide, qui docet invenire
orthogonium triangulum equalem circulo cuius area
probata est, ergo et area circuli.   10

Quod sic probatur. Concesso quod linea curva
sit equalis recte, fiat circulus, et protrahatur
semidyameter a circumferentia, et vocetur b.
Item protrahatur linea recta contingens circumfer-

---

4 38: ?F / post 38 add. pedum M / dimidii: dimidum ?M /
    pedum: pedis FH   om.M
5 ducatur dyameter tr.EF / ducatur om.H
5-6 multiplicatur per 11: per 11 multiplicatur FH
6 dividatur per 14: per 14 dividatur FH
8 ex: in FH
10 ergo: ?F
11 curva: recta E
14 item om.EFH / recta om.FH
14-15 circumferentiam om.FH

I, 22. TO DEDUCE THE AREA OF A CIRCLE.[67]

Let half the diameter be multiplied by half the circumference. The product will give the area (fig. 28). So that if
the diameter is 7 feet, the area will be 38 1/2 feet. The
same thing can be done in another way.[68] Let the diameter be
multiplied by itself. Let the product be multiplied by 11;
let the result be divided by 14. The quotient gives the area
of the circle. The proof of this proposition will be had
from Archimedes,[69] who shows how to find a right triangle
equal to a circle. The area of this is shown, therefore so
is the area of the circle.

Fig. 28   Fig. 29

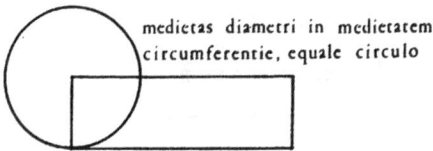

medietas diametri in medietatem
circumferentie, equale circulo

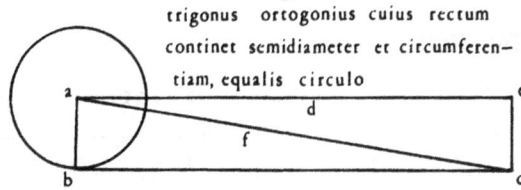

trigonus ortogonius cuius rectum
continet semidiameter et circumferentiam, equalis circulo

---

[67]"Circuli . . . 38 et dimidii pedum" (I, 22.1-4). This
method of finding the area of a circle is equivalent to the
formula $A = \frac{d}{2} \cdot \frac{c}{2}$. The same procedure may be found in "Boethius" Geom. II, ed. Folkerts, 1970: 166.885-888, Geom. incert.,
Ep. et Vit. Ruf., and Adelbold's letter to Pope Sylvester II,
ed. Bubnov, 1899: 346.15-16, 546.10-14 and 304.19-305.3. The
method is adequate as it stands and does not require the use
of π; in the calculation the author has used the value of π
from the previous proposition (I, 21.3-5) to compute the circumference.

[68]"Item . . . circuli" (I, 22.4-7). This procedure may
be expressed as $A = \frac{11}{14}d^2$; it assumes $\frac{22}{7}$ as the value of π.
The same algorithm may be found in "Boethius" Geom. II, ed.
Folkerts, 1970: 166.880-884, Geom. incert. and Ep. et Vit. Ruf.,
ed. Bubnov, 1899: 356.10-12 and 546.5-9.

[69]"Probatio . . . area circuli" (I, 22.7-10). The source
of the author's knowledge of Archimedes must remain a mystery

entiam a termino semidyametri equalis circumfer-   15

entie b̲ et vocetur c̲.   A centro circuli scil. ter-

mino [illius semidyametri] a̲ protrahatur linea f̲

ad terminum [illius] c̲.   Illud orthogonium est equale

circulo secundum Archimenidem.   Quod sic probatur.

Protrahatur equedistans [ad lineam] a̲ a termino   20

[illius] c̲ et vocetur e̲, et alia equedistans [ad

lineam] c̲ et vocetur d̲, a termino e̲ et a̲.   Et habes

quadrangulum duplum ad circulum.   Quod sic probatur.

Quadrangulum cuius unum latus est a̲; oppositum e̲, 3^m

c̲, 4^m d̲ fit ex ductu a̲ in c̲ ut premissa docent, sed   25

a̲ est trium pedum et dimidii pedis, c̲ 22.   Ergo

quadrangulum continet pedes 77.   Sed area circuli

fit ex ductu a̲ in medietatem c̲, quia in medietatem

---

15 p̲o̲s̲t̲ semidyametri a̲d̲d̲. c̲ H̲

16 et vocetur c̲ o̲m̲.F̲H̲ / p̲o̲s̲t̲ circuli a̲d̲d̲. in termino M̲ /
   scil.: e H̲

17 protrahatur i̲t̲e̲r̲.M̲

20 a̲ o̲m̲.H̲ / a o̲m̲.M̲F̲

21 c̲ o̲m̲.F̲

22 et a̲ o̲m̲.H̲

25-26 sed a̲ est: scil. a̲ F̲H̲

26 dimidii pedis: dimidium E̲ / 22: 11 M̲E̲ patet ?̲F̲

27 sed: et F̲H̲

28 c̲: b̲c̲ ?̲M̲

This is proved as follows.[70] Having conceded that a
curved line may be equal to a straight line, let a circle
be made, and let a radius be drawn to the circumference, and
let [the circumference] be called b. Then let a straight
line be drawn touching the circumference at the end of the
radius equal to the circumference b and let it be called c.
From the center, of the circle, that is from the end [of
radius] a let a line f be drawn to the end of c. That right
triangle is equal to a circle according to Archimedes. This
is proved as follows.[71] Let a parallel to a be drawn from
the end of c; and let it be called e. And another parallel
to c from the ends of e and a; and let it be called d. And
you have a rectangle double the circle. This is proved as
follows. [In] a rectangle of which one side is a, the

---

until further volumes of Clagett's Archimedes become available. These lines seem to have been adopted word for word
in the fourteenth-century Practica geometrie found in Vat.
Reg. suev. 1261, f.294v, as cited in Clagett, 1964: p. 36,
n. 13.

[70]"Quod . . . Archimenidem" (I, 22.11-19). The author
once again appeals to the rectification postulate (see I,
21.5-8 and n. 65, above). The triangle constructed here
(fig. 29) has an easily calculated area. The procedure
seems to have been taken over and improved upon slightly by
Gordanus in his Compilacio quorundam canonum in practicis
astronomie et geometrie (Vat. Pal. lat. 1389, f.107r) as
given in Clagett, 1964: p. 143, n. 4.

[71]"Quod . . . per secundam" (I, 22.19-34). The author is
unnecessarily going through a proof of the area of a triangle
using the method of I, 2.32-49. Because of the way in which
the triangle was constructed, its area will be equal to that
of the circle.

b. Ergo area circuli est subdupla ad quadrangulum
continens 38 pedes et dimidium. Ergo medietas quad-   30
ranguli est equalis circulo. Ergo orthogonium acf est
equale circulo. Probatio quod sit medietas eius,
quia paralellogramum orthogonaliter dividitur per
equalia. Ergo trianguli sunt equales per secundam.
Et hoc erat propositum probare, scil. aream circuli   35
invenire.

---

30 post continens add. scil. EF / pedes: pedis E
31 acf: acb MH
32 probatio: probo FH
33 quia om.MF / orthogonaliter om.E
33-34 per equalia om.FH
34 secundam: ?F
35-36 aream circuli invenire: are que circuli est invenire
    quantitatem H

opposite side e, the third c, and the fourth d, [the area]
is produced by the product a by c, as the preceding [propo-
sitions] teach. But a is three and a half feet, c 22.
Therefore the rectangle contains 77 feet. But the area of
the circle is produced by multiplying a by half of c because
[it is produced by multiplying a] by half of [the circumfer-
ence], b. Therefore the area of the circle, containing 38
and a half feet, is subdouble to the rectangle. Therefore
half the rectangle is equal to the circle. Therefore the
right triangle acf is equal to the circle. The proof that
it is its half, [is had] because the parallelogram is divided orthogonally [? into right triangles] by equals.
Thus the triangles are equal by the second proposition. And
this is what it was proposed to prove, namely to find the
area of a circle.

[I, 23]   CIRCULI INAURATURAM INVENIRE.

Metiatur dyameter se. Productum multiplicetur in
22, productum dividatur in 7.   Denominatio dabit circuli
inauraturam, que semper quadrupla est ad eius aream.
Eamdem habebis si dyameter metiatur periferiam.   Quod   5
probatur ratione numeri.   Ut si 7 pedum est dyameter,
154 erit inauratura.

---

1 inauraturam: inanraturam M̲
2 multiplicetur: feriatur F̲H̲
3 in 7: per 7 E̲
4 eius aream t̲r̲.̲F̲H̲
5 habebis: habebit F̲ / metiatur: ducatur per E̲
6 numeri: manifesta ex proxima H̲
7 154: et 54 M̲

I, 23. TO FIND THE SURFACE OF A CIRCLE [I.E. SPHERE].[72]

Let the diameter be multiplied by itself. Let the pro-

duct be multiplied by 22. Let the product be divided by 7.

The quotient gives the surface of the circle [i.e. sphere],

which is always quadruple its area [i.e. of a circle with

the same diameter]. You will have the same if the diameter

be multiplied by the periphery. This is proved by argument

from number. So that if the diameter is 7 feet, the surface

will be 154.

---

[72]"Circuli . . . inauratura" (I, 23.1-7). The methods
employed for finding the surface area of a sphere can be ex-
pressed as $S = \frac{22}{7} d^2$ and $S = dc$; they are equivalent if $\frac{22}{7}$
is taken as the value of π. Both of these rules are in
Geom. incert., ed. Bubnov, 1899: 360.1-4. The statement that
the surface is four times the area of a circle with the same
diameter is not found in Geom. incert. Different rules for
finding the surface appear in Ep. et Vit. Ruf., ed. Bubnov,
1899: 545.13-546.3. Proof "by argument from number" is not
adequate from the point of view of rigorous mathematics, but
for the purposes of practical geometry, it seems quite fit-
ting. "Argument from number" shows that the various expressions produce the same results; that is all the author is
claiming here. Gordanus has adopted our author's text in
his Compilacio (Vat. Pal. lat. 1389, f.107v) as printed by
Clagett, 1964: p. 145, n. 5.

[I, 24]   DYAMETRUM DUPLUM AD ALIUM, AREAM QUADRUPLAM AD
ALIAM CONCLUDERE.

   Hic docet quod si dyametrum circuli unius est

duplum ad diametrum alterius circuli, area maioris

circuli erit quadrupla ad aream minoris.   Hoc probatur   5

per precedentem.

1-6 In codices F et H, haec propositio antecessit propositionem I, 23.
1 alium: aliud F
3 hic: hoc E   hec FH / circuli unius tr.FH
3-4 est duplum: est dupla EM   om.H
4 post diametrum add. duplum H
5 circuli om.M / erit: exit M   est FH / minoris: mi H /
   hoc: hec E   quod FH / ante probatur add. sic F

I, 24. TO SHOW [THAT WHEN] ONE DIAMETER [IS] DOUBLE
ANOTHER, THE AREA [IS] QUADRUPLE THE OTHER.

This shows that if the diameter of one circle is double
the diameter of another circle (fig. 30), the area of the
greater circle will be quadruple the area of the smaller.
This is proved by means of the preceding.[73]

Fig. 30

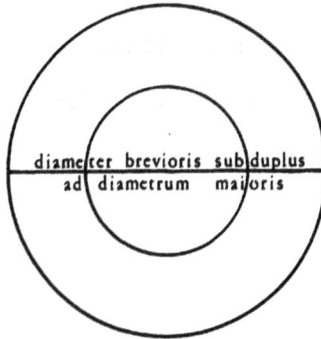

diameter brevioris subduplus
ad diametrum maioris

---

[73]"Hoc . . . precedentem" (I, 24.5-6). Using the procedure for the area of a circle from I, 22.1-7, the desired
result is easily obtained. The author would probably have
worked by applying specific numerical values rather than
any sort of algebraic expression.

[I, 25] DATIS TRIBUS PUNCTIS IN LINEA NON RECTA,
CENTRUM INVENIRE.

Esto primum datum punctum a, secundum b, tertium

c. Ponatur pes circini in a. Et contacto alio pede

super b, fiat quedam sectio. Positoque eodem pede cir-   5

cini in b, fiat sectio intersecans primam sectionem.

Similiter fiat in inferiori parte. Et ducatur linea a

sectione in sectionem. Similiter fiat inter b et c

ita quod linea protracta a sectione in sectionem

intersecet aliam protractam a sectione in sectionem.   10

Et necessario centrum erit in intersectione. Et

---

1 punctis: om.M

2 centrum: circumferentiam illa tangentum E

3 datum: arcum M

4 pes circini tr.F / circini om.E

5 super b: super a et b MF ad medietatem spatii b E /
   fiat: fiatque H

5-11 fiat . . . intersectione om.E

6 b: ?F

7 inferiori: inferiore H

8 fiat: fiet M

10 intersecet: interseceret F

11 centrum om.M / in om.H / intersectione: intersecatione FH /
   post intersectione add. Altrinsecus ducantur due linee.
   Positoque pede in b; ducatur pes ita quod lineas factas
   intersecet. A sectione in sectionem, ducatur linea recta
   f. Similiter fiat inter b et c ut altrinsecus signentur.
   Posito pede (hic add. in b intersecantes signa altrinsecus posito pede E) in a, a sectione in sectionem ducatur linea recta e ita quod intersecet f. In intersectione erit centrum circuli. EM

I, 25.   GIVEN THREE POINTS NOT IN A STRAIGHT LINE, TO FIND
THE CENTER [OF A CIRCLE TOUCHING THEM].[74]

Let the first given point be a, the second b, the third

c (fig. 31). Let the foot of the compass be put at a. And

having touched b with the other foot, let there be made some

arc. Having put the same [first] foot of the compass at b,

let an arc crossing the first arc be made. Let the same be

done below. And let a line be drawn from crossing to cross-

ing. Similarly let [arcs] be made between b and c so that

the line drawn from crossing to crossing intersects the

other drawn from crossing to crossing. And the center will

necessarily be at the intersection. And the circumference

Fig. 31

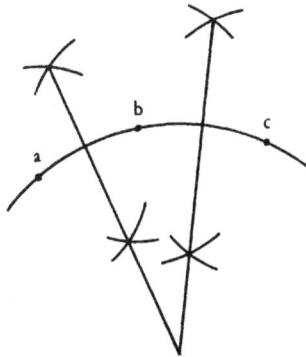

---

[74]"Datis . . . equales" (I, 25.1-14). This is the only
construction taught as such in this whole treatise. All of
the others are auxiliary to some proof, and none of them
give detailed instructions about placing the compass points.

circumferentia tanget a̲, b̲, c̲. Quod probatur quia

linee exeuntes a centro, in intersectione scil., ad

tria puncta erunt equales.

Et hec propositio valet ad horas in astrolabio   15

inveniendas.

---

12 a̲, b̲, c̲: ab eo F̲ / quia o̲m̲.F

13 p̲o̲s̲t̲ linee a̲d̲d̲. intersecantes se F̲ / a centro . . . scil.:
   ab intersecatione F̲H̲   (v̲a̲r̲.) s̲u̲p̲e̲r̲ s̲c̲r̲. ultima, scil.
   lenearum [sic] predictarum H̲$_a$

15-16 in astrolabio inveniendas: in astrolabio faciendas E̲
   inveniendas in astrolabio F̲H̲

will touch <u>a</u>, <u>b</u>, <u>c</u>. This is proved because the lines going
out from the center, that is, the intersection, to the
three points, will be equal.

And this proposition is of value for finding the hours
in an astrolabe.[75]

---

[75]"Et hec . . . inveniendas" (I, 25.15-16). To construct
the lines used in finding the unequal hours in an astrolabe,
one must draw several circles, each passing through three
given points. The unequal hours divide the periods of day-
light and night each into 12 equal parts, no matter what
the time of the year and resulting length of the day.

[I, 26]   CIRCULI QUADRATURAM REPERIRE.

Inveniatur area circuli per precedentem antepenultimam.
Illius aree radix erit latus quadrati equalis circulo.
Quod probatur quia latus illud, scil. radix, ductum in
se, facit aream quadrati equalem aree circuli.   Ergo   5
circulus quadratus.   Ut si dyameter circuli sit 7 pedum,
radix aree, scil. latus quadrati, erit 6 pedum et 5te
pedis, quod ductum in se reddet aream utriusque.

Hoc notato quod si numerus aree est surdus, per
minutias sumas radicem, scil. 6 pedum, 12 minuta, 16   10
secunda, 12 tercia in radice predicte aree, que ducta

---

2 per precedentem: presentati $\underline{E}$ / antepenultimam:
  penultimam $\underline{F}$ ⸱per antepenultimam $\underline{ME}$
5 aree circuli $\underline{tr.E}$
6 quadratus: quadratur $\underline{FH}$
7 erit $\underline{om.FH}$
8 pedis $\underline{super\ scr.}$ et paulo plus $\underline{H}_a$
10 sumas: sumes $\underline{E}$ capies $\underline{FH}$ / pedum: pedes $\underline{FH}$ / 12: 10 $\underline{MEF}$
11 12: 14 $\underline{F}$ / ducta: ducte $\underline{H}$

I, 26. TO ASCERTAIN THE QUADRATURE OF A CIRCLE.

Let the area of the circle be found as in the last proposition but three [i.e. I, 22]. The root of that area will
be the side of a square equal to the circle. This is proved
because that side, namely the root, multiplied by itself,
makes the area of the square equal to the area of the circle.
Thus [it makes] the circle square. So that if the diameter
of the circle is 7 feet (fig. 32), the root of the area,
namely the side of the square, will be 6 1/5 feet, which
multiplied by itself will give the area of both.[76]

But if the number of the area is a surd, once this has
been noted, you will take the root by means of fractions,
namely 6 feet, 12 minutes, 16 seconds, and 12 thirds in

Fig. 32

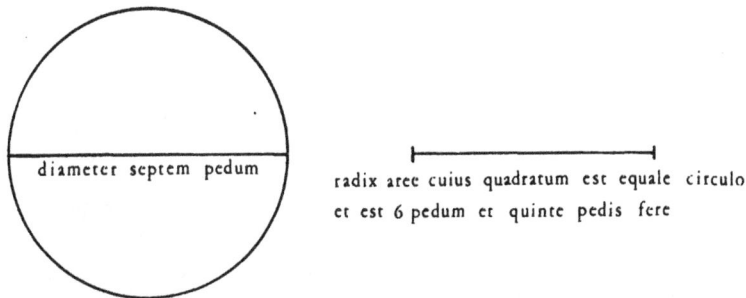

diameter septem pedum

radix aree cuius quadratum est equale circulo
et est 6 pedum et quinte pedis fere

---

[76]"6 pedum . . . utriusque" (I, 26.7-8). Six and one-
fifth feet squared is 38 11/25 square feet, a value not very
far from the 38 1/2 square feet found in I, 22.4 for the area
of a circle.

in se aream reddet.   Si autem insensibilis est excessus

differentie, quasi neglectum preterit geometria.

---

12 reddet: reddunt FH
13 neglectum: lac.F / preterit: poterit ?F

the root of the stated area, which root, multiplied by itself gives the area.[77] If, however, the excess of the difference cannot be sensed, geometry passes over it as if
disregarded.[78]

---

[77]"Hoc notato . . . reddet" (I, 26.9-12). The square of
6; 12, 16, 12 feet is 38; 29, 44, 57, 10, 26, 24 square feet,
where 6; 12, 16, 12 feet is to be read 6 feet, 12 minutes,
16 seconds and 12 thirds and is equal to
$6 + 12 \times 60^{-1} + 16 \times 60^{-2} + 12 \times 60^{-3}$. We will normally use
this notation for sexagesimal fractions. The square is of
course very close to the required 38 1/2 square feet. The
procedure for extracting roots in sexagesimal fractions is
given below in IV, 26-28.

[78]"Si autem . . . geometria" (I, 26.12-13). The author
seems to be saying that the difference between the squares
of 6 and one-fifth and of 6; 12, 16, 12 is insignificant.
Not being sensible is a very curious definition of equality
in geometry; however, it seems quite apt in practical geometry.

[I, 27]   QUADRATI CIRCULATIONEM INVENIRE.

Metiatur latus quadrati se. Productum multiplicetur
per 14; summa dividatur per 11. Denominationis radix
erit dyameter circuli equalis quadrato predicto. Ut
si latus quadrati est 6 pedum et 5te pedis, dyameter   5
circuli erit 7 pedum. Quod probatur per primam et per
illam qua probatur area circuli.

---

3 denominationis: denominatio <u>ME</u>
4 quadrato predicto <u>tr.FH</u>
5 est <u>om.FH</u> / pedis <u>super scr.</u> et paulo plus <u>H</u>$_a$
6 primam: ipsam <u>H</u>
7 qua: <u>?F</u>

I, 27. TO FIND THE CIRCLING OF A SQUARE.[79]

Let one side of a square be multiplied by itself.[80] Let

the product be multiplied by 14, the result divided by 11.

The root of the quotient will be the diameter of a circle

equal to the aforesaid square. So that if the side of a

square is 6 1/5 feet (fig. 32), the diameter of the circle

will be 7 feet. This is proved by the first proposition

[before this] and by that by which is proved the area of a

circle.[81]

---

[79]"Quadrati . . . circuli" (I, 27.1-7). Once again Gordanus seems to have taken over our author's text in his Compilacio (Vat. Pal. lat. 1389, f. 111v) as found in Clagett, 1964: p. 145, n. 5.

[80]"Metiatur . . . predicto" (I, 27.2-4). The procedure may be expressed by the formula $d = \sqrt{\frac{14\,s^2}{11}}$, where d is the diameter of the required circle and s the side of the given square.

[81]"Quod . . . circuli" (I, 27.6-7). Proposition I, 26 provides the means of finding a square equal to a given circle. The area of a circle is shown in Proposition I, 22.

[I, 28] CIRCULI QUADRATO INSCRIPTI EXCESSUM REPERIRE.

Per proximas enim nota est area circuli, que si subtrahatur ab area quadrati inscribentis nota per precedentes, notus erit excessus. Ut si circuli inscripti dyameter est 7 pedum, excessus quadrati est 10 pedum et dimidii, ut precedentes demonstrant.   5

---

1 excessum mg. hab. active intelligatur et passive, scil. excessus ad [sic] circuli ad quadratum inscriptum et inscribentem H$_a$
2 ante per add. Hec patet ?H / proximas: primas FH / enim om.E / si om.E
4 post erit add. et E
5 inscripti tr.post pedum F / est om.FH / 7: ?F / quadrati est tr.FH / post est (2) super scr. inscribentis circulum ad circulum inscriptum H$_a$
6 demonstrant: monstrant EF

I, 28. TO ASCERTAIN THE EXCESS OF A CIRCLE INSCRIBED IN
A SQUARE.[82]

The area of a circle is, in fact, known by the previous

propositions. If the area be subtracted from the area of

the inscribing square, known by the preceding propositions,

the excess will be known. So that if the diameter of the

circle is 7 feet the excess of the square is 10 1/2 feet,

as the preceding propositions demonstrate.[83]

[Fig. 33]

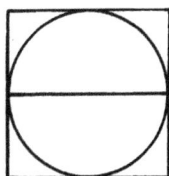

---

[82]"Circuli . . . demonstrant" (I, 28.1-6). A similar
proposition is found in Geom. incert., ed. Bubnov, 1899:
349.12-22.

[83]"Ut si . . . demonstrant" (I, 28.4-6). From I, 22.3-4
we know that the area of a circle of a diameter 7 feet (fig.
33) is 38 1/2 square feet. From I, 10 we may conclude that
the area of a square of side 7 feet is 49 square feet. Ten
and a half [square] feet is thus the difference.

[I, 29] QUADRATI CIRCULO INSCRIPTI EXCESSUM HABERE.

Ab area circuli nota subtrahantur 4 14^e [aree quadrati inscribentis], residuum erit area quadrati inscripti circulo. Ut si 14 pedum est dyameter circuli,
probatum est quod 154 erit area. Subtractis quatuor   5
14^{is} [aree, scil. 196, quadrati inscribentis], 98 erit
area quadrati [inscripti], ut probatur per secundam.

---

1 quadrati . . . habere om.MH
2 4 14^e: 14 E 4 undecime H (corr. ex 4 quatuordecime)
4 ante circulo add. in E / dyameter circuli tr.F
5 154: 158 ME
6 14^{is}: undecimis H (corr. ex quatuordecimis)

I, 29. TO HAVE THE EXCESS OF A SQUARE INSCRIBED IN A
CIRCLE.[84]

From the known area of a circle, let four fourteenths
[of the area of the inscribing square] be subtracted, the
remainder will be the area of the square inscribed in the
circle.[85] So that if the diameter of the circle is 14 feet
(fig. 34), it has been proved that the area will be 154.
Having subtracted four fourteenths [of the area, 196, of
the inscribing square], the area of the inscribed square will

Fig. 34

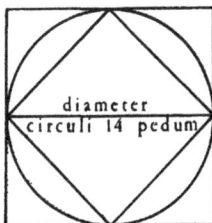

84"Quadrati . . . propositum" (I, 29.1-10). This proposition, intelligible only after a major emendation, has as its
source Geom. incert., ed. Bubnov, 1899: 349.12-350.9. The
figure accompanying the proposition in Geom. incert., ed.
Bubnov, 1899, fig. 87, is like fig. 34 here, but is is marked
with numerical values for the area of the circle and its inscribed and circumscribed squares. The revision in MS H
offers different emendations which also make the text coherent. However, mine are supported by reference to the author's
probable source.

85"Ab area . . . circulo" (I, 29.2-4). The procedures
given here are equivalent to the application of the following formulas, where $\underline{d}$ is the diameter of the circle, its
area $A_o = (11/14)d^2$. The area of the larger square, $A_s = d^2$,
and the area of the inscribed square, $A_i = A_o - (4/14)d^2 = d^2/2$.

Si latus [minoris quadrati] se metiatur, cui si addas

illas 4 14as [illius 196], <154> fient et restauratur

area <circuli>.   Et hoc erat propositum.   10

---

8 <u>ante</u> si (1) <u>add.</u> ut <u>F</u> / latus <u>super scr.</u> inscripti, scil.
   quadrati <u>H</u>$_a$ / se metiatur <u>tr.FH</u>
9 illas: istas <u>E</u> / 154 (<u>ed.</u>): 158 <u>ME</u>   scil. 56 <u>F</u>   scil. 26 <u>H</u> /
   fient et <u>om.FH</u>
10 circuli (<u>ed.</u>): quadrati <u>MEFH</u>

be 98, as is proved by the second proposition.[86] If the
side of the smaller square is multiplied by itself, and if
you add to it those four fourteenths [of the same 196], 154
is made, and the area of the circle is restored. And this
was to be proved.

---

[86]"area . . . secundam" (I, 29.7). The area of a square
is shown in I, 10, not in I, 2.

[I, 30] QUADRATI CIRCULUM INSCRIBENTIS AD QUADRATUM
INSCRIPTUM CIRCULO, EXCESSUM RIMARI.

Inscribatur circulus quadrato et eidem circulo
minus quadratum. Quia nota est quantitas maioris quadrati
et circuli inscripti, et nota est quantitas minoris   5
quadrati per precedentem, notus est ergo excessus maioris
quadrati ad minus quia est duplum ad ipsum. Est enim area
maioris 196 cuius radix, scil. 14, est latus eius et
dyameter circuli. Et hoc probatur per precedentia.

---

3 eidem: in eodem $\underline{E}$
4 est $\underline{om.FH}$
5-6 et circuli . . . minoris quadrati $\underline{om.E}$
5 inscripti, et . . . quantitas minoris $\underline{om.H}$
6 $\underline{ante}$ quadrati $\underline{add.}$ inscripti $\underline{F}$ / ergo excessus $\underline{tr.FH}$ /
   maioris: minoris $\underline{M}$ $\underline{super\ scr.}$ scil. quantitas $\underline{H}_a$
7 ad minus: ad ipsum $\underline{M}$ / quia: que $\underline{FH}$ / duplum ad ipsum:
   duplum ad minus $\underline{M}$ dupla ad ipsam $\underline{FH}$
8 196: $\underline{?F}$ / 14: 4 $\underline{MF}$ / eius $\underline{om.FH}$
9 Et $\underline{om.E}$

I, 30.   TO DISCOVER THE EXCESS OF A SQUARE INSCRIBING A
CIRCLE OVER A SQUARE INSCRIBED IN A CIRCLE.[87]

Let a circle be inscribed in a square and in the same

circle, a small square (fig. 34). Since the quantity of

the larger square and of the inscribed circle is known, and

the quantity of the smaller square is known by the preceding

proposition, thus the excess of the greater square over the

smaller is known, for it is double the latter. Indeed the

area of the larger is 196, the root of which, namely 14, is

its side and the diameter of the circle. And this is proved

by the preceding.[88]

---

[87]"Quadrati . . . ipsum" (I, 30.1-7). This proposition,
a continuation of I, 28 and 29, also has as its source Geom.
incert., ed. Bubnov, 1899: 349.12-350.9.

[88]"Quia . . . precedentia" (I, 30.4-9). The area of the
larger square was necessary for I, 29 but was given only in
the emendations. As shown in n. 85, $A_s = d^2 = 2A_i$. The side
of A_s is of course $\underline{d}$, the diameter.

[I, 31]   SUPERFICIEI NOTE CAPACITATEM LATERCULORUM
COMPREHENDERE.

Habeat superficies in longitudine pedes 200, in

latitudine, 100.   Teneatque laterculus 5 pedes in longum

et 4 in latum.   Numerum eorum sic scies.   Accipe 5   5

partem 200, scil. 40, et quartam partem 100, scil. 25.

Ducas 40 in 25, productum dabit numerum laterculorum.

Hoc probatur ratione numeri.

---

1 laterculorum om.F
1-2 laterculorum comprehendere tr.H
3 pedes 200: pedes 300 M   200 pedes FH
5 et om.FH
6 quartam: 5 M

I, 31. TO COUNT THE NUMBER OF SMALL RECTANGLES IN A KNOWN
SURFACE.[89]

Let the surface have 200 feet in length, 100 in width.

Let the small rectangle contain 5 feet in length and 4 in

width (fig. 35). You will know the number of them as fol-

lows. Take the fifth part of 200, namely 40, and the fourth

part of 100, namely 25. Multiply 40 by 25; the product gives

the number of small rectangles. This is proved by argument

from number.[90]

Fig. 35

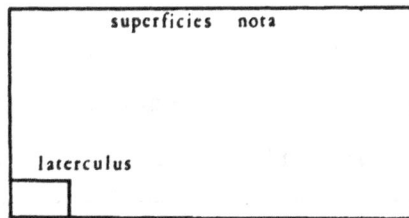

---

[89]"Superficiei . . . laterculorum" (I, 31.1-7). Instead
of the number of small rectangles, the probable source of
this proposition, Geom. incert., ed. Bubnov, 1899: 352.15-21,
finds the number of sheep which can be quartered in a field
of the same size as this one. Elsewhere in Geom. incert.,
ed. cit.: 355.17-27, the number of "laterculi" or tiles in
the floor of a basilica is computed. "Laterculus" is also
the name of a surface-area measure, having dimensions 12 x
23 inches, in Gerbert, Geom., ed. Bubnov, 1899: 59.9-15.

[90]"Hoc . . . numeri" (I, 31.8). One can easily find the
number N of small rectangles of dimensions a and b in a
field of dimensions A and B from the formula
$N = \frac{AB}{ab} = \frac{A}{a} \cdot \frac{B}{b}$. Without this technique for generalizing,
the computation using numerical examples offers a fairly
convincing argument.

[I, 32] CAMPI FASTIGIOSI SUPERFICIEM INVENIRE.

Si frons campi fuerit 50 perticarum, medium 60,
utrumque latus 100. Iungatur frons medio. Summe
medietatem scil. 55, ferias in 100, id est in
longitudinem, et habes numerum perticarum. Quas divide   5
per perticas unius arpenti, scil. 144, secundum nos.
Denominatio dabit numerum arpentorum, scil. 38. Vel
divide per 72, secundum quosdam tot enim perticas
continet arpentus. Secundum eos denominatio dabit
numerum arpentorum.   10

Hoc notato quod 6 pedes faciunt trabuscum et
trabusci 2, perticam; 4 pertice <quadratum includentes>
tabulam; 12 tabule, sextarium; 8 sextarii, modium.

---

1 superficiem invenire: noti capacitatem arpennorum concludere E
2 si frons campi fuerit: sit frons FH
3 utrumque: ut quantum M / latus mg. hab. Hic dico latus non
  latera continentia fastigiosum, sed longitudines ipsius
  fastigiosi. H_a
4 ferias: multiplicet E / id est: scil. H
6 arpenti: arpenni EF apemni H / 144: 184 M / nos: hoc E
7 arpentorum: arpennorum EF appemniorum H / 38: ?F super
  scr. fere H_a
9 arpentus: arpennus E arpemnus F arpemnium H / eos:
  hoc M
10 arpentorum: om.E arpennorum F arpemniorum H
11 trabuscum: trabucum FH / et: 2 FH
12 trabusci: trabuci FH / 2 om.FH
13 12: 13 FH

I, 32. TO FIND THE SURFACE OF A SLOPE-SIDED FIELD.[91]

Let the frontage of the field be 50 perches, the inner

boundary 60 and each side 100 (fig. 36). Let the frontage

be added to the inner boundary. Multiply half the sum, that

is to say 55, by 100, that is by the length, and you have the

number of [square] perches. Divide them by the perches in

one arpent, which is 144 according to us. The quotient gives

the number of arpents, that is 38. Or divide by 72, for ac-

cording to certain persons, one arpent contains that many

perches. According to them, the quotient gives the number

of arpents.

Once this has been noted [we shall observe] that 6 feet

make a trabuscum,[92] 2 trabusci a perch; 4 perches enclosing

a square make a tabula, 12 tabula a sextarius, 8 sextarii

a modius. Twelve perches on each side make an arpent

Fig. 36

[incorrect in H, ab should be parallel to cd]

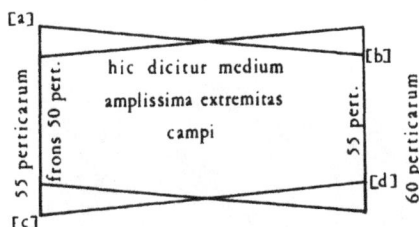

_____

[91]"Campi . . . arpentorum" (I, 32.1-10). This proposition is very close to Geom. incert., ed. Bubnov, 1899: 352.22-
353.12. The procedure yields only an approximation of the
area because the slant height is used as if it were the altitude in computing the area of this trapezoidal field. The
remainder is ignored in the conversion to arpents.

[92]"Hoc notato . . . arpentum" (I, 32.11-15). Cf. IV,
1.8-12, where 4 feet make a trabuscum, and the square measures are similarly defined. Some of these measures are defined by Gerbert in his geometry, some with the same equivalences, some differently. He gives "passus duos," or two

12 pertice in utroque latere, arpentum; secundum alios

12 in longo, 6 in lato faciunt arpentum.   15

---

14 pertice om.E / utroque: unoquoque FH / post latere add.
   pertice E / arpentum: arpennum E arpennium FH
15 post longo add. et FH / faciunt arpentum tr.H /
   arpentum: arpennum E arpenninium ?F arpennium H

according to us, according to others 12 in length and 6 in

width.

---

paces, each of 5 feet, as the value of the "pertica" or
perch (ed. Bubnov, 1899: 60.1 and 60.6). The "tabula" is
described as a fourth of a "jugerum" (ibid.: 61.11). Gerbert gives 144 square rods ("pertice constrate") as the
value of an arpent (ibid.: 60.22-61.3). His term for arpent
is "agripennus" or "aripennus." The "sextarius" and "modius"
are commonly measures of capacity; they may also mean the
amount of land that may be sowed with that much grain. The
"tabula" is also a measure of surface. It is not clear
whether a tabula is 4 square perches or 4 perches on each
side.

[I, 33] CAMPI QUADRANGULI NOTI NUMERUM ARPENNORUM CAPERE.

Esto unum latus 30, oppositum 32, 3^m 34, quartum 36.

Iunge a, b, scil. opposita latera, hoc est longitudines,
et sume medietatem. Iunge latera latitudinis et sume
medietatem. Et metiatur una medietas aliam. Productum   5
divide per 144. Denominatio dabit propositum, quod
probatur ratione numeri.

---

1 arpennorum: arpenninorum F arpenniorum H / capere mg. hab.
  Quod autem hic dicitur et quod dictum est prius in 12^a.
  Non autem intelligo quomodo fiat, nisi [bis] per noticiam
  angulorum et laterum, aut laterum et diametrum, aut nisi
  intelligantur duo latera equidistancia, reliqua autem duo
  quomodocumque et latitudines sumantur proprie que tunc
  sunt equalia. H_a
2 post 3^m add. metiatur una medietas reliquum uni M metiatur medietas una reliqua uni F / 34: 33 E / quartum:
  quartam M / 36: 34 E
3 a, b, scil. om.FH / hoc est: scil. FH
4-5 iunge latera . . . aliam om.FH
4 sume (1): summe MH / post medietatem add. in mg. de hinc
  latitudines et sume medietatem. Metiaturque una medietas
  reliquam H_a / sume (2): summe M
5 et metiatur: metiaturque FH / aliam: reliquam FH
6 divide om.F / ante denominatio add. secundum nos H
6-7 quod probatur ratione: quod probari ratione potest F
  et potest probari ratione H

I, 33.   TO OBTAIN THE NUMBER OF ARPENTS IN A KNOWN
QUADRANGULAR FIELD.[93]

Let one side be 30, the opposite 32, the third 34, the

fourth 36 (fig. 37).   Add the opposite sides, a, b, that is

the lengths and take half.   Add the sides of the width and

take half.   Let one half be multiplied by the other.   Divide

the product by 144.   The quotient gives what was proposed.

This is proved by argument from number.[94]

Fig. 37

---

[93]"Campi . . . per 144" (I, 33.1-6).   Geom. incert., ed.
Bubnov, 1899: 353.20-29, gives an almost identical proposition; except where our text has 36, it has 32.   Cf. I, 12
and n. 44 above where the area of a comparable quadrilateral
is considered using a similar technique.

[94]"quod . . . numeri" (I, 33.6-7).   Argument from number
may demonstrate the number of arpents in a fixed area, but
an area of the sort considered here is not generally computable.

[I, 34]   SUPERFICIEI NOTE NUMERUM DOMUUM CAPIENDARUM
COMPREHENDERE.

Esto superficies in una longitudine 1000 pedum,

in alia 1100, in utroque latere 600. Habeatque domus

40 pedes longitudinis, 30 latitudinis. Iunctarum   5

longitudinum medietas dividatur per 40, iunctarum

latitudinum medietas dividatur per 30. Una denominatio

metiatur aliam, scil. 20, 26. In hoc exemplo denominatio

dabit numerum domuum comprehensarum, quod probatur

ratione numeri.   10

---

2 comprehendere mg. hab. Hic dico sicut prius, latera non
  que continent superficiem cum longitudinibus, sed equidistantes ortogonales ductas a terminis minoris longitudinis super maiorem longitudinem. Longitudines vero
  duas, dico duo latera inequalia equidistancia, superficiem cum duabus declinatis continencia, que declinate hic
  non ponuntur 600 pedum, sed linee perpendiculares que
  subtenduntur angulo quem continent declinate cum maiore
  longitudine. $\underline{H}_a$
3 una om.F tr.post longitudine H
4 1100: 1000 F 1080 H / habeatque: habeat FH
5 pedes: pedum E / 30 latitudinis: latitudinis 20 M
6 post per add. in mg. 40, medietas latitudinum per $\underline{H}_a$ /
  40: 50 ?F
6-7 40 . . . per om.H
8 aliam: reliquam FH / scil. bis F / denominatio om.F
  tr.post numerum H
9 comprehensarum: comprehendarum F comprehendendarum H
9-10 quod . . . numeri om.FH

I, 34. TO COUNT THE NUMBER OF HOUSES TO BE OBTAINED FROM
A KNOWN SURFACE.[95]

Let a surface be 1000 feet on one length, 1100 on the

other, 600 feet on each side (fig. 38). Then let a house

have 40 feet in length, 30 in width. Let half the combined

lengths be divided by 40; let half the combined widths be

divided by 30. Let one quotient be multiplied by the other,

namely 20 by 26. In this example, the result will give the

number of counted houses, which is proved by argument from

number.

---

[95]"Superficiei . . . comprehensarum" (I, 34.1-9). The
probable source of this proposition, Geom. incert., ed. Bubnov, 1899: 354.18-29, has "a quadrangular city" instead of
the less explicit "known surface" found here. The numerical
values in the example are the same in the two texts. As in
I, 32, slant height is used instead of altitude in the area
computation (cf. n. 91 above). Note that the remainder is
ignored in the division by 40. The emendation of H changes
the value of one of the lengths to eliminate the remainder.

Et hac arte cuiuslibet note superficiei sumendo
denominationes contentorum in longitudine et latitudine
certissime invenies numerum pavimenti.

---

12 contentorum: contentarum F / longitudine: latitudine F
   latitudinem H / latitudine: in longitudine F longitudinem H
13 certissime invenies tr.FH

And with this art, you will find with certitude the number of small areas in any known surface by taking the product
of the quotients contained in the length and width.

Fig. 38

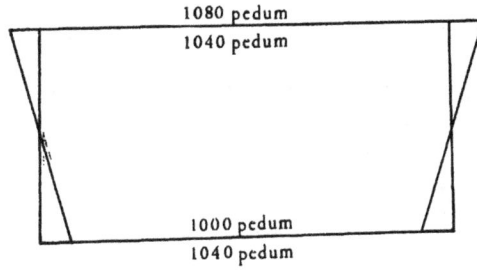

1080 pedum
1040 pedum

1000 pedum
1040 pedum

[I, 35] SUPERFICIEI TRIANGULE NUMERUM LATERCULORUM
COMPREHENDERE.

Esto laterculus 20 pedum longitudinis et 10

latitudinis. Medietas laterum, iunctorum superficiei

scil., dividatur per 20. Medietas frontis, scil. basis   5

medietas, dividatur per 10. Denominationes ducte in se

dabunt numerum. Ut si 100 pedum est utrumque latus,

frons 90, 20 erunt laterculi. Quod probatur ratione

numeri.

---

2 comprehendere: concludere <u>FH</u> comprehendere vel concludere <u>E</u>
4 <u>ante</u> medietas <u>add.</u> et <u>F</u> / medietas: medietatem <u>E</u> /
   superficiei: superficies <u>E</u> <u>super scr.</u> triangule <u>H</u>$_a$
5 dividatur <u>mg. hab.</u> In hac demonstratione subponitur aream
   trianguli esse equale ei quod fit ex ductu unius laterum
   medietas trianguli, quod non est verum nisi computemus
   mensuras semiplenas juxta regulam Boetii in principio
   positam de invenienda area trianguli. Possunt laterculi
   trianguli facile inveniri. Si [?] inventa cateto, per
   ipsam et medietatem basis operemur iuxta presentem
   regulam. <u>H</u>$_a$
5-6 20 medietas . . . per <u>om.</u>E
6 10: 20 <u>M</u> / in se <u>super scr.</u> scil. una scil. in alteram <u>H</u>$_a$
7 100: 1000 <u>MH</u> / est <u>om.FH</u>
8 <u>ante</u> frons <u>add.</u> et <u>H</u> / frons: frontis <u>F</u> <u>super scr.</u> basis,
   scil. <u>H</u>$_a$ / 90: 80 <u>H</u> / 20: 200 <u>H</u>
9 <u>post</u> numeri <u>add.</u> explicit planimetria <u>H</u>

I, 35. TO COUNT THE NUMBER OF SMALL RECTANGLES IN A
TRIANGULAR SURFACE.[96]

Let the small rectangle be 20 feet in length and 10 in
width. Let half of the sides, namely those combined [to
form] the surface be divided by 20. Let half the frontage,
that is half the base, be divided by 10. The quotients multiplied by themselves give the number. So that if each side
is 100 feet, and the frontage 90 (fig. 39), there will be
20 small rectangles. This is proved by argument from number.

Fig. 39

----

[96]"Superficiei . . . laterculi" (I, 35.1-8). Once again
our source, Geom. incert., ed. Bubnov, 1899: 354.30-355.8,
specifies "city" and "house" where our text has "surface"
and "small rectangle." Geom. incert. gives the same numerical values. When half the base, 45 feet, is divided by the
width of the small rectangle, 10 feet, the remainder is dis-
carded. Not keeping the remainder probably compensates, to
some extent, for the use of slant height instead of altitude.

[I, 36] CIVITATIS ROTUNDE NUMERUM DOMUUM COLLIGERE.

Esto domus quelibet 30 pedum in longitudine 20 in
latitudine. Ex ductu 20 in 30 surgunt 600, per quem
numerum divides pedes aree civitatis. Denominatio dabit
numerum domuum. Aream invenies ex ductu semidyametri   5
in medietatem circumferentie, ut probatum est.

Esto sic planimetriam instruximus. Et operis
huius prima partitio terminatur.

Restat ut in secundo volumine altimetriam et
profundimetram instruamus.   10

1-6 civitatis . . . est om.FH
2-3 20 in latitudine add.mg.E   om.M
7 esto: et E
7-8 esto . . . terminatur om.FH
9-10 restat . . . instruamus om.H

I, 36. TO INFER THE NUMBER OF HOUSES IN A ROUND CITY.[97]

Let some house be 30 feet in length, 20 in width. From

taking 20 times 30, 600 arises, by which number you divide

the area of the city in [square] feet. The quotient will

give the number of houses. You will find the area from the

product of half the diameter by half the circumference, as

has been proved.[98]

Let us thus have taught planimetry. And the first section

of this work is ended.

It remains for us to instruct the altimeter and profundi-

meter in the second volume.

---

[97]"Civitatis . . . est" (I, 36.1-6). The source of this
proposition seems to be Geom. incert., ed. Bubnov, 1899:
355.8-16, which uses the same values for the size of the
house. Geom. incert. goes into greater detail, computing
the area of a city of given size.

[98]"Aream . . . est" (I, 36.5-6). The method for finding
the area of a circle is given in I, 22.

[II, 1] SOLIS ALTITUDINEM PER UMBRAM METIRI.

Metiatur umbra se ipsam. Producto addantur 144,
totalis summe radix sumatur. Item, umbra extendatur
in 60, productum dividatur per radicem inventam.
Denominatio erit corda, cuius sumatur arcus qui a 90   5
subtrahatur, residuum erit altifa solis, i.e. altitudo.
Ut si 12 digitorum sit umbra, 45 graduum erit altitudo.
Probationem huius propositionis in Almagesti inveniens,
quam quia nimium habet prolixitatis ut probatam transcurrimus.   10

Hoc notato quod de umbra plana loquimur, que sumitur

---

1 ante solis add. incipit liber secundus E incipit
   altimetria H
2 metiatur tr.post ipsam H
3 totalis: talis M / umbra extendatur: feriatur H
5 a 90: 790 H
6 altifa: altipha E arcipha H
7 12: 13 [corr. in mg.] H
8 probationem: probatio H / in om.H / inveniens: invenies H
9 prolixitatis: prolixitates H
9-10 transcurrimus: transeimus H

II, 1. TO MEASURE THE ALTITUDE OF THE SUN BY MEANS OF A
SHADOW.[1]

Let the shadow be multiplied by itself. Let 144 be added

to the product. Let the root of the whole sum be taken.

And then let the shadow be multiplied by 60. Let the pro-

duct be divided by the root found [above]. The result will

be the Sine; let its arc be found. Let [the arc] be sub-

tracted from 90, the remainder will be the altifa,[2] that is,

the altitude, of the sun. So if the shadow be 12 digits,

---

[1]"Solis . . . altitudo" (II, 1.1-6). The procedure here
is equivalent to solving the forumula Sin (90 - a) =
$(60 \cdot s)/(s^2+144)^{\frac{1}{2}}$, where the shadow of a vertical gnomon of
12 digits is s and the altitude of the sun is a. We shall
use Sin x to mean 60·sin x. Note that the author uses corda
for Sine; if he had intended to use chords, the formula
would have to be emended so that Sin (90 - a) =
½ Crd (180 - 2A). Al-Battani also uses chorda for the half-
chord of the double arc; see Nallino, 1903-1907, v. 1: 10.40-
11.2. Ibn Ezra alternates between corda and algeib, the
Latin form of the Arabic word for sine; cf. Millás-Vallicrosa,
1947: p. 55. A table of arcs and sines is needed to find
the value of a. This method for finding the solar altitude
is frequently found in medieval astronomical works. Most of
them take the radius of the unit circle as 150 parts; cf.
Millás-Vendrell, 1963: 153.1-6, Suter, 1914: pp. 21-22, Neugebauer, 1962: p. 56 and Curtze, 1900: 342.13-24. The author
takes the Ptolemaic value of 60 parts (Almagest, I, 10 and 11),
as does al-Battani; cf. Nallino, 1903-1907, v. 1: 22.21-38.

[2]"altifa . . . altitudo" (II, 1.6). Altifa is a corrupted
transcription of al-irtifaᶜ, meaning altitude; cf. Biruni,
1934: par. 225. Other Latin forms are found in other astronomical works; cf. Millás-Vendrell, 1963: 154.1 (alertifa),
Gerbert, Astrolabe, ed. Bubnov, 1899: 128.13 (ertifa) and
Suter, 1914: pp. 21-22 (artifa).

secundum proportionem ipsius ad suum cathetum qui semper
est 12 digitorum, que si est dupla ad cathetum, erit 24;
si equalis, 12; si subdupla, sex, etc. Et invenitur
per regulam divisam secundum quantitatem sui catheti   15
orthogonaliter in ea fixi et per divisionem cuiuslibet
sectionis facte in 12.

Qualiter etiam altitudo solis per astrolabium et
quadrantem invenitur patet.

---

12 ipsius om.H / ad suum om.E / cathetum: catheti E
13 12: 13 H
17 sectionis: lectionis H / facte tr.ante cuiuslibet H /
    12: 13 H
18 qualiter: quantum H / etiam: sit H / solis: celi H
18-19 et quadrantem: per quadrans H tr.post invenitur E /
    invenitur: inveniri potest ut H

the altitude will be 45 degrees. Finding the proof of this proposition in the <u>Almagest</u>,[3] we hurry by it as proved since it is too tedious.

This having been noted, we speak about the plane shadow[4] which is taken according to the ratio to its vertical, which is always 12 digits. If the shadow is double the vertical, it will be 24; if equal, 12; if half, 6, etc. And it is found by a ruler divided according to the size of the vertical fixed at right angles to it, and by dividing any section made [equal to the vertical] into 12 parts.

And how the altitude of the sun is found by means of the astrolabe and quadrant is clear.[5]

---

[3]"probationem . . . inveniens" (II, 1.8). The proof of this proposition is not given directly in the <u>Almagest</u>, but may perhaps be derived from part of the proof in <u>Almagest</u>, II, 5. The proof is found in Ibn-al-Muthanna, ed. Millás-Vendrell, 1963: 153.13-154.14.

[4]"umbra . . . in 12" (II, 1.11-17). The plane shadow differs from the shadow in a quadrant in that it may be any number; the shadow in a quadrant or astrolabe has 12 as its maximum value (see I, 1). The instrument described here will yield a value for which the formula given in n. 1 above may be applied directly for any value of the shadow. A more detailed description of the instrument is given in II, 38.

[5]"Qualiter . . . patet" (II, 1.18-19). The astrolabe and quadrant are normally marked with a scale of degrees so that altitude can be read directly. They also have shadow quadrants which could be used after a minor adaptation of the method above.

[II, 2] STELLE ALTITUDINEM PER TRIANGULUM ORTHOGONIUM
INVENIRE.

Fiat orthogonium sic, quod basis sit virga recta in

directo pupille procedens. Erigaturque cathetus scil.

virga recta orthogonaliter. Et tantum contrahatur basis   5

predicta quod per summitatem catheti videatur stella.

Que est proportio basis ad cathetum prefatum, eadem

est proportio umbre ad totam altitudinem solarem. Sed

nota est basis 12 digitorum si est equalis catheto. Et

ita secundum proportionem, ergo nota est umbra. Ergo   10

nota est altitudo stelle, quod probatur per precedentem,

que per umbram solis altitudinem docet invenire.

Qualiter per astrolabium et quadrantem altitudo

stelle invenitur patet.

---

4 scil.: in H
7 que est: et quantumcumque H
8 proportio om.H / totam: tantam H
9 post basis add. sic H / 12: 13 H
11 quod: hoc H
12 que: et H / solis altitudinem tr.H / altitudinem docet
    tr.M
13 quadrantem: quadrans H / altitudo stelle om.E

II, 2. TO FIND THE ALTITUDE OF A STAR BY MEANS OF A RIGHT TRIANGLE.[6]

Let a right triangle be made whose base is a [horizontal] straight stick pointing towards the eye. And then let a vertical be set up perpendicularly to the straight stick. And let the aforementioned base be shortened [or lengthened] so that the star may be seen across the top of the vertical. And the ratio of the base to the aforementioned vertical is the same as the ratio of the shadow to the whole solar altitude [i.e. 12]. But the base is known, [that is,] 12 digits if it is equal to the vertical. And thus the shadow is known according to the ratio. Therefore the altitude of the star is known. This is proved by the preceding,[7] which teaches how to find the altitude of the sun by means of a shadow.

How the altitude of a star is found by means of an astrolabe and a quadrant is clear.[8]

[Fig. 40]

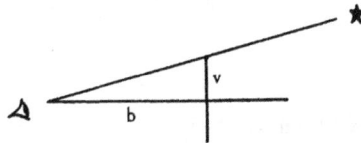

---

[6]"Stelle . . . umbra" (II, 2.1-8). The instrument described here, fig. 40, contains the essential elements of the "Jacob's staff" and may be considered an early version of it. The shadow, s, is found by applying $\frac{b}{v} = \frac{s}{12}$, where b and v are the known base and vertical. The "whole solar altitude" I take to mean a vertical of 12 digits like that in an astrolabe or quadrant.

[7]"Ergo . . . invenire" (II, 2.10-12). The method given in II, 1.3-8 is perfectly applicable here; see n. 1.

[8]"Qualiter . . . patet" (II, 2.13-14). Cf. II, 1.20-21 and n. 5.

[II, 3] MERIDIONALIS ALTITUDINIS LINEAM INVENIRE.

Fiat circulus in plana superficie, et erigatur
cathetus in centro ita quod ante meridiem umbra intret
circumferentiam. Et ubi umbra intrabit circumferentiam
fiat punctum a. Item, ubi umbra post meridiem exibit a   5
circumferentia fiat b punctum. In ea arcus que est inter
a et b item arcus inter b et <a> dividatur in duo equalia
ex utraque parte per 2 puncta, e, d. Protrahatur linea
a d in e. Illa erit linea meridionalis quam tanget umbra
sole existente in meridie. Post meridiem autem vel ante   10
declinabit ab illa.

Et hec omnibus astrologis est ubique necessaria.
Probaturque quod sit meridionalis linea, quia tunc est
minima umbra tangens eam.

---

3 post centro add. a E / ita quod: in quam H
4 ubi om.H
5 fiat: intrabit H / punctum: predictum M / a (1): b E /
    item: iterum E / a om.H
6 circumferentia: circumferentiam H / b punctum: punctum c E
    punctum b H
6-7 in . . . b om.E
7 arcus om.H / item: iterum E / item . . . a om.H /
    a (2,ed.): c EM / in: inter M / equalia: equa H
8 per om.H / e: c H / post d add. et H
9 a d in e: ad in e M a c in d H / illa: ante a H
10 autem om.H
11 ab illa om.H
12 astrologis: astronomicis H / est ubique tr.H
14 tangens eam: eius tangens H

II, 3.   TO FIND THE LINE OF MERIDIONAL ALTITUDE.[9]

Let a circle be made in a plane surface and let a vertical
be erected in the center so that the shadow may enter the circumference before noon. And where the shadow enters the circumference, let a point a be made. Likewise where the shadow
leaves the circumference after noon, let point b be made.
Let that arc which is between a and b as well as that between
b and a be divided into two equals in either direction by two
points, e, d. Let a line be extended from d to e. That line
which the sun's shadow touches at noon will be the meridional
line. However, after noon or before, it will deviate from it.

And this is necessary to all astronomers everywhere. And
it is proved that it is the meridional line, because at noon
the shadow is the smallest one touching it.

[Fig. 41]

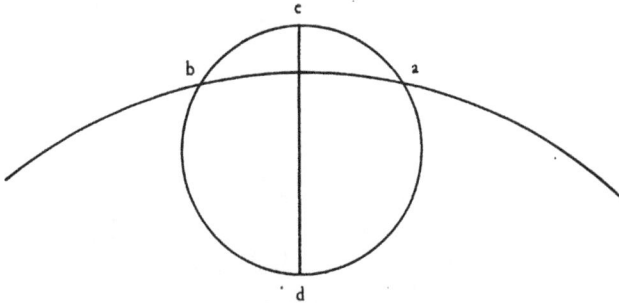

<hr>

[9]"Meridionalis . . . illa" (II, 3.1-11). This device,
fig. 41, for finding the north-south line is described in
Geom. incert., ed. Bubnov, 1899: 363.5-364.9, where it is
taken from the Roman agrimensores who used it to find the
cardo and decuman, or north-south and east-west lines for
laying out colonial cities. Al-Battani also explains its construction; see Nallino, 1903-1907, v. 1: 24.1-46. In the Arab
world it is known as the Indian Circle; cf. Biruni, 1934:
par. 131.

[II, 4]   MERIDIANAM SOLIS ALTITUDINEM INVENIRE.

Hec monstratur per precedentem. Quando enim umbra
prefatam tangit lineam in predicto circulo, tunc est
meridies. Et quia tunc est minima umbra per primam de
altitudine umbre eius inveniatur altitudo. Et illa erit   5
meridiana.

Et aliter invenire eam docebimus inferius.

---

1 meridianam: meridionalis H̲
2 hec: bis H̲ / enim om̲.H
3 prefatam: predeterminata E̲ / tangit lineam tr̲.H /
   predicto: dicto E̲
4 est om̲.M / umbra om̲.H
7 et om̲.M / invenire eam tr̲.M / docebimus: docebitur H̲

II, 4. TO FIND THE MERIDIAN ALTITUDE OF THE SUN

This is shown by the preceding.[10] When the shadow touches the aforemade line in the aforementioned circle, then it is noon. And since then it is the minimum shadow, the altitude may be found from the height [? length] of its shadow, by means of the first proposition. And that will be the meridian altitude.[11]

And below we will teach how to find it in another way.[12]

---

[10]"Meridianam . . . precedentem" (II, 4.1-2). This proposition may be considered an extension of II, 3.

[11]"per primam . . . meridiana" (II, 4.4-6). Applying II, 1.3-8 to the noonday shadow, we find the meridian altitude.

[12]"Et . . . inferius" (II, 4.7). See II, 5.11-13 and n. 15 below.

[II, 5] EQUINOCTIALEM SOLIS ALTITUDINEM INVESTIGARE.

Hec sic ostenditur. Quando sol est in primo gradu
arietis vel libre circa 17 kalendas aprilis vel 17
kalendas octobris, sumatur altitudo meridiana per
precedentem et illa erit equinoctialis.   5

Aliter, si sol est ab ariete ad libram, remotio
solis subtrahatur ab altitudine meridiana. Residuum
erit equinoctialis altitudo. Si vero est a libra ad
arietem, remotio addatur altitudini meridiane. Et erit
equinoctialis altitudo, que in climate parisiensi est   10
42 graduum. Per equinoctialem etiam addendo vel sub-
trahendo remotionem solis, ut dictum est, habebis
altitudinem meridionalem.

---

1 investigare: invenire H
2 est: erit M
4 kalendas octobris: septembris HM
7 ab . . . meridiana: a meridiana altitudine M / residuum
    erit tr.H
8 vero om.H
9 altitudini meridiane tr.M
10 altitudo om.E
11 graduum: gradus E
12 est: et E
13 altitudinem meridionalem tr.H / post meridionalem add.
    ipsius M

II, 5. TO INVESTIGATE THE EQUINOCTIAL ALTITUDE OF THE SUN.[13]

This is shown as follows. When the sun is in the first degree of Aries or of Libra, around March 16 or September 15, let the meridian altitude be taken by the preceding and that will be the equinoctial altitude.

In another way, if the sun is moving from Aries to Libra, let the declination of the sun be subtracted from the meridian altitude. The remainder will be the equinoctial altitude. If, however, it is moving from Libra to Aries, let the declination be added to the meridian altitude. And the result will be the equinoctial altitude, which is 42 degrees in the Parisian zone.[14] Also by adding or subtracting the declination of the sun to the equinoctial altitude[15], as has been said, you will have the meridional altitude.

---

[13]"Equinoctialem . . . altitudo" (II, 5.1-10). When the sun is at the first degree of Aries or Libra, that is at the vernal or autumnal equinox, it has no declination, and the meridian altitude of the sun, h_n, equals the equinoctial altitude or co-latitude, $\bar{\phi}$, of the place where the observation is made. In general, $\bar{\phi} = h_n - \delta$ where δ is the declination, considered positive to the north, i.e. when the sun is moving from Libra to Aries, and negative to the south, i.e. sun moving from Aries to Libra. Cf. Millás-Vendrell, 1963: 133.30-134.26, Suter, 1914: p. 18, Neugebauer, 1962: p. 46, Millás-Vallicrosa, 1947: 155.23-156.6, Nallino, 1903-1907, v. 1: 30.13-17, Millás-Vallicrosa, 1940: 16.1-18 and Trac. quadrantis, ed. Britt, 1972: pp. 149-150.

[14]"Et erit . . . 42 graduum" (II, 5.9-11). This would indicate that the author was writing in or near Paris. Cf. II, 6.7, II, 7.5-6, II, 9.8, II, 10.6 and II, 18.6.

[15]"Per . . . meridionalem" (II, 5.11-13). This is simply the inverse of the procedure in II, 5.1-10. The formula in n. 13 above may be rewritten $h_n = h_e + \delta$. Cf. Millás-Vallicrosa, 1947: 17.6-13.

[II, 6] SOLSTICIALEM SOLIS ALTITUDINEM MERIDIANAM
INVENIRE.

Altitudini circuli equinoctialis per precedentem
invente addatur maxima remotio solis, scil. 23 gradus et
33 minuta, habebis solsticialem estivalem, scil. 66 gradus 5
fere. Eadem subtrahatur ab equinoctiali, et habebis
hyemalem, scil. 19 gradus in climate parisiensi.

---

3 altitudini: altitudinis E / circuli equinoctialis om.H
4 invente tr.ante per H / post remotio add. sit E /
    23: 25 MH / et om.MH
5 33: 23 M lac. H / ante habebis add. et H / solsticialem:
    solsticium E
6 ab equinoctiali: equinoctii H / habebis: habes M
7 scil. om.H

II, 6. TO FIND THE SOLSTITIAL ALTITUDE OF THE SUN.[16]

Let the greatest declination of the sun, namely 23 degrees
and 33 minutes,[17] be added to the altitude of the equinoctial
circle found by the preceding. And you will have the summer
solstitial altitude, namely about 66 degrees. Let the same
[declination] be subtracted from the equinoctial altitude,
and you will have the winter solstitial altitude, namely 19
degrees in the Parisian zone.

---

[16]"Solsticialem . . . parisiensi" (II, 6.1-7). Applying
the formula of n. 15 (see II, 5.11-13) for the maximum value
of δ, which is the obliquity of the ecliptic, ε, we have h_n =
$\bar{\phi} + \varepsilon$. When ε is taken as a positive declination, the result
is the meridian altitude of the sun at the summer solstice.
When taken as negative, we get the altitude at the winter
solstice.

[17]"maxima . . . 33 minuta" (II, 6.4-5). The value of ε
given here is that of the Mumtaḥan zīj, or tabule probate,
found by Yaḥiā b. Abī Manṣūr who worked for the Caliph
al-Ma'mūn. The Ptolemaic value is 23° 51' 20". See Millás-
Vallicrosa, 1943-1950: 39.1-2 and n. 1 and p. 45, Millas-
Vendrell, 1963: 130.35-38, and pp. 63-64, and Millás-
Vallicrosa, 1947: 77.12-18. Adding and subtracting this
value of ε, 23° 33', to the value of h_e, 42° for Paris, as
given in II, 5.9-11, we get approximately 66° and 19° as
the summer and winter solstitial altitudes of the sun at
noon.

[II, 7] PER ALTITUDINEM EQUINOCTIALEM LATITUDINEM
REGIONIS INVENIRE.

Altitudo equinoctialis inventa ut docuimus subtrahatur

a 90 gradibus, residuum est latitudo regionis et poli

altitudo. Secundum hoc 48 graduum est latitudo regionis   5

parisiensis, scil. distantia ab equinoctiali linea.

Quod tanta sit altitudo poli per secundam huius voluminis

probabis. Quod etiam aliter invenire docebimus.

1-8 In codice H haec propositio cum illa praecedenti con-
    iuncta est.
1 per altitudinem: ex altitudine H / equinoctialem:
    equinoctialis E / latitudinem: latitudine M
4 gradibus om.E
5 48: 46 M / graduum: gradus E / est om.H
7 altitudo poli tr.H / secundam: 3^{am} E
8 etiam ?E

II, 7. TO FIND THE LATITUDE OF A REGION BY MEANS OF THE
EQUINOCTIAL ALTITUDE.[18]

Let the equinoctial altitude, found as we have taught,

be subtracted from 90 degrees, the result is the latitude

of the region, and the altitude of the pole. According to

that, 48 degrees is the latitude of the Parisian region,

that is the distance from the equinoctial line, [i.e. the

equator]. That the altitude of the pole is that much, you

will prove by the second proposition of this volume. We

will also teach how to find it in another way.[19]

---

[18]"Per altitudinem . . . linea" (II, 7.1-6). The equinoctial altitude, $\bar{\phi}$, is also called the co-latitude because
it equals 90° - ϕ, where ϕ is the latitude of the place.
Cf. Millás-Vendrell, 1963: 133.30-134.26, Millás-Vallicrosa,
1943-1950: 38.22, Millás-Vallicrosa, 1947: 155.26-27, Suter,
1914: p. 18, Neugebauer, 1962: p. 46, Millás-Vallicrosa, 1940:
16.18-20 and Trac. quadrantis, ed. Britt, 1972: pp. 148-149.
For Paris $\bar{\phi}$ is 42°, hence ϕ equals 48°.

[19]"quod . . . docebimus" (II, 7.7-8). Proposition II, 2
teaches how to find the altitude of any star, including the
pole star required here. Using any circumpolar star, we can
find the altitude of the pole by applying II, 8.

[II, 8] POLI ALTITUDINEM INQUIRERE.

Sumatur alicuius stelle vicine polo maxima
altitudo per secundam et per meridionalem lineam,
quando erit in directo eius. Similiter sumatur minima
altitudo eius que subtrahatur a maiori. Excessus   5
medietas addatur minori. Et illa erit altitudo poli.

---

1 inquirere: invenire H
3 per (2) om.E
5 maiori: maioris E

II, 8. TO INQUIRE INTO THE ALTITUDE OF THE POLE.[20]

Let the maximum altitude of any star near the pole be taken using the second proposition and by means of the meridional line when the star is along it. Similarly let its minimum altitude be taken. Let that minimum be subtracted from the maximum. Let half the excess be added to the smaller, and that will be the altitude of the pole.

---

[20]"Poli . . . poli" (II, 8.1-6). The maximum and minimum altitudes of a star will occur when the star is along the north-south line. For the purposes of this proposition, one must choose a star which never sets, i.e. a circumpolar star. The average of the maximum and minimum values of the star's altitude will yield the polar altitude or the latitude of the place of observation. Cf. Millás-Vendrell, 1963: 135.1-31, Millás-Vallicrosa, 1943-1950: 38.23, Suter, 1914: p. 18, Neugebauer, 1962: p. 46, Millás-Vallicrosa, 1940: 20.8-10 and Trac. quadrantis, ed. Britt, 1972: pp. 150-151.

[II, 9] PER POLI ALTITUDINEM SUMMAM IMPERFECTORUM
AZIMUTH ET ALMUCANTARATH IN QUODLIBET CLIMATE INVENIRE.

Poli altitudo sive regionis latitudo subtrahatur a

66. Residuum dividatur per 5 si fiat walzagora per

quinquenas. Denominatio dabit imperfectos circulos qui   5

subtrahantur a 18; residuum dabit perfectos. Secundum

hoc in primo climate latitudinis graduum 15, 10 sunt

inperfecti. In parisiensi 4 fere.

---

2 azimuth et om.H / invenire: tr.ante in E   reperire H
5-6 circulos . . . perfectos om.H
7 primo: proprio H / graduum 15 tr.E / post 15 add. igitur M
8 fere om.H

II, 9.   TO FIND THE SUM OF INCOMPLETE AZIMUTHS AND ALMUCANTARS
IN ANY ZONE BY MEANS OF THE ALTITUDE OF THE POLE.[21]

Let the altitude of the pole or latitude of the region be
subtracted from 66. Let the remainder be divided by 5 if
the astrolabe is made by fives [i.e., the azimuths and almucantars are drawn for every fifth degree]. The quotient will
give the [number of] circles. Let them be subtracted from
18; the remainder will give the [number of] complete circles.
According to that, in the first zone, of latitude 15 degrees,
there are 10 incomplete circles. In the Parisian [zone],
about 4.

---

[21]"Per poli . . . perfectos" (II, 9.1-6). The lines called
almucantars on the walzagora (the face of the astrolabe, or
even the astrolabe in general) are the planispheric projections
of the circles parallel to the horizon; the azimuths are projections of the great circles perpendicular to the horizon.
On an astrolabe all the azimuths are incomplete circles; I
do not see why they were included in the enunciation of this
proposition. Only some of the almucantars are complete circles; others are cut off by the projection of the Tropic of
Capricorn. This proposition gives us a procedure for finding
the number of incomplete circles, n_i, and complete circles,
n_c. We may write the procedure as the formula
$n_i = [(90 - \varepsilon) - \phi]/c$ where c is the number of the class of
astrolabe, that is, the almucantars are drawn for every cth
degree. The rest of the procedure is a simple subtraction
$(90/c) - n_i = n_c$.

[II, 10] PER POLI ALTITUDINEM DISTANTIAM CENITH AB
IPSO ET PARALELLIS INVENIRE.

Subtrahatur poli atitudo a 90. Residuum est

distantia cenith ab ipso. A quo subtrahantur 36.

Residuum est distantia ab artico paralello. Secundum   5

hoc, 6 gradibus distat cenith parisius ab eo. Et ab

equinoctiali linea distantia est tanta quanta est poli

altitudo a solsticiali estivali, fere 24. Et dico

24 fere quia revera non sunt ibi nisi 23 gradus et 33

minuta et 8 secunda.   10

---

2 paralellis: paralellus H
4 ab ipso: ?E
5 post distantia add. cenith E
6 gradibus: gradus E / distat cenith: ab eo distamus H /
    ab eo om.H
7 distantia: distamur ?M distamus H / est tanta quanta:
    quantum H
8 24: 34 H / et: ideo H
8-10 et . . . secunda om.E
9 fere om.H / 23: 33 M

II, 10. BY MEANS OF THE ALTITUDE OF THE POLE TO FIND THE
DISTANCE OF THE ZENITH FROM IT [I.E. THE POLE] AND THE
PARALLELS.[22]

Let the altitude of the pole be subtracted from 90. The
remainder is the distance of the zenith from the pole. From
the remainder let 36 be subtracted; the remainder is the
distance [of the zenith] from the arctic parallel. According to that, the zenith at Paris is 6 degrees from it. And
from the equinoctial line [i.e. the equator] the distance
[of the zenith] is as much as that of the altitude of the
pole from the summer solstice [i.e. the Tropic of Cancer],
about 24. I say about because it is not actually 24 in
that place but is 23 degrees and 33 minutes and 8 seconds.[23]

---

[22]"Per poli . . . ab eo" (II, 10.1-6). The author finds
the co-latitude of a location and then attempts to find the
distance from that place to the Arctic Circle. The value
he uses for the radius of the Arctic Circle, 36°, is incorrect; it should be the inclination of the ecliptic, about
24°, the value given a few lines down. However, 36° is the
size of the frigid zone, according to Macrobius, Somnium II,
vi, 2, tr. Stahl, 1952: p. 207. In other words, the author
confused the Arctic Circle with the frigid zone; Paris is
18° from the Arctic Circle but only 6° from the frigid zone.

[23]"Et ab equinoctiali . . . secunda" (II, 10.6-10). In
other words, the sun at the summer solstice is about half-
way between the equator and the latitude of Paris. The value
for ε given here is an unusual one not found in Battani's
table, v. Nallino, ed., 1903-1907, v. 1: p. 160.

[II, 11]   PER SOLIS ALTITUDINEM HORAM DIURNAM REPERIRE.

Altitudinis corda per 60 multiplicetur, productum
dividatur per cordam meridiane altitudinis eiusdem diei.
Eius que provenit corde arcus dividatur per 15. Denominatio dabit horas finitas si est altitudo sumpta ante   5
meridiem, si vero post, finiendas. Probationes huius
et aliorum invenies in Almagesti, quas ideo hic non
ponimus ne opus longum agamus et in alterius vicula aremus.

---

2 <u>post</u> altitudinis <u>add.</u> solis <u>H</u> / per 60 multiplicetur:
  multiplicetur per 60 <u>E</u> feriatur in 60 <u>H</u>
3 per: in <u>H</u> / eiusdem diei <u>om.</u>H
4 provenit: pervenit <u>E</u> proveniet <u>H</u>
6 si vero: sed <u>H</u> / finiendas: finiendo <u>M</u> / <u>ante</u> probationes
  <u>add.</u> has <u>M</u>
7 aliorum: aliarum <u>H</u> / invenies <u>tr.post</u> Almagesti <u>H</u>
8 longum: actum <u>H</u> / et . . . aremus <u>om.</u>E

II, 11. TO ASCERTAIN THE HOUR OF THE DAY FROM THE ALTITUDE OF THE SUN.[24]

Let the Sine of the altitude [of the sun] be multiplied by 60; let the product be divided by the Sine of the meridian altitude of the same day. Let the arc of the Sine which is produced be divided by 15. The quotient will give the hours completed [since sunrise] if the altitude is taken before noon, and those to be completed [before sunset] if after. You will find in the Almagest the proofs of this and of other things[25] which we do not set forth here lest we make a long work and hoe another man's row.[26]

[Fig. 42]

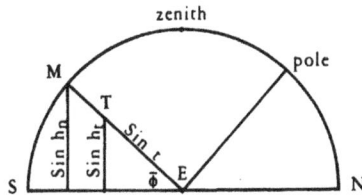

---

[24]"Per solis . . . finiendas" (II, 11.1-6). The procedure here, may be restated as $\text{Sin } t = (60 \cdot \text{Sin } h_t)/\text{Sin } h_n$, where t is the arc of the day already completed or to be completed, h_t the altitude of the sun at the time, h_n the meridian altitude of the sun, as in fig. 42, a projection on the plane of the meridian. M and T are the positions of the sun at the meridian and at t. Since 15° are traversed in an hour, dividing t by 15 yields the number of hours. The formula is an approximation, accurate only at the equinox; see Goldstein, 1967: pp. 207-208 and Millás-Vallicrosa, 1943-1950: pp. 46-47. As in II, 1.3-8 and n. 1 thereto, corda must be taken to mean Sine, and the unit circle has 60 parts. In other works on astronomical tables similar propositions, but with a 150-part unit circle, may be found. Cf. Millás-Vendrell, 1963: 147.4-14 and Millás-Vallicrosa, 1943-1950: 39.17.

[25]"Probationes . . . Almagesti" (II, 11.6-7). I have not found this proposition given explicitly in the Almagest, but it could probably be derived from the material in Books I and II of the Almagest.

[26]"quas . . . aremus" (II, 11.7-8). Vicula seems to be a transformed diminutive of vicus, meaning furrow.

[II, 12]   PER SOLIS ALTITUDINEM ASCENDENS INVENIRE.

Hore preterite diei invente per precedentem sumpta
altitudine ante meridiem multiplicetur in gradibus et
minuta corde hore. Productum addatur orientalibus
gradibus. Signum sub quo invenitur, erit ascendens   5
per tot gradus quot habet latitudo. Finiende post meridiem
feriatur similiter. Productum dematur ab occidentalibus,
et habes gradum <descendentem>.

---

3 altitudine: altitudo H̲ / multiplicetur: multiplicentur E̲
   feriatur H̲ / gradibus: gradus H̲ / ante et add. hore
   corte ?E̲
4 corde hore: corda hore E̲ om.H̲ / addatur: addaturque ?E̲
6 finiende: finiente M̲ / post post add. etiam M̲
7 feriatur: feriantur E̲ om.M̲
8 habes: habet ?E̲ / descendentem: ascendentem M̲E̲H̲

II, 12. TO FIND THE ASCENDENT FROM THE ALTITUDE OF THE SUN.[27]

Let the hours of the day [already] passed by, found by
the preceding proposition when the altitude was taken before
noon, be multiplied by the degrees and minutes of the [arc]
of an hour.[28] Let the product be added to the degrees [of
longitude] of the rising sun. The sign under which [that
sum] is found will be the ascendent for that latitude. Let
the [hours] to be finished after noon be multiplied similarly; let the product be taken from the degrees [of longi-
tude] of the setting sun and you have the descendent degree.

---

[27]"Per solis . . . ascendentem" (II, 12.1-8). The ascendent is merely the point of the ecliptic crossing the eastern horizon at a particular time or along with a certain
heavenly body. The procedure seems to be a simple conversion of the seasonal hours since sunrise into equatorial
degrees, which are then added to the solar longitude, or
position of the sun along the ecliptic. The result should
be the ascendent for the particular date, time and latitude.
The hours before sunset are converted to degrees by the same
procedure. That angle subtracted from the sun's longitude
gives the descendent, the point at the western horizon.
Whether these operations are done before or after noon makes
no difference--the ascendent and descendent can be calculated at any time during the day by these methods. Cf.
Millás-Vallicrosa, 1943-1950: p. 47, Nallino, 1903-1907, v. 1:
30.36-38, and Goldstein, 1967: p. 209.

[28]"in gradibus . . . hore" (II, 12.3-4). To make sense,
corda in the Latin should be omitted or changed to arca.

[II, 13]   PER ALTITUDINEM STELLE, HORAM NOCTIS INVENIRE.

Sumatur altitudo stelle note. Et ponatur acumen
eius super altitudinem tantam in eadem parte. Gradus
solis dabit horam; gradus super orizonta ascendens in
astrolabio.   5

---

3 altitudinem tantam tr.H / eadem: ea H

II, 13. TO FIND THE TIME OF NIGHT FROM THE ALTITUDE OF A
STAR.[29]

Let the altitude of a known star be taken. And let its

point [on the rete] be put over its altitude in the same

direction. The degree of the sun will give the time; the

degree at the horizon [will give] the ascendent in the

astrolabe.

---

[29]"Per altitudinem . . . astrolabio" (II, 13.1-5). The
procedure here is the standard one for finding the time at
night using an astrolabe. Cf. Bubnov, 1899: 130.17-131.3,
Llobet de Barcelona (tr.), "Tractat d'us de l'astrolabi,"
in Millás-Vallicrosa, 1931: 282.169-184 and Gunther, 1929:
p. 220.

[II, 14]   PER SOLIS ALTITUDINEM UMBRAM ALTITUDINIS INVENIRE.

Sumatur corda et servetur.   Item altitudo subtrahatur
a 90.   Residui sumatur corda, que multiplicetur in 12,
et dividatur per cordam prius servatam.   Denominatio
dabit digitos umbre.   Quod remanebit similiter multipli-   5
cetur in 60, et similiter dividatur per cordam ut prius
et habebis minuta.   Hoc probatur in Almagesti.

---

1 altitudinis om.EH
2 ante sumatur add. altitudinis H / item: iterum ?EM
3 multiplicetur: feriatur H
5 similiter om.H
5-6 multiplicetur: feriatur H
6-7 per . . . minuta om.H
7 probatur tr.post Almagesti H

II, 14. TO FIND THE SHADOW OF THE ALTITUDE FROM THE
ALTITUDE OF THE SUN.[30]

Let the Sine of the altitude be taken and put aside. Let
the same altitude be subtracted from 90. Let the Sine of
the remainder be taken, multiplied by 12, and divided by
the previously put aside Sine. The result will give the
digits of the shadow. Let what remains similarly be multiplied by 60 and similarly divided by the ·Sine as before,
and you will have the minutes of the shadow.[31] This is
proved in the Almagest.[32]

[Fig. 43]

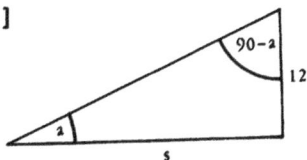

[30]"Per solis . . . : umbre" (II, 14.1-5). The sine rule
for a right triangle, fig. 43, yields the result, equivalent
to s = [12·Sin (90 - a)]/Sin a, where s is the shadow and
a the altitude of the sun. The right triangle has known
angles and one side equal to a gnomon of 12 digits. The
shadow is the other side, adjacent to the right angle.
Corda is the author's word for Sine; cf. II, 1 and II, 11
and notes 1 and 24. Similar propositions may be found in
Millás-Vendrell, 1963: 151.34-38, Suter, 1914: p. 21,
Neugebauer, 1962: p. 55, and Curtze, 1900: 342.5-12.

[31]"Quod remanebit . . . minuta" (II, 14.5-7). The author
tells us how to convert a remainder into minutes. Cf. Suter,
1914: p. 21 and Neugebauer, 1962: p. 55.

[32]"Hoc . . . Almagesti" (II, 14.7). The proof is not
given directly in the Almagest but is implicit in Almagest
II, 5.

[II, 15]   PER UMBRAM ALTITUDINIS 29 GRADUUM ASCENSUM
ARIETIS IN SPERA RECTA REPERIRE.

Umbram illius altitudinis 21 digitorum et 38
minutorum multiplices in 40, et dividas illam altitudinem
29 <graduum> per 31. Denominatio dabit ascensum arietis   5
in circulo recto, qui est 27 gradus et 40 minuta.

---

2 reperire: invenire M
3 21: 29 E
4 multiplices: ferias H / illam: in H
5 graduum (ed.): digitorum MEH / per 31: si 21 H
6 gradus: graduum H / et om.H / minuta: minutorum H

II, 15. TO ASCERTAIN THE ASCENSION OF ARIES IN THE RIGHT
SPHERE USING THE SHADOW OF AN ALTITUDE OF 29 DEGREES.[33]

You should multiply by 40 the shadow of that altitude,

which is 21 digits and 38 minutes, and divide by 31 that

[product of the shadow of the] altitude of 29 degrees.

The quotient will give the ascension of Aries in the right

sphere, which is 27 degrees, 40 minutes.

---

[33]"Per umbram . . . 40 minuta" (II, 15.1-6). The calculation here is easily performed, given a table of shadows
for various angles. The value for umb 29° given in the text
is nearly correct for a gnomon of 12 digits; Battani gives
21 digits and 40 minutes (ed. Nallino, 1903-1907, v. 2: p. 60)
and Khwarizmi gives 21 digits and 39 minutes (ed. Suter,
1914: p. 179). The source and meaning of the 40/31 is un-
clear, as is that of 29°. Most puzzling of all is the reason the author chose to give this procedure (and those of
II, 16 and 17) without any justification; a table of the
resulting values would have told us no less. Prof. O.
Neugebauer suggested to me that the technique is most probably rooted in Indian astronomical methods; see n. 35 below.
The value for the right ascension of the 30° of Aries is
27° 50', according to the Almagest, I, 16. Computing as
required by the text yields 27° 55', a value different from
either Ptolemy's or the author's. Different procedures,
generally more satisfactory, may be found in other treatises; cf. Millás-Vendrell, 1963: 136.13-140.10, Millás-
Vallicrosa, 1947: 147.17-153.16, Suter, 1914: pp. 18-19,
Neugebauer, 1962: pp. 46-48, Nallino, 1903-1907, v. 1: 13.31-
14.3, and Millás-Vallicrosa, 1943-1950: 39.1-5 and pp. 45-46.

[II, 16]   PER UMBRAM ALTITUDINIS 27 GRADUUM ASCENSUM
TAURI IN SPERA RECTA COLLIGERE.

Umbra illa, scil. 23 digitorum 33 minutorum

multiplicetur in 40, et dividatur per 31.   Denominatio

dabit ascensum tauri ibi qui est 29 graduum et 54   5

minutorum.

---

3 umbra illa: umbram illam E / 23: 33 H / digitorum:
   digiti H / minutorum: minuta H
4 multiplicetur: feriatur H / 31: 21 H
5 post ibi add. est E

II, 16. TO INFER THE ASCENSION OF TAURUS IN THE RIGHT
SPHERE USING THE SHADOW OF AN ALTITUDE OF 27 DEGREES.[34]

Let that shadow, namely 23 digits, 33 minutes, be multiplied by 40 and divided by 31. The quotient will give the
ascension of Taurus for that place, which is 29 degrees and
54 minutes.

---

[34]"Per umbram . . . 54 minutorum" (II, 16.1-6). This
procedure is similar to II, 15, except that 40/31 times the
shadow of 27° is used to find the right ascension of the
30° of Taurus. Battani (ed. Nallino, 1903-1907, v. 2: p. 60)
and Khwarizmi (ed. Suter, 1914: p. 174) give the same value
for the shadow. The author's result coincides with that of
Ptolemy in the Almagest, I, 16. My calculation following
the technique of the text yields 30° 23'.

[II, 17]   PER UMBRAM ALTITUDINIS 26 GRADUUM ASCENSUM
GEMINORUM IN SPERA RECTA REPERIRE.

    Illa umbra, scil. 24 digitorum 36 minutorum

multiplicetur in 40, et dividatur per 31.   Denominatio

dabit ascensum geminorum qui est 32 gradus et 34   5

minuta.   Hec in Almagesti probatur.

---

4 multiplicetur: feriatur H
5 gradus: graduum H
6 minuta: minutorum H

II, 17.   TO ASCERTAIN THE ASCENSION OF GEMINI IN THE RIGHT
SPHERE USING THE SHADOW OF AN ALTITUDE OF 26 DEGREES.[35]

Let that shadow, namely 24 digits, 36 minutes, be multiplied by 40 and divided by 31. The quotient will give the
ascension of Gemini which is 32 degrees and 34 minutes.
This is proved in the Almagest.

---

[35]"Per umbram . . . probatur" (II, 17.1-6). The shadow
of 26° and the puzzling 40/31 found in II, 15 is used here
to find the right ascension of the 30° of Gemini. The same
value for the shadow of 26° is found in Battani (ed. Nallino,
1903-1907, v. 2: p. 60) and Khwarizmi (ed. Suter, 1914: p. 174)
Ptolemy in the Almagest, I, 16 gives 32° 16'; my computation,
using the method of the text, produced 31° 44'.
    If we tabulate the computed values for the right ascen-
sions along with those given as results in the text of II,
15-17 and those of Ptolemy, some interesting patterns appear.

	Ptolemy	Given in text	Computed
Aries	27° 50'	27°40'	27°55'
Taurus	29° 54'	29°54'	30°23'
Gemini	32° 16'	32°34'	31°44'
sum	90° 0'	90° 8'	90° 2'

The right ascensions for the first quadrant of the zodiac
should add up to 90°, as do Ptolemy's values. The values
given in II, 15-17 are each very close to Ptolemy's and add
up very nearly to 90°. The computed values, although considerably different from those of Ptolemy, have a sum very
close to 90°. In other words, the computations produce a
consistent set of values. The author of the treatise or its
immediate source seems to have replaced the computed values
of the text with values like the correct ones of Ptolemy,
without correcting the method of computation. Scribal error
can account for the remaining small differences.
    We are still left with making sense of the computational
method described in II, 15-17. Prof. Otto Neugebauer kindly
looked at this material and suggested it is the survival of
an Indian mnemonic technique. Since the computed values have
a sum very close to 90°, they may be considered acceptable,
consistent values for the right ascensions. Thus, the computations produce numbers within the realm of possible values.
They do not, however, make astronomical sense; that is, the
numbers involved in the computation are not related to astronomical parameters. They are numbers arbitrarily chosen from
some table, in this case a table of shadows, and then multiplied by some constant, 40/31 here, to produce values close
to some known values. In this instance, the constant was
probably determined by dividing 90° by the sum of the chosen
shadows. For this to have been an effective mnemonic tech-

Idem enim est ibi ascensus oppositorum signorum.

Idem etiam est ascensus piscium qui arietis; aquarii

qui tauri; capricorni qui geminorum; virginis qui libre;

leonis qui scorpii; cancri qui sagittarii.   10

---

7 enim om.H

8 est om.H

There is in fact the same ascension of the opposite signs.[36] Also the ascension is the same for Pisces and Aries, Aquarius and Taurus, Capricorn and Gemini, Virgo and Libra, Leo and Scorpio, Cancer and Sagittarius.

[Table 1]
Signs and Right Ascensions

Sign	Right Ascension	Opposite Sign
Aries	27° 40'	Libra
Taurus	29° 54'	Scorpio
Gemini	32° 34'	Sagittarius
Cancer	32° 34'	Capricorn
Leo	29° 54'	Aquarius
Virgo	27° 40'	Pisces

nique, the table of shadows must also have been memorized, not an unlikely possibility for Indian astronomers who usually memorized not only astronomical parameters but also trigonometric tables, often using poems in which certain syllables stood for certain numbers, independent of the meaning of the poem.

The intermediary between the original Indian source and the author would have been a set of Arabic astronomical tables known as a zīj. Many of these Arabic astronomical works include some Indian material. Indeed, the method of finding oblique ascension in II, 18 is also an Indian procedure (see below, n. 37), and Indian astronomical notions are explicitly mentioned in II, 27.

[36]"Idem enim . . . sagittarii" (II, 17.7-10). The signs opposite the ones for which the right ascensions have been found have the same right ascensions; the next quarter of the ecliptic has the same ascensions as those found but in opposite order. The signs opposite these also have the same right ascensions. Table 1, based on the values given in II, 15-17 should make this clear. Cf. Suter, 1914: p. 19, Neugebauer, 1962: p. 47, Millás-Vallicrosa, 1947: 153.17-25 and Nallino, 1903-1907, v. 1: 14.7-13.

[II, 18] PER EQUINOCTIALIS ALTITUDINIS UMBRAM,
ASCENSUM SIGNORUM IN OMNI SPERA DECLINI INVENIRE.

Illa umbra multiplicetur in 15 et dividatur per 16.
Denominatio subtrahatur ab ascensu arietis in circulo
directo. Reliquum est ascensus arietis in illa terra,   5
que in climate parisiensi est 15 gradus 36 minuta.

Si eiusdem umbre equinoctialis sumatur media pars,
scil. 13 digitorum 18 minutorum. Eadem umbra ducatur
in 12, et dividatur in 16. Denominatio subtrahatur

---

2 ascensum: ascensus H̲
3 illa umbra t̲r̲.H̲ / multiplicetur: feriatur H̲ / per: in H̲
5 directo: recto H̲ / reliquum: residuum H̲ / illa: ipsa H̲
6 que: qui H̲ / gradus: graduum H̲ / a̲n̲t̲e̲ 36 a̲d̲d̲. et H̲ /
   minuta: minutorum H̲
7 si eiusdem umbre: cum umbra H̲ / sumatur media pars:
   meridiana H̲
8 p̲o̲s̲t̲ scil. a̲d̲d̲. cum sit H̲ / p̲o̲s̲t̲ digitorum a̲d̲d̲. et H̲ /
   eadem: eandem E̲
9 et o̲m̲.H̲

II, 18. TO FIND THE ASCENSION OF THE SIGNS IN EVERY OBLIQUE
SPHERE USING THE SHADOW OF THE EQUINOCTIAL ALTITUDE.[37]

Let that shadow be multiplied by 15 and divided by 16.

Let the result be subtracted from the ascension of Aries in

the right circle. The remainder is the ascension of Aries

in that land, which in the zone of Paris is 15 degrees and

36 minutes.

Let half of the same equinoctial shadow, namely 13 digits

and 18 minutes, be taken. Let the same shadow be multiplied

by 12, and divided by 16. Let the result be subtracted from

---

[37]"Per equinoctialis . . . minutorum" (II, 18.1-16). The
procedure for finding the ascensions in an oblique sphere,
or the oblique ascensions, ρ, is, in origin, an Indian one.
It rests on computing the difference between the oblique and
right ascensions, α. The difference, $\gamma = \alpha \mp \rho$, depends only on
the latitude, ϕ, of the place in question and the longitude
λ of the point on the ecliptic under consideration. This
difference, in the Indian technique, can in fact be reduced
to a simple product of the shadow of the equinoctial altitude, $\underline{u}$, by a factor which is a function of λ, that is
$\gamma = c \cdot u$. This factor, $\underline{c}$, is very closely approximated by the
formula $c = (R \sin \lambda)/(g \cos \lambda)$, where R is the radius of
the unit circle and g is the number of digits in the gnomon
used to find $\underline{u}$. The original astronomical tables, or zīj,
of al-Khwarizmī, seem to have contained a table computing
this function (see Millás-Vendrell, 1963: pp. 70-71 and
Neugebauer, 1962: pp. 52-53). Similar tables are found in
Azarquiel (see Millás-Vallicrosa, 1943-1950: p. 63). The
function and its use are described in Millás-Vendrell, 1963:
141.18-144.23 and 145.6-35, Millás-Vallicrosa, 1943-1950:
39.5 and pp. 45-46, Suter, 1914: pp. 20-21, Neugebauer,
1962: pp. 48-55, and Nallino, 1903-1907, v. 1: 27.1-28.25
and pp. 187-189. The author also uses the coefficient on $\underline{u}$,
to compute the difference of ascensions; unfortunately, he
gives few values for $\underline{c}$, which are unexplained and inaccurate (see table 2). Khwarizmi's values (see Suter, 1914:
pp. 20-21 and Neugebauer, 1962: pp. 48-55) for these three
coefficients are 114/115, 113/116, 1/3; perhaps a scribal
error can explain the differences in the first two values.

ab ascensu tauri in circulo recto. Reliquum est ascensus   10

tauri ibi, qui in climate est 20 gradus 15 minuta ut

probatum est.

Eiusdem umbre 3^a pars sumatur, et subtrahatur ab

ascensu geminorum in circulo recto. Reliquum est

ascensus geminorum in terra. Et est in hac terra 24   15

graduum 17 minutorum.

Idem vero est ascensus piscium qui arietis; aquarii

qui tauri; capricorni qui geminorum. Ascensus libre

tantum superat 30 quantum arietis superatur. Scorpii

---

10 reliquum: residuum H̲
11 ibi o̲m̲.M̲ / minuta: minutorum H̲
12 probatum est: probavimus H̲ / p̲o̲s̲t̲ est a̲d̲d̲. aliter x (?̲)
     si umbra equinoctialis meridiana sit 13 digitorum, 18
     minutorum eadem umbra ducatur in 12 M̲
13 subtrahatur o̲m̲.M̲
14 p̲o̲s̲t̲ recto a̲d̲d̲. subtrahatur M̲ / reliquum: residuum H̲
15 24: 14 E̲
17 vero o̲m̲.H̲ / ascensus o̲m̲.H̲ / piscium: piscis H̲ /
     p̲o̲s̲t̲ qui a̲d̲d̲. est H̲
19 30 o̲m̲.E

the ascension of Taurus in the right circle. Then the remainder is the ascension of Taurus in that place, which in
[this] zone is 20 degrees, 15 minutes as has been proved.

Let the third part of the same shadow be taken and sub-
tracted from the ascension of Gemini in the right circle.
The remainder is the ascension of Gemini in [that] land.
And in this land it is 24 degrees, 17 minutes.

Indeed the ascension is the same[38] for Pisces and Aries,
Aquarius and Taurus, Capricorn and Gemini. The ascension
of Libra exceeds 30 as much as 30 is exceeded [by the
ascension] of Aries. [The ascension] of Scorpio exceeds

---

[38]"Idem vero . . . sagittarii" (II, 18.17-23). The symmetries of the ecliptic and the equator facilitate the computation of the remaining oblique ascensions, once one has
determined those for the first quarter of the ecliptic. We
have seen how the difference between ascensions, $\gamma = c \cdot u$ is
computed (n. 37 above); this γ need only be added to or subtracted from α to produce the required oblique ascensions.
The author, in effect, adds to and subtracts from 30° instead
of α. Other than that, the symmetries he describes are correct. Cf. Millás-Vendrell, 1963: 144.23-34, Suter, 1914:
pp. 20-21, Neugebauer, 1962: pp. 48-50, 52, and Nallino,
1903-1907, v. I: 27.27-30.

tantum superat quantum tauri superatur. Sagittarii   20

superat tantum quantum geminorum superatur.   Idem vero

est virginis qui libre; leonis qui scorpii; cancri qui

sagittarii.

---

21 superat tantum <u>om.</u>H / vero est <u>om.</u>H

that [30] as much as 30 is exceeded [by that] of Taurus.
[The ascension] of Sagittarius exceeds that [30] as much as
30 is exceeded [by that] of Gemini. [The ascension] is the
same for Virgo and Libra, Leo and Scorpio, Cancer and Sagittarius.

[Table 2]
Signs and Oblique Ascensions
(in Degrees and Minutes)

Sign	Oblique Ascension Formula from Text	Oblique Ascension for Paris		
		Given in Text[a]	Computed[b]	Ptolemy[c] II, 8
Aries	$\alpha(\Upsilon) - (15/16) \cdot u$	15 36	15 1	14 20
Taurus	$\alpha(\vartheta) - (12/32) \cdot u$	20 15	24 49	18 23
Gemini	$\alpha(\mathbb{I}) - (1/3) \cdot u$	24 17	28 8	27 17
Cancer	$\rho(\mathbb{Z})$	35 43	31 52	37 15
Leo	$\rho(\mathfrak{m})$	39 45	35 11	41 25
Virgo	$\rho(\triangle)$	44 24	44 59	41 20
Libra	$60 - \rho(\Upsilon)$	44 24	44 59	41 20
Scorpio	$60 - \rho(\vartheta)$	39 45	35 11	41 25
Sagittarius	$60 - \rho(\mathbb{I})$	35 43	31 52	37 15
Capricorn	$\rho(\mathbb{I})$	24 17	28 8	27 17
Aquarius	$\rho(\vartheta)$	20 15	24 49	18 23
Pisces	$\rho(\Upsilon)$	15 36	15 1	14 20

[a]The first three values are given explicitly in the text;
the others are computed using the formulas of the second
column.

[b]These values are computed using the formulas in the
second column, the right ascensions given in II, 15-17, and
the value for u correctly given in II, 18.8.

[c]One of the columns of the Table of Ascensions by 10° in
the Almagest, II, 8 is for the latitude 48°. Since Ptolemy's
table is by 10° sections of the ecliptic, one must add three
successive values to obtain the values given here. They may
then be compared with the values in the other columns.

[II, 19] PER MERIDIANAM SOLIS ALTITUDINEM, LOCUM EIUS
SCIRE.

Sit ignotum in quo gradu cuius signi sit sol.
Sumatur altitudo meridiana per precedentes. Necessario 2
gradus zodiaci cadunt in meridionali linea in tanta   5
altitudine in astrolabio in quanta est sol nisi sit
solsticium. Necessario igitur sol erit in altero
illorum graduum. Cum autem scias quis mensis sit, et
quod signum cui mensi debetur, necessario scies in quo
gradu erit sol. Si est solsticium, necessario scies in   10
quo gradu erit per parvam vel magnam altitudinem.

---

3 cuius om.M / post sol add. et cuius signi M
6 in quanta est sol om.H
7 igitur om.H / sol erit tr.M
8 illorum: eorum M / scias: sciat H / sit: sciat H om.M
10 necessario: tunc H
11 parvam: ?M / post vel add. per H / post magnam add. eius H

II, 19. TO KNOW THE PLACE OF THE SUN BY MEANS OF ITS
MERIDIAN ALTITUDE.[39]

Let it be unknown in which degree of which sign the sun

is. Let the meridian altitude be taken by the preceding.

Two degrees of the zodiac necessarily fall on the meridional

line at as great an altitude as is the sun in an astrolabe,

unless it is the solstice. Therefore the sun will neces-

sarily be at one of those two degrees. Since, however, you

know what month it is, and what sign is appropriate for that

month, you necessarily know in what degree the sun will be.

If it is the solstice, you know necessarily at what degree

it will be by means of its small or large altitude.

---

[39]"Per meridianam . . . altitudinem" (II, 19.1-11). This
proposition shows how to use an astrolabe to determine the
place of the sun on the ecliptic. The technique, a common
one, relies on the fact that any circle parallel to the
horizon intersects the ecliptic at two places at the most.
Knowing the approximate date will eliminate one of the pos-
sibilities. The solstitial altitudes are on circles tangent
to the ecliptic; the sun can be simultaneously on the ecliptic and one of the solstitial circles at only one time of
the year. The greater altitude indicates the summer solstice; the lesser, the winter solstice. The method for finding the meridian altitude is described in II, 4. Cf. Gunther,
1929: p. 221 and Millás-Vallicrosa, 1940: 13.1-9.

[II, 20] PER EQUINOCTIALEM ALTITUDINEM SOLIS ET
MERIDIANAM IN QUOVIS DIE REMOTIONEM EIUS SCIRE.

Nota erit altitudo meridiana diei que si est
maior quam equinoctialis que similiter nota est, remotio
erit septemtrionalis. Et quot gradibus est maior tanta   5
est remotio. Si est minor ea quot gradibus est minor,
tanta est remotio meridiana.

Similiter per altitudinem planete meridianam vel
cuiuslibet alterius stelle gradus scies que et quanta
sit eius declinatio, i.e. remotio ab equinoctiali.   10

3-4 est maior <u>tr.</u>H
4 que <u>om.</u>H / est <u>om.</u>H
5 erit: est <u>E</u> enim <u>H</u> / gradibus: graduum <u>H</u>
9 cuiuslibet alterius stelle <u>om.</u>H / <u>ante</u> gradus <u>add.</u> eius <u>H</u> /
    gradus <u>om.</u><u>E</u> / <u>post</u> que <u>add.</u> etiam <u>M</u>

II, 20. TO KNOW THE DECLINATION OF THE SUN BY MEANS OF ITS
EQUINOCTIAL ALTITUDE AND MERIDIAN ALTITUDE IN ANY DAY.[40]

The meridian altitude of the day will be known; if it is

greater than that of the equinoctial, similarly known, the

declination will be northern. And the declination is as

many degrees as [the meridian altitude] is greater [than the

equinoctial altitude]. If it is smaller, the declination is

as many degrees as the meridian is smaller than [the equi-

noctial].

Similarly by the meridian altitude of a planet or of the

degree of any other star[41] you will know which and how much

is its declination, that is, its distance from the equinoc-

tial.

---

[40]"Per equinoctialem . . . meridiana" (II, 20.1-7). In
II, 5.6-10, we saw how to find the equinoctial altitude of
the sun, given the meridian altitude and declination; no
technique for finding the declination was given, though it
was implicit in II, 5.11-13. Here we are simply told the
declination is the difference between the meridional and
equinoctial altitudes, both of which may be found using an
astrolabe or quadrant (see II, 4.1-6 and II, 5.1-5). Cf.
Britt, 1972: pp. 145-146, Gunther, 1929: p. 223 and Millás-
Vallicrosa, 1940: 16.5-18.

[41]"Similiter . . . equinoctiali" (II, 20.8-10). The declination of a star or planet is its angular distance from
the celestial equator; that of the sun is its angular distance from the plane of the ecliptic. Thus determining the
declination of a star or planet is a simple measurement of
altitude. Cf. Gunther, 1929: p. 223.

[II, 21] PER MAXIMAM STELLE ALTITUDINEM, GRADUM EIUS
INVENIRE.

Videatur quando stella fixa est in meridionali
linea supra inventa. Tunc est in meridie. Sumptoque
ascendente illa hora per precedentes videatur in   5
walzagora qui gradus est in meridie. In illo necessario
erit prefata stella.

---

1 gradum: graduum E
2 invenire: scire H
4 in meridie: meridionalis E
5 videatur: videat M
7 prefata om.H

II, 21. TO FIND THE DEGREE OF A STAR BY MEANS OF ITS
GREATEST ALTITUDE.[42]

Let the star be seen when it is positioned on the meridional line found above. Then it is at the meridian. The
ascendent for that time having been found by the preceding
propositions, let the degree [of the ecliptic] at the meridian be seen in the walzagora. The aforementioned star will
necessarily be in that degree.

---

[42]"Per maximam . . . stella" (II, 21.1-7). The degree of
a star is the point of the ecliptic which reaches the meridian together with the star. The rising degree of a star
is that point of the ecliptic which rises at the same time
as the star; the setting degree is that point which sets
with it. The meridional degree is correctly determined
using the meridional line found in II, 3 and an astrolabe.
Why the ascendent is mentioned here is not clear; II, 12
gives the procedure for finding it. Cf. Gunther, 1929:
p. 224.

[II. 22]   PER ALTITUDINEM STELLE NOTE ET ALTITUDINEM
PLANETE, LOCUM EIUS INVENIRE.

Sumatur altitudo planete et hora per stellam notam.
In astrolabio videatur quis gradus zodiaci sit in sumpta
altitudine planete in oriente, si sumpta fuit altitudo   5
orientalis. Vel in occidente, si sumpta fuit altitudo
occidentalis. Nam in illo erit planeta. Hoc est juxta
veritatem astrolabii. Hec tamen ecentricitas circuli
planete et latitudo planete quandoque impediunt.

---

4 zodiaci sit <u>tr.</u>H
5 fuit: fuerit H
6 in <u>om.</u>E
7 nam: et H / illo: illo gradu H / erit planeta <u>tr.</u>H
8 astrolabii <u>om.</u>H / hec tamen <u>om.</u>H / ecentricitas <u>om.</u>E /
   <u>post</u> ecentricitas <u>add.</u> enim H
9 impediunt: impediant H

II, 22. TO FIND THE PLACE OF A PLANET BY MEANS OF THE
ALTITUDE OF A KNOWN STAR AND THE ALTITUDE OF THE PLANET.[43]

Let the altitude of the planet and the time be taken by

means of the known star. If the eastern altitude was taken,

see which degree of the zodiac in the astrolabe [corresponds

to] the altitude of the planet taken in the east, or in the

west, if the western altitude was taken. Now the planet

will be in that degree. This is according to the truth of

the astrolabe, though sometimes the eccentricity of the

circle of the planet and the latitude of the planet may

impede this.

_____

[43]"Per altitudinem . . . impediunt" (II, 22.1-9). In
this proposition the time is used to arrange the astrolabe
so that when the latitude of a planet is found, the corresponding degree of the ecliptic may be read on the astrolabe. The procedure is not very satisfactory; the inaccu-
racies noted in the text can be significant. Cf. Gunther,
1929: p. 227 and Millás-Vallicrosa, 1940: 21.11-20.

[II, 23]   PER ALTITUDINEM STELLE ET PLANETE, DIRECTIONEM
EIUS INVENIRE.

Sumatur altitudo stelle note in astrolabio et hora
et altitudo planete cuiusvis.   Tunc notetur per 3 dies
post secundum quod planeta est levis vel gravis.   In   5
eadem hora noctis sumatur eadem altitudo stelle prefate
et eiusdem planete.   Si altitudo planete sumpta ultimo
est minor quam prius sumpta, et planeta est orientalis,
tunc est directus.   Si est maior, tunc est retrogradus.
Si tanta, stationarius.   Econtrario erit si planeta est   10
occidentalis.

4-5 et altitudo . . . in: noctis et planete illius in aliqua
   hora noctis deinde post cres<cit 3> dies   vel 4 in mg.E
4 planete cuiusvis tr.H / tunc: tunc et H
5 post om.H / gravis: ponderosus H
6 eadem hora noctis in mg.E / post eadem (2) add. habitudo H
7 ultimo: ultima H
8 quam om.E / planeta: planete M

II, 23. TO FIND THE DIRECTION OF A PLANET BY MEANS OF THE
ALTITUDE OF A STAR AND THE PLANET.[44]

Let the altitude of a known star and the time and the

altitude of any planet be taken with the astrolabe. Then

let [the altitude] be noted three days later during which

the planet is light or heavy [? ahead of or behind the sun].

At the same time of night let the same altitude of the

aforementioned star and of the same planet be taken. If

the last-taken altitude of the planet is less than that

previously taken, and the planet is in the east, then it

is direct. If it is greater, then it is retrograde; if as

much, stationary. It will be the opposite if the planet is

in the west.

---

[44]"Per altitudinem . . . occidentalis" (II, 23.1-11).
The uniform relation between the latitude and longitude of
points of the ecliptic at the same time of day permits the
use of planetary altitude instead of planetary longitude in
finding whether a planet is direct, retrograde or stationary.
Here as in II, 22 the latitude of a planet may cause diffi-
culties. Cf. Gunther, 1929: p. 227 and Millás-Vallicrosa,
1940: 21.30-22.15.

[II, 24]   PER PLANETE ALTITUDINEM ET ALTITUDINEM GRADUS
IN QUO EST, LATITUDINEM PLANETE INVENIRE.

Sumatur altitudo planete quando erit in meridionali

linea de nocte. Gradus eius notus per precedentia ponatur

in meridionali linea in astrolabio. Et computetur   5

altitudo eius ab orizonte. Si altitudo planete est

maior illa, latitudo est septemtrionalis secundum maius

et minus. Si est minor, meridiana. Si est equalis est

in nodo.

---

1 planete altitudinem tr.H
4 notus: motus H
5 in (2) om.H
6 planete: eius M
7 maius: magis H / post maius add. similiter M
8 post minor add. est H / est (3) om.H

II, 24.   TO FIND THE LATITUDE OF A PLANET BY MEANS OF THE
ALTITUDE OF THE PLANET AND THE ALTITUDE OF THE DEGREE IN
WHICH IT IS.[45]

Let the altitude of the planet be taken when it is at

the meridional line at night.   Let its known degree be put

in the meridional line of the astrolabe.   And let its alti-

tude from the horizon be computed.   If the altitude of the

planet is greater than that, the latitude is northern by

more and less.   If it is less, it is southern.   If it is

equal, it is in the node.

---

[45]"Per planete . . . in nodo" (II, 24.1-9).   In this
proposition, the altitude of a planet found by calculation
or in an almanac is compared with that measured with an
astrolabe.   The difference between the two values determines
the latitude.   Cf. Gunther, 1929: p. 227.

[II, 25]   PER DISTANTIAM PLANETE AB ALTO LOCO ECENTRICI,
EQUATIONEM EIUS INVENIRE.

Si planeta est in alto loco ecentrici vel ymo, tunc
est idem medius motus eius et verus. Et nulla equatio.
Quod probatur per lineas ductas a centro terre et centro   5
ecentrici per centrum planete ad zodiacum, quia sunt una
sola ille linee. Sed in maxima distantia planete alti
loci et ymi, maxima est linearum diversitas secundum quas
attenditur equatio.

---

1 loco om.M
4 est idem tr.H / equatio: adequatio H
5 ductas: directas M / centro: a centro E
6 sunt om.H
7 ille linee: iste linee E linea est illa que ducitur ab
   utroque H
8 et ymi: inim ?H
9 attenditur: ascenditur E   om.H

II, 25. TO FIND THE EQUATION OF·A PLANET FROM ITS
DISTANCE FROM THE APOGEE OF THE ECCENTRIC.[46]

If the planet is in the apogee or perigee of the eccen-

tric, then its mean motion and its true motion are the same,

and there is no equation. This is proved by means of the

lines drawn to the center of the earth and from the center

of the eccentric through the center of the planet to the

ecliptic because those lines are one [line] only. But at

the greatest distance of the planet from the apogee and

perigee, the diversity of the lines, in accordance with

which the equation is marked, is greatest.

---

[46]"Per distantiam . . . equatio" (II, 25.1-9). The
author begins his brief version of planetary theory with a
discussion of the equation of center. Note that the author
is describing the motion of the center of the planet's epicycle although he refers to it as planeta or centrum planete. (The moon and sun do not have epicycles.) This is
the correction required because the centers of the epicycles
Q for the various planets, fig. 44, are taken to revolve
uniformly around a center E different from the center of
the earth O. The point E is usually called the equant
point. The mean motion is measured around E, the true motion around O. The author is correct in stating that no
correction is required when the center of the epicycle Q
is at apogee A or perigee P, because then Q, E and O are all
on the same line. The equation of center d, the angle EQO,
reaches a maximum when QC is perpendicular to EO.

Ergo tunc est maxima equatio cum centrum planete   10
est 3 signorum vel 9. Ergo secundum distantiam planete
ab auge usque ad 3 signa crescit equatio proportionaliter.
Et eadem proportione usque ad ymum locum descrescit.
Iterum, crescit usque ad medium augem, et post descrescit.

Tantaque est maxima equatio planete quanta ecentri-   15
citas circuli eius; in sole, gradus unus, 59 minuta, 8

---

10 centrum: punctum, scil. H
12 crescit: curret H / equatio: equilatio H
13 ymum: unum ?H
14 iterum: item H / augem: augis H
15 est om.H / maxima equatio: adequatio maxima H /
     post quanta add. est H
15-16 ecentricitas lac.H
16 unus: unius et H / post minuta add. et H

Thus the equation is greatest when the center of the planet is three signs or nine.[47] Therefore along the distance of the planet from the aux [apogee] up to the third sign, the equation increases proportionally. And it decreases by the same proportion up to the perigee. Likewise it increases up to the mean aux and afterwards decreases.

Then the eccentricity of the circle of the planet is as much as its greatest equation.[48] For the sun it is one

---

[47]"Ergo tunc . . . decrescit" (II, 25.10-14). The author is here involved in a confusion arising from his failure to distinguish mean motion and anomaly, measured around the equant E, from true motion and anomaly, measured around the point of observation O. (See fig. 44.) For all planets with equants, the maximum equation of center, d_{max}, occurs when QC is perpendicular to ACO, at which time angle AEQ, the mean anomaly, is somewhat greater than 90° (three signs) or somewhat less than 270° (nine signs).

The author's comments are correct for the simple eccentric model, fig. 44a, in which C and E coincide, if he is referring to angle AOS, the true anomaly. In that case, d acquires its maximum value when angle AOS is 90° or 270°.

[48]"Tantaque . . . probantur" (II, 25.15-20). The eccentricity $2e = EO$ in fig. 44 can be derived from the maximum equation of center d_{max} because $R \cdot \tan \frac{1}{2} d_{max} = e$, where R is the radius of the deferent. Thus the eccentricity is a function of, but is not identical with, the maximum equation of center. For the model with no equant, fig. 44a, eccentricity $e = OC$ is again a function of maximum equation of center since $e = R \cdot \sin d_{max}$. The author's values are not precisely Ptolemy's values in the Almagest, as may be seen in table 3 below. He probably derived his values from some other source, perhaps Ibn Ezra, ed. Millás-Vallicrosa, 1947: 78.14-15 for the sun, 108.32-34 for the moon, 145.33-146.9 for the planets, or the Toledan Tables, see Toomer, 1968: pp. 56-67. Cf. Nallino, 1903-1907, v. 2: p. 80.

secunda; in luna, 13 gradus 4 minuta; in saturno, 6

gradus, 31 minuta; in jove 5 gradus, 15 minuta; in marte

11 gradus, 24 minuta; in venere, ut in sole; in mercurio

3 gradus, 1 minuta. Hec in Almagesti probantur.   20

---

18 31: 13 H̲
19 11: 2 E̲
20 3: est H̲ / 1: 21 H̲ / p̲o̲s̲t̲ hec a̲d̲d̲.̲ autem M̲ /
    probantur: probatur E̲

degree, 59 minutes, 8 seconds; for the moon, 13 degrees
4 minutes; for Saturn, 6 degrees, 31 minutes; for Jupiter,
5 degrees, 15 minutes; for Mars, 11 degrees, 24 minutes;
for Venus as for the sun; for Mercury, 3 degrees, 1 minute.
These are proved in the Almagest.

[Fig. 44]

[Fig. 44a]

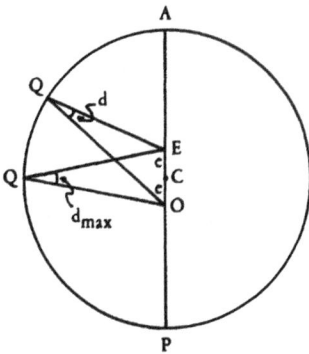

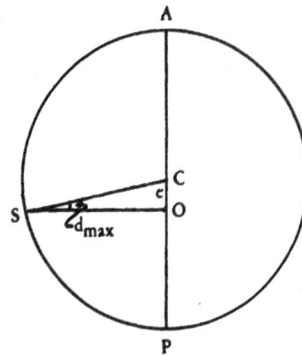

[Table 3]
Maximum Equations of Center

Planet	Maximum equation of center		
	Ptolemy[a]	ACC	Toledan Tables[b]
Sun	2;23°	1;59,8°	1;59,10°
Moon	13;9	13;4	13;9
Saturn	6;31	6;31	6;31
Jupiter	5;15	5;15	5;15
Mars	11;25	11;24	11;24
Venus	2;24	1;59,8	1;59
Mercury	3;2	3;1	3;2

---

[a]Ptolemy's values are given in the Almagest, III, 6 for
the sun, V, 8 for the moon, and XI, 11 for the other planets.
The form in which the values in this table are to be read
is a convenience when dealing with sexagesimal parts. Thus
1;59,8° is to be read as 1 degree, 59 minutes and 8 seconds
or $1 + (59 \times 60^{-1}) + (8 \times 60^{-2})$ degrees.

[b]These values are from Toomer, 1968: pp. 56-67.

[II, 26]   ALTA LOCA 6 PLANETARUM SECUNDUM MOTUM STELLARUM
FIXARUM INVESTIGARE.

Probatum est in Almagesti quod quibuslibet 100 annis

movetur quelibet stella fixa uno gradu ab occidente in

orientem.   Eodem modo altus locus cuiuslibet planete   5

preter quam lune.   Est autem hodierno die ipse altus

locus solis 2 signa 27 gradus 40 minuta, scil. anno domini

1193; saturni 8 signa 10 gradus 49 minuta; jovis 5 signa

22 gradus 58 minuta; martis 4 signa 11 gradus 48 minuta;

veneris qui et solis; mercurii 6 signa 26 gradus 40   10

minuta.

---

1 stellarum om.H

2 investigare: reperire H

3 probatum: probata H

3-4 quibuslibet . . . gradu: quinquaginta unius secundi
    movetur stella in anno H

5 ante eodem add. et H

5-6 eodem . . . lune om.E

5 cuiuslibet: cuiusvis H

6 die ipse: tempore H

7 2 signa: duo secundi M 3 signa H / 27: 37 H / 40: 45 H /
    scil. anno domini: anno verbi incarnati H

8 1193: 1188 E / gradus: signa H / 49: 47 H

9 22: 37 H / 58: 54 E 57 H

10 26: 35 H / gradus om.H

II, 26. TO INVESTIGATE THE APOGEES OF THE SIX PLANETS
ACCORDING TO THE MOTION OF THE FIXED STARS.[49]

It is proved in the Almagest that any fixed star moves
one degree from west to east in 100 years. The apogee of
any planet but the moon moves in the same way. But at this
present time, in the year of the Lord 1193,[50] the apogee
of the sun is 2 signs, 27 degrees, 40 minutes; of Saturn
8 signs, 10 degrees, 49 minutes; of Jupiter 5 signs, 22
degrees, 58 minutes; of Mars 4 signs, 11 degrees, 48 minutes; of Venus as of the sun; of Mercury 6 signs, 26 degrees,
40 minutes.

---

[49]"Alta loca . . . lune" (II, 26.1-6). The precessional
changes in the values of the apogees are considered here.
Ibn Ezra insisted that the apogees move with the fixed stars
even though Thabit ibn Qurra and Azarquiel had claimed a
different motion (see ed. Millás-Vallicrosa, 1947: 80.8-20).
Ptolemy's value for precession (Almagest, VII, 3) . ⅈ one
degree for 100 years.

[50]"Est autem . . . 40 minuta" (II, 26.6-11). Here the
author gives an unequivocal indication of the time at which
he is writing. The source of the apogees given here may be
Ibn Ezra (see table 4). The values in our text differ from
those of Ibn Ezra by less than a degree except for the sun
and Jupiter. They are not as close to the values given in
Millás-Vallicrosa, 1943-1950: pp. 65-66, Suter, 1914: pp. 14-
15, Neugebauer, 1962: p. 41, or Toomer, 1968: p. 45.

Et si vis scire ubi erunt post multos annos, pro
singulis annis distantie addes 51 secunda. Si vis scire
ubi fuerunt ante multos annos, pro singulis annis
demes ab altis locis prefata 51 secunda. Fuit autem   15
tempore Ptholemei aux solis 15 gradus geminorum.

---

13 addes: adde H̲ / secunda: secundum E̲H̲
15 prefata: prefatis H̲ / 51: 5 H̲
16 geminorum: terminorum H̲

And if you want to know where they will be after several
years, you will add to the distance 51 seconds for each of
the years.[51] If you want to know where they were several
years ago, take from the apogees the above-mentioned 51
seconds. At the time of Ptolemy, the aux of the sun was
at 15 degrees of Gemini.

[Table 4]
Apogees of the Six Planets

Planet	Apogees		Reference in ed. Millás-Vallicrosa, 1947
	From II, 26 (1193 A.D.)	From Ibn Ezra (1154 A.D.)	
Sun	2^s 27° 40'	2^s 17° +	78.4-5
Saturn	8 10 49	8 11	110.16-17
Jupiter	5 22 58	= sun +90°	118.8-9
Mars	4 11 48	4 11	120.22-23
Venus	= sun	= sun	121.1
Mercury	6 26 40	[6 25][a]	123.22-23

[a]The apogee of Mercury is given as opposite the jauzahar,
which is 0^s 25°; cf. II, 28.7 and n. 56.

---

[51]"Et si . . . geminorum" (II, 26.12-16). The Ptolemaic
value for precession is equivalent to 36 seconds per year,
not 51. Ibn Ezra used al-Sufi's value, one degree in 70
years, equivalent to about 51 seconds per year; cf. Millás-
Vallicrosa, 1947: 78.9-10 and 83.15-17. A common medieval
value for precession is one degree in 66 years, equivalent
to 54 seconds per year; cf. Thabit ibn Qurra, Liber de motu
octave sphere, ed. Millás-Vallicrosa, 1943-1950: p. 500 and
Millás-Vallicrosa, 1947: 78.9.

[II, 27]   SECUNDUM REDDITUM SOLIS AD ALTUM LOCUM, ANNUM
COMPUTARI.

Indi autem annum solis computant secundum reditum
eius a stella fixa ad eamdem. Et continet 365 dies et
6 horas et centesimam et 20^{am} partem diei. Ptholomeus   5
autem a iunctura ad iuncture reditum metitur annum et
continet 365 dies et 6 horas, 30 parte unius diei minus.
Latini vero ponunt integras 6 horas tantum cum 365 diebus
in anno solis grossam facientes computationem.

Cum ergo secundum Ptholomeum sol intravit arietem   10
i.e. iuncturam, i.e. lineam equinoctialem in 23 die
martii, sed secundum Yndos in 13 die, sed hodierno die
fit econtrario, intrat enim arietem 13 die martii

---

3 autem om.H / post computant add. aut M
4 eius: solis M
5 et centesimam: sive tricentesimam M
5-7 Ptholomeus . . . unius diei om.H
7 unius diei tr.E
8 ponunt om.H / integras 6 horas: horas 6 integras M /
    post horas add. punctum ?E / cum 365 diebus om.H
9 solis om.E / post solis add. cum 365 diebus H /
    grossam: crassam H ?E
10 secundum Ptholomeum: tempore Ptholomei H / intravit:
    intraverit M / arietem: radix ?H
11 i.e. (1) om.E / iuncturam: iunctura ?H / lineam om.E
12 sed (2) om.H / die (2): tempore H
13 arietem: modo radix ?H

II, 27. TO ASSESS THE YEAR BY MEANS OF THE RETURN OF THE
SUN TO THE APOGEE.[52]

The Indians compute the solar year by means of the sun's

return to a certain fixed star, and it contains 365 days

and 6 hours and a hundred-and-twentieth part of a day.

But Ptolemy measures the year from juncture to juncture,

and it contains 365 days and 6 hours, less the thirtieth

part of a day. The Latins in fact assume just 6 whole hours

with 365 days in a year of the sun making a gross computa-

tion.

Therefore, according to Ptolemy the sun entered Aries,[53]

that is the juncture or the equinoctial line, on the twenty-

third day of March, but according to the Indians, on the

thirteenth; at the present time it happens conversely, for

it now enters Aries on the thirteenth day of March according

---

[52]"Secundum . . . computationem" (II, 27.1-9). The length
of the solar year is considered in this proposition. The
Indian value of the length of the year is often found in
Arabic treatises, cf. Neugebauer, 1962: p. 131, Millás-
Vendrell, 1963: 200.9-11, Millás-Vallicrosa, 1947: 75.1-3
and Nallino, 1903-1907, v. 1: 40.27-28. The author errs in
reporting Ptolemy's value; Ptolemy gives (Almagest, III, 1)
365¼ days less one three hundredth part of a day. The author
distinguishes the sidereal year, marked by the sun's revolu-
tion through the fixed stars, from the tropical year, determined by the sun's successive crossings through the point of
the vernal equinox. Cf. ed. Millás-Vallicrosa, 1947: 79.21-
25, where Ibn Ezra cites the lost work of Azarquiel on the
solar year.

[53]"Cum ergo . . . 13 die" (II, 27.10-12). Ptolemy did
not use the Julian calendar, but the Egyptian one, which
has no month called March. The vernal equinox observed by
Ptolemy on 7 Pachom, 463 years after the death of Alexander
(Almagest, III, 1) is the same day as 23 March 140 A.D., by
my calculations. The author may have used a source which
computed this equivalence, unless he computed it himself.

secundum Ptholomeum, que est 23 die secundum Yndos.

Et est idem dies, 13us dies martii secundum Ptholomeum   15

qui est 23us secundum Yndos. Estque moderno tempore

ipsa iunctura, scil. equinoctialis linea, scil. initium

arietis in 22° gradu piscium secundum Yndos. Et ibi

est equinoctium revera, similiter in 22° gradu virginis.

Indi enim signa secundum ymagines computant. Et hoc   20

este valde attendendum.

---

14 qui est 23 die: sed E om.H / post Yndos add. in 23 die
    E 23 die H

16 est om.E / moderno: hodierno H

17 ipsa om.H

18 arietis: radix ?H / 22°: 21 H

18-19 ibi est tr.H

19 22°: 21 ?H

20 et hoc: quod notandum M

21 est tr.post attendendum E / post valde add. et M /
    attendendum: attendum H

to Ptolemy, which is the twenty-third according to the Indians.[54] And the thirteenth day of March following Ptolemy is the same day as is the twenty-third following the Indians. And in modern times, this juncture, namely of the equinoctial line, or the beginning of Aries, is at the twenty-second degree of Pisces following the Indians.[55] In fact the [autumnal] equinox is similarly at the twenty-second degree of Virgo. Indeed the Indians compute the signs according to the images. And this is certainly to be considered.

---

[54]"sed hodierno . . . Yndos" (II, 27.12-16). Because the Julian calendar uses a year that is a little too long, the date of the equinox had shifted from Ptolemy's March 23 to March 13 in the author's time. Ibn Ezra (ed. Millás-Vallicrosa, 1947: 81.29-82.3) gives an explanation (taken from the magistri ymaginum, whom I take to be the same as the Indi of II, 27.20) that at the time of Ptolemy it was the end of the motion of the ascension of the circle and that at Ibn Ezra's time it is the end of the motion of descension; cf. ibid.: 86.17-19. The explanation seems to rely on trepidation theory.

[55]"Estque . . . attendendum" (II, 27.16-21). The author claims one of the intersections of the ecliptic and equator occurs in the constellation Pisces, the other in the con-stellation Virgo. This position with respect to the constel-lations of fixed stars should not be confused with the con-ventional zodiacal signs. In Western astronomy, each sign is one-twelfth of the ecliptic and need not correspond to the visible constellation with the same name. In Indian astronomy, the zodiacal signs seem to correspond to the con-stellations. Precession causes the equinoxes to shift with respect to the fixed stars; therefore, the equinoxes in Indian astronomy will not necessarily occur at the beginning of the visible Aries and Libra. Cf. ibid.: 84.9-18 and 85.5-15.

[II, 28]   PER ALTA LOCA PLANETARUM 5 GENZAHAR EORUM
INVENIRE.

Subtrahantur enim ab alto loco saturni 180 gradus.

Erit que ibi suum caput vel caput sui drachonis, i.e.

genzahar, in opposito cauda.   Ab alto iovis 70 gradus,   5

reliquum dat genzahar eius.   Ab alto loco martis, 90;

mercurii, 180; caput drachonis veneris est ubi

altus locus eius est, que capita moventur sicut auges.

---

1 genzahar: yenzar E yelizar ?H
3 subtrahantur: subtrahatur H / enim om.H / 180: 140 ?H
4 erit que: erit E / post erit add. reliquum dat yenzar
    eius ab alto martis que E / suum caput vel om.H /
    i.e. om.H
5 genzahar: yenzar E yelizar ?H / opposito: opposita E /
    cauda tr.ante in H / gradus om.H
6 reliquum: residuum H / genzahar: yenzar EH / loco om.E
7 veneris est tr.H
8 est om.H / auges: angues H

II, 28. TO FIND THE 5 JAUZAHARS [I.E. NODES] OF THE
PLANETS BY MEANS OF THEIR APOGEES.[56]

Let 180 degrees be subtracted from the apogee of Saturn.

And there will be its head or the head of its dragon, that

is, the jauzahar, in opposition will be the tail. From the

apogee of Jupiter [let] 70 degrees [be subtracted]; the re-

mainder gives its jauzahar. From the apogee of Mars, 90;

from that of Mercury, 180; the head of the dragon of Venus

is where its apogee is; the heads move just as the auges.

---

[56]"Per alta loca . . . auges" (II, 28.1-8). The nodes,
or jauzahars, are used to find the planetary latitudes.
Our text does not give us sufficient information to compute
the latitude of a planet. The instructions for finding the
nodes, as given here, are identical with those of Ibn Ezra
except for Saturn, where Ibn Ezra gives 140°, instead of
180°, to be subtracted from the apogee. The Arabic numeral
4 is easily misread for 8 in their medieval Latin forms.
Misreading the number could account for the difference. Ibn
Ezra's values are found in various places in ed. Millás-
Vallicrosa, 1947: 116.11-13 for Saturn, 118.19-20 for Jupiter, 120.28-29 for Mars, 122.24-25 for Venus and 123.22-23
for Mercury. The word jauzahar does not, however, appear
in Ibn Ezra's work; it occurs as geuzahar, a form very close
to ours, in BM Arundel 377, 56v - 63r, another redaction of
the work of Ibn Ezra giving very similar expressions for the
nodes; see Millás-Vallicrosa, 1947: pp. 63-64. Cf. the different values for the jauzahars in Suter, 1914: p. 15, Neugebauer, 1962: p. 42 and Toomer, 1968: p. 46. The Ptolemaic
values for the nodes are equivalent to those of II, 28, except that 90° must be subtracted from the values for Venus
and Mercury; cf. Almagest, XIII, 6 and Toomer, 1968: p. 46,
n. 10.

[II, 29] PER ALTITUDINEM POLI ORIZONTIS DEPRESSIONEM
ET DIERUM VARIATIONEM DESIGNARE.

Quanta est poli altitudo super orizonta in uno
climate, tanta est depressio orizontis. Et quanto est
maior in uno climate quam in alio, tanto est orizontis   5
depressio maior. Unde dierum inequalitas provenit, quia
ubi magna est altitudo poli, scil. magna latitudo regionis,
ibi est magna dierum prolixitas estivalium. Citius que
est dies in magna poli altitudine sole ente in signis
sinistris ab ariete, scil. ab libram.   10

Ubi est minor altitudo poli citius est dies sole
ente in dextris signis. Ubi maxima altitudo poli,
continuus est dies sole ente in sinistris; continua est
nox sole ente in dextris. Et est equinoctialis linea
ubi orizon. Hoc per astrolabium probatur.   15

---

3 poli altitudo tr.M / super orizonta om.H
3-4 in uno climate om.E
4 depressio orizontis tr.H
5 maior: minor E / climate om.H
6 maior: minor E / inequalitas provenit tr.H
7 ubi: illa M / magna est tr.H
8 que: quod H
9-10 in signis sinistris: a sinistris signis H
11 est (1) om.E / altitudo poli tr.H
12 post ubi add. est M
13 continuus: cum citius E / post in add. signis M /
   est om.H
14 et est om.H
14-15 linea ubi: linea ibi M minima est ubi H

II, 29. TO DESCRIBE THE DEPRESSION OF THE HORIZON AND THE
VARIATION OF DAYS BY MEANS OF THE ALTITUDE OF THE POLE.[57]

The depression of the horizon is as much as the altitude
of the pole above the horizon in a given climate. And the
depression of the horizon is as much greater in one climate
than in another as [the altitude of the pole] is greater.
Hence the inequality of the days results,[58] since wherever
there is a great polar altitude, that is the latitude of
the region is large, there the summer days are long. And
where the polar altitude is great, the day [breaks] sooner
when the sun is in the northern signs, namely from Aries to
Libra.

Wherever the polar altitude is small, the day [breaks]
sooner when the sun is in the southern signs. Where the
polar altitude is greatest, the day is continuous when the
sun is in the northern signs; the night is continuous when
the sun is in the southern signs. And the equator is where
the horizon is. This is proved by means of the astrolabe.

---

[57]"Per altitudinem . . . depressio maior" (II, 29.1-6).
The depression of the equatorial horizon seems to mean the
terrestrial latitude. The inclination of the equator is
thus equal to the polar altitude. Cf. Millás-Vendrell, 1963:
130.19-32 and 134.10-13.

[58]"Unde dierum . . . probatur" (II, 29.6-15). The author
is simply explaining that the further north one goes, the
days are longer and the sunrise is earlier during the spring
and summer but shorter during the fall and winter. The maximum altitude of the pole occurs at the north pole, where
the day and night are each six months long. By describing
longer days when the sun is going from Aries to Libra, the
author has neglected to describe the situation in the southern hemisphere.

[II, 30]   TURRIS ALTITUDINEM METIRI.

Expecta quod altitudo solis sit 45 graduum ut
docuimus, tunc erit umbra cuiuslibet corporis in plano
siti equalis suo corpori.   Nota ergo erit umbra turris.
Mensura eam, qualiter autem metiatur, patet eius altitudo. 5

Sic proportio cuiuslibet umbre ad suum corpus in
quavis altitudine solis, eadem est que umbre turris ad
turrem.   Si ergo nota est proportio, nota erit umbra,
ergo nota erit altitudo.

Item per astrolabium, pone allidadam in 45 gradu   10
altitudinis.   Et per eam aspice turris summitatem.   Et
adice staturam tuam retro ab oculis infra.   Quantum erit
spatium usque ad pedem turris, tanta erit altitudo
turris.   Quod probatur per triangulos orthogonos, qui

---

2 altitudo solis <u>tr.</u>H / sit <u>om.</u>H
3 cuiuslibet: cuiusvis <u>H</u>
4 <u>post</u> ergo <u>add.</u> quod <u>E</u>
5 <u>post</u> mensura <u>add.</u> ergo <u>M</u> / qualiter . . . altitudo:
   aliter autem metiatur eius altitudo <u>E</u>   ergo altitudo
   eius <u>H</u>
6 sic: patet sic <u>M</u>   aliter videatur <u>H</u> / cuiuslibet: eius
   vis <u>H</u>
7 quavis: quacumque <u>E</u> / est que: erit <u>H</u>
9 erit: est <u>H</u>
10 item: iterum <u>E</u> / 45: 85 <u>?H</u>
11 summitatem: summitate <u>M</u>
12 adice: addite <u>H</u> / <u>ante</u> quantum <u>add.</u> et <u>M</u>
13 pedem: pede <u>H</u>

II, 30. TO MEASURE THE HEIGHT OF A TOWER.[59]

Wait until the altitude of the Sun is 45 degrees as we
have taught, then the shadow lying in the plane of any body
will be equal to its body. Therefore the shadow of the
tower will be known. As the length [of the shadow] measures
it, its height is clear.

The ratio of any shadow to its body[60] for any altitude
of the sun is the same as the [ratio] of the shadow of the
tower to the tower. Therefore if the ratio is known, the
shadow will be known, and therefore the height will be
known.

The same by means of the astrolabe:[61] put the allidad
at 45 degrees of altitude, and through it look at the summit of the tower. And [going] to the rear increase [the
distance to the tower] by your height up to the eyes. The

---

[59]"Turris . . . altitudo" (II, 30.1-5). Measuring the
height of a tower is one of the most traditional problems
in altimetry; it occurs in many variations. The method
given here requires the use of some device to measure 45°
as well as an instrument to measure length. Proposition
II, 1 taught how to measure solar altitudes. Cf. Geom.
incert. and Ep. et Vit. Ruf., ed. Bubnov, 1899: 327.20-328.8,
and 550.23-551.4, Hugh of St. Victor, Practica geometriae,
ed. Baron, 1966: 30.167-171 and Trac. quadrantis, ed. Britt,
1972: p. 167.

[60]"Sic proportio . . . altitudo" (II, 30.6-9). One may
use the shadow of any vertical object having a directly mea-
sureable height to determine the height of the tower. This
method eliminates the need for an angle-measuring device.
It demands an understanding of proportion which the preceding method did not. Cf. Geom. incert., ed. Bubnov, 1899:
322.15-28, Hugh, ed. Baron, 1966: 22.170-179, Adelard, De
eodem et diverso, ed. Willner, 1903: 29.30-30.8 and Trac.
quadrantis, ed. Britt, 1972: pp. 167-169.

[61]"Item . . . laterum" (II, 30.10-16). Using this method, one need not wait for the sun to shine at any particular
angle. The right isosceles triangles, fig. 45, ABC, AED and

super cathetum ypothenuse sunt constituti, eamdem enim   15

habent proportionem laterum.

---

15 cathetum ypothenuse: eandem ypothenusam H̲ / post
   constituti add. sive super eamdem ypothenusam M̲ /
   enim om.H̲
16 habent: habebit H̲

height of the tower will be as much as will be the [new]
distance to the foot of the tower. This is proved by right
triangles which are disposed on the same hypotenuse; they
thus have the same ratio of sides.

[Fig. 45]

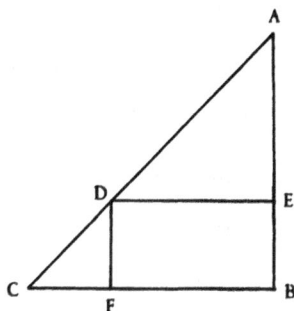

DFC all have the same line as a hypotenuse; therefore, adding FC, the distance the measurer moves backwards, to BF,
the original distance to the tower, is equivalent to adding
EB to AE. Cf. Geom. incert., ed. Bubnov, 1899: 317.24-
318.7, Hugh's practical geometry, ed. Baron, 1966: 30.176-
31.201 and Trac. quadrantis, ed. Britt, 1972: pp. 161-162.

[II, 31]   TURRIS INACCESSIBILIS ALTITUDINEM METIRI.

Stet altimetra in metiende turris altitudine.
Aspiciatque per mediclivium donec videat cacumen turris,
et signetur locus, et numerentur digiti quos capit
allidada in catheto, qui verbi gratia sint 4.   Tunc ab   5
inaccessimetra retro pergatur donec cacumen iterum
videatur per mediclivium, et signetur locus.   Et sit
intervallum stationum 12 pedum.   Notenturque digiti quos
capit allidada in catheto, qui verbi gratia sint 2.
Binarius autem in latere quadrati, scil. 12, sexcies   10
continetur.   Prefatus vero quaternarius, ter, vel prefatus ternarius, quater.   Minus continens de maiori
continenti subtrahatur semel, scil. quaternarius de senario,
remanet binarius.   Ergo secundum hoc, intervallum
stationum est duplum ad altitudinem.   Si unum remaneret, 15
esset equale ei.

---

2 altimetra: planimetra H / altitudine: spatio H
3 aspiciatque . . . turris om.E
5 in: de H / post 4 add. vel 3 E
6 inaccessimetra: ?M incessimetra H / pergatur: ?E
   purgatur H
7 et om.E
8 12: 15 E / post digiti add. catheti H / quos: quod E
9 in catheto om.EH
11-12 vero . . . prefatus om.H
12 de: a H
13 subtrahatur semel tr.H / quaternarius de senario:
   4 de 6 ?H
14 binarius: 2 H
15 si unum: summum H / remaneret: remanet H
16 equale ei tr.H

II, 31. TO MEASURE THE HEIGHT OF AN INACCESSIBLE TOWER.[62]

Let the planimeter stand before the altitude of the
tower to be measured. Let him look through the mediclivium
until he sees the peak of the tower, and let the place be
marked. And let the digits in the vertical side [of the
shadow quadrant] which the allidad reaches be counted, which,
for example, may be 4. Then let him proceed back from the
inaccessible tower until he sees at the peak once again
through the mediclivium, and let the place be marked. And
let the interval of the stations be 12 feet. Then let the
digits which the allidad reaches in the vertical side be
noted, for example, let them be 2. But in the side of the
square, that is 12 [digits], 2 is contained six times. The
aforementioned 4 [is contained] three times, or the aforementioned 3 [is contained] four times. Let the smaller quotient be subtracted from the larger quotient once, that is,
4 from 6; 2 remains. Therefore, according to that, the interval of the stations is double the height. If one remains,
[the interval] would be equal to [the height].

---

[62]"Turris . . . triangulorum" (II, 31.1-20). The technique given here is best understood by reference to fig. 46,
where the tower to be measured is DE, and A and B are the
two positions from which sightings are made. Let q = AC/CD
and r = BC/CD, then q is also the first or smaller quotient
and r the larger quotient. Then q·CD = AC and r·CD = AB +
AC. Subtracting, CD·(r-q) = AB. Since AB is the interval
of the stations, we have what was to be proved. Although
the text does not mention it, the height of the measurer,
equal to CE, must be added. Cf. Geom. incert., ed. Bubnov,
1899: 318.13-320.14, which is very close to the text. Cf.
also Hugh's practical geometry, ed. Baron, 1966: 32.212-
33.253 and Trac. quadrantis, ed. Britt, 1972: pp. 171-174.

Et ita est generalis regula quod perfecta subtractione
continentium si unum remanet, intervallum est equale
altitudini, si duo duplum, etc. quod probatur per
proportionem triangulorum.   20

Per speculum et per filum et sagittam, et per aquam,
et per orthogonium, et per cannas mensuratur altitudo quod
quasi notum preterimus.

---

17 quod om.E / perfecta: facta H
18 continentium om.E / si unum: summum H / remanet bis E /
   est equale: erit H
19 post altitudini add. equale H / si duo: scil. 2 H /
   etc.: et est H
21 et (1): etiam H / et sagittam om.E / et per aquam
   tr.post orthogonium E / et (3) om.H
22 et (1) om.H / et (2) om.H / post cannas add. etiam H /
   mensuratur: mensurantur E / altitudo om.H
23 preterimus: pretermitto H

And this is the general rule: when a subtraction of the
quotients has been completed, if one remains, the interval
is equal to the height, if 2, [it is] double, etc. This is
proved by the ratio of triangles.

The height may be measured also by means of a mirror,
an arrow and string, [a dish of] water, a right triangle,
or reeds, which we pass by as if they were known.[63]

[Fig. 46]

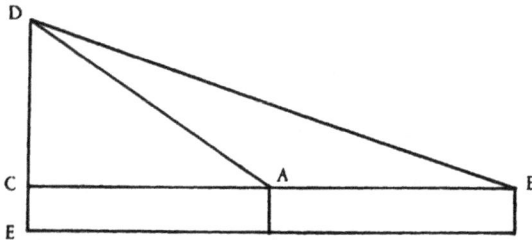

---

[63]"Per speculum . . . preterimus" (II, 31.21-23). The
different techniques mentioned here are common in practical
geometries. The author includes some of them in II, 32 and
II, 33, even after claiming to omit them. For the mirror,
see Geom. incert., ed. Bubnov, 1899: 323.1-10 and 333.3-14,
Hugh of St. Victor, ed. Baron, 1966: 37.384-391 and Trac.
quadrantis, ed. Britt, 1972: pp. 181-183. For the arrow and
string, see Geom. incert., ed. Bubnov, 1899: 334.19-335.65
and II, 33 below. For the dish of water, see Geom. incert.,
ed. Bubnov, 1899: 323.1-10 and 333.3-14 and Hugh, ed. Baron,
1966: 37.373-383. For the right triangle, see Geom. incert.,
ed. Bubnov, 1899: 327.20-328.21, Hugh, ed. Baron, 1966:
34.290-35.317 and II, 32. For the reeds or sticks, see Geom.
incert., ed. Bubnov, 1899: 323.11-325.10 and 326.14-327.2,
Hugh, ed. Baron, 1966: 35.318-37.372, Trac. quadrantis, ed.
Britt, 1972: pp. 177-179 and II, 32.

[II, 32]   ALTITUDINEM PER ORTHOGONIUM ET CANNAS MENSURARE.

Fiat orthogonium cuius cathetus sit equalis basi et
in tali loco ponatur in plano scil. quod per summitatem
katheti summitas rei videatur oculo posito in termino
basis. Quanta erit linea ab oculo ad pedem altitudinis   5
tanta erit altitudo. Hoc probatur quia illi 2 trianguli,
abc, efg, habent eandem proportionem laterum super eandem
enim ypothenusa sunt constituti. Si stans mensuraveris
staturam tuam adde.

Item ad quantitatem tui ab oculo infra sume virgam.   10
Et iacens per summitatem virge erecte super pedes,
cacumen respice. Tanta erit altitudo quanta est linea
ab oculo ad pedem rei. Hec probatur per precedentem.

---

3 plano: planos H / scil. om.H
5 post basis add. et M / quanta: quota H / linea: linia H
7 abc, efg om.E
8 eandem (2): eadem M
9 enim om.H / ypothenusa: ypothenusam H / sunt om.H /
   si: sed H / mensuraveris: mensurabis H
11 item: iterum E vel H

II, 32. TO MEASURE A HEIGHT BY MEANS OF A RIGHT TRIANGLE AND REEDS.[64]

Let there be a right triangle of which the altitude is equal to the base, and let it be put in a place such that the summit of the thing may be seen by the eye put at the end of the base. The height will be as much as the line from the eye to the foot of the altitude. This is proved because those two triangles, abc, efg have the same ratio of sides for they are constituted on the same hypotenuse. But if you measured standing, add your height.

Or take a stick [equal] to your size from your eye down. And lying down, look at the peak across the top of the stick erected above your feet.[65] The height will be as much as the line from the eye to the foot of the thing. This is proved by means of the preceding.

---

[64]"Altitudinem . . . adde" (II, 32.1-9). This method is so simple that the right isosceles triangle constructed for it could be replaced by a simple stick, which is exactly what the author has done in II, 32.10-13. The two triangles are obviously similar. Cf. Geom. incert., ed. Bubnov, 1899: 327.20-328.8 and Hugh, ed. Baron, 1966: 35.305-317.

[65]"Item ad quantitatem . . . precedentem" (II, 32.10-13). Here the prone body of the measurer performs the function of the base of a right isosceles triangle. A pole, stuck in the ground and equal to the height of the measurer up to his eyes, serves as the altitude of the triangle. This method is more notable for its whimsy than its mathematics.

Item sit altitudo metienda, ab. Sumatur canna maior

statura mensoris, ef, statura hominis, cd, linea recta   15

ducta ab oculo per arundinem orthogonaliter, scil. usque

ad altitudinem, dgh. Ita quod per summitatem, ef,

videatur summitas rei. Et sit radius afd. Que erit

comparatio dg ad gf in triangulo parvo, eadem erit dh ad

ha. Et prior nota ergo secunda. Hoc per proportionem   20

triangulorum probatur per primas.

Si inaccessibilis est altitudo, sumatur planities

retro ci. Esto statura metientis ik. Sit linea ortho-

gonaliter ducta ab oculo ad altitudinem kh. Sit arundo

---

14 item: iterum E vel H / metienda: metianda H
15 cd: ed H / recta tr.post ducta E om.H
17 dgh: dg H / ita: hoc itaque H
19 gf: gef EM
20 secunda: secundum M et secundam ?H
21 per primas: ?H
23 esto tr.post metientis H / ik: ak MH
24 kh: kn ?E ?H

Or let the height to be measured be ab.[66] Let a stick,

ef, be taken greater than the height of the measurer; let

the height of the man be cd, let the straight line, dgh, be

drawn from the eye through the stick perpendicularly, namely

as far as the height, so that the summit of the thing is

seen across the top [of] ef. And let the [visual] ray be

afd. The ratio [of] dg to gf in the smaller triangle will

be the same as that of dh to ha. And the former is known,

therefore the latter. This is proved by the ratio of tri-

angles in the first propositions.

If the height is inaccessible,[67] let the [measurer]

move back [the distance] ci. Let the height of the measur-

er be ik. Let the line drawn perpendicular from the eye to

the height be kh. Let a stick equal to the one above be lm,

---

[66]"Item sit altitudo . . . primas" (II, 32.14-21). This
method presupposes the ability to solve similar triangles
using their proportional sides. See fig. 47, where triangles
fgd and ahd are similar. The procedure has the advantage
of requiring very little equipment. The author has changed
the proposition very little from its form in Geom. incert.,
ed. Bubnov, 1899: 325.11-326.2. Cf. also Hugh, ed. Baron,
1966: 35.327-36.335, Adelard, ed. Willner, 1903: 29.13-29
and Trac. quadrantis, ed. Britt, 1972: pp. 177-179.

[67]"Si inaccessibilis . . . mensuratum" (II, 32.22-30).
An extension of the technique of II, 32.14-21 to the problem
of the inaccessible tower, this method has much in common
with II, 31. This technique substitutes triangles of which
one side is a known part of the measuring stick for the triangles in the square of the shadow quadrant in II, 31. See
fig. 48. The proposition is very similar to Geom. incert.,
ed. Bubnov, 1899: 326.3-13. Cf. Hugh, ed. Baron, 1966:
36.336-356.

equalis superiori <u>lm</u> <transiens <u>kh</u> in <u>n</u>>. Videatur   25

proportio <u>kn</u> ad <u>nl</u>, eadem erit <u>kh</u> ad <u>ha</u>. Si <u>dh</u> fuerit

duplum ad <u>ha</u>, <u>kh</u> ad <u>ha</u> erit quadruplum. Subtrahatur

ergo <u>dh</u> de <u>kh</u>; remanet <u>kd</u>, quod est duplum ad <u>ha</u>. Sed

<u>ki</u> est equalis <u>mn</u> et <u>dc</u> et <u>ge</u> et <u>hb</u>. Ergo totum <u>ba</u> est

mensuratum.   30

---

26 <u>nl</u>: <u>ml</u> <u>E</u>
26-28 si <u>dh</u> . . . ad <u>ha</u> <u>om.H</u>
28 <u>kd</u>: <u>ha</u> <u>E</u> / quod est duplum: quadruplum <u>M</u>
29 <u>ki</u>: <u>ka</u> <u>?M</u> / <u>dc</u>: <u>de</u> <u>M</u> / <u>ba</u>: <u>ha</u> <u>M</u>

crossing $\underline{kh}$ at $\underline{n}$. Let the ratio of $\underline{kn}$ to $\underline{nl}$ be seen; the
same will be that of $\underline{kh}$ to $\underline{ha}$. If [for example] $\underline{dh}$ was
double $\underline{ha}$, [and] $\underline{kh}$ [was] quadruple $\underline{ha}$, then let $\underline{dh}$ be subtracted from $\underline{kh}$. $\underline{kd}$, which is double $\underline{ha}$, remains. But $\underline{ki}$
is equal to $\underline{mn}$ and $\underline{dc}$ and $\underline{ge}$ and $\underline{hb}$. Therefore all of $\underline{ba}$
is measured.

[Fig. 47]   [Fig. 48]

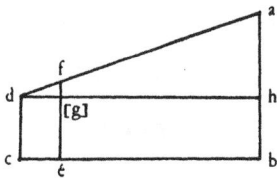
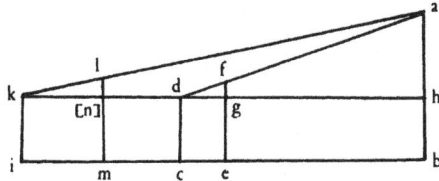

[II, 33]   ALTITUDINEM CUM FILO ET SAGITTA COLLIGERE.

Ferias summitatem rei cum sagitta filo ligato.
Similiter pes altitudinis feriatur cum sagitta filo
ligato. Tunc mensuretur utrumque filum quot pedum sit.
Et utrumque numerum multiplica in se. Et utriusque note-   5
tur summa. Minus de maiori subtrahatur. Et illius quod
remanserit sumatur radix, et tot pedum erit altitudo.
Ut si altitudo sit _ab_ et filum primum scil. _ac_, sit 5
pedum; aliud filum _cb_, 4 pedum. Producto quaternarii
subtracto a producto quinarii remanent 9 cuius radix,   10
scil. ternarius, dat altitudinem _ab_. Hoc probatur
ratione numeri.

---

2 ligato: ligata H
3 cum sagitta om.H
6 et om.H / illius: illis ?H / quod: qui H
7 sumatur radix tr.H / pedum erit tr.M
8 ac: at M / sit (2) tr.ante scil. H
9 aliud filum om.M / cb: bc E / quaternarii: quadrati sive
    quaternarii M 4^{ti} H
10 quinarii: 5^{ti} H
11 scil. ternarius om.H

II, 33. TO INFER A HEIGHT WITH A STRING AND ARROW.[68]

Hit the summit of the thing with an arrow with a string
attached. Similarly let the foot of the altitude be hit by
an arrow with a string attached. Then let how many feet
there are in each string be measured. And multiply each
number by itself. And let the product of each be noted.
Let the smaller be subtracted from the larger. And let the
root of what remains be taken, and the altitude will be that
many feet. As for example, if the altitude is ab, and the
first string, namely ac, is 5 feet; the other string, cb is
4 feet. The product of 4 [by itself] having been subtracted
from the product of 5 [by itself], there remain 9, the root
of which, namely 3, gives the altitude ab. This is proved
by reason of number.

[Fig. 49]

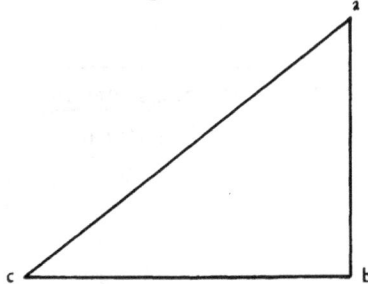

---

[68]"Altitudinem . . . numeri" (II, 33.1-12). A simple
application of the Pythagorean theorem will solve this problem, in which the hypotenuse and base are known, with an
accuracy dependent on the skill of the archer. See fig.
49. The proposition is close to Geom. incert., ed. Bubnov,
1899: 334.19-335.25.

[II, 34]   PUTEI PROFUNDITATEM METIRI.

Stet profundimetra, tenens exterius astrolabium vel
quadrans, aspiciatque per ambo foramina terminum oppositi
lateris in fundo putei, ita quod circulatio sit perpendiculariter perpensa.   Et quantitas dyametri notetur.   5
Videaturque quot digitos dyametri capit allidida in
quadrato.   Et fiat proportio ad 12, et eadem erit dyametri
et profunditatis cum ipsius statura, qua remota de
profunditate, quod superest est putei profunditas.   Sit
igitur putei profunditas ab.   Sit dyameter ac, 3 pedum.   10
Sit statura cd, 4 pedum.   Sint digiti in quadrato 3; et
sit radius visualis db.   Quia igitur ternarius est
subquadruplus ad 12, est igitur dyameter ac subquadrupla
ad de.   Ergo subtracto dc statura, scil. 4 pedum,
remanet ce.   Est ergo altitudo ce, 8 pedum.   Hoc per   15
proportionem triangulorum probatur.

---

2 ante tenens add. et H / exterius om.H
3 quadrans: quadrantem M / aspiciatque: respiciat H /
   terminum: tantum H
4 fundo: profundo H
6 dyametri om.H
9 est om.H
11 cd: ad E
13 est igitur om.H / subquadrupla: est subquadruplus H
14 ergo subtracto tr.E / scil. om.M   tr.ante statura E
15 8: 4 MH
16 ante probatur add. scietur vel E

II, 34. TO MEASURE THE DEPTH OF A WELL.[69]

Let the depth measurer stand [on the edge of the well],
holding the astrolabe or quadrant outside [the well]. And
through both holes [of the allidad] let him look at the end
of the opposite side in the bottom of the well, so that its
circular surface is considered perpendicularly. And let the
size of the diameter be noted. And how many digits the allidad marks in the shadow square may be seen. And let
[their] ratio to 12 be taken, and it will be the same as the
ratio of the diameter to the depth plus his own height. When
that height is taken away from the depth, what is left over
is the depth of the well. Therefore let the depth of the
well be ab. Let the diameter ac be three feet. Let the
height [of the measurer] cd be 4 feet. Let there be 3
digits [marked] in the shadow square; and let the visual
ray be db. Therefore since 3 is subquadruple to 12, the
diameter ac is subquadruple to de. Therefore, having subtracted the height [of the measurer] dc, namely 4 feet, ce
remains. Therefore the altitude ce is 8 feet. This is
proved by means of the ratio of triangles.

---

[69]"Putei . . . probatur" (II, 34.1-16). The fact that
the diameter of the well is the same at top and bottom permits this determination. The technique is a simple use of
the quadrant or of the shadow square in an astrolabe. See
fig. 50. The text seems to have been taken from Geom. incert., ed. Bubnov, 1899: 320.26-321.21; it is closest to
the readings of Bubnov's group E. Cf. ibid.: 331.6-16,
Hugh, ed. Baron, 1966: 43.541-45.570, and Trac. quadrantis,
ed. Britt, 1972: p. 200.

Idem scies per cannam sic. Stans super puteum, pone
virgam sub pedibus donec per summitatem virge putei
profunditatem in opposita parte videas. Et pars illius
virge notetur. Et comparetur cum statura tua. Nam que   20
est proportio illius partis ad staturam, tota erit stature
et profunditatis ad dyametrum. Sit ergo putei profunditas
ab, dyameter ac, statura cd, arundo ce, altera pars cf.
cd est quadruplum ad ce. Ergo df est quadruplum ad ac.
Subtracto dc, remanet cf. Igitur nota est putei profun-   25
ditas, scil. ab.

---

18 virge: eius H
19 ante pars add. alia M   illa H / illius om.H
20 cum om.H
22 et . . . dyametrum: ad profunditatem E
23 cf: ef M ?E
24 cd: om.?E   cdt H / ce ergo . . . quadruplum ad om.H /
    df: ef M
25 subtracto: sed subtracto E   sublato H / igitur nota est:
    ergo H
25-26 putei profunditas tr.H

You will know the same thing by means of a stick as follows.[70] Standing above the well, put the stick under [your]
feet until you look across the top of the stick at the opposite side of the bottom of the well. And let the part of
that stick be noted, and let it be compared with your
height. Now the ratio of that part to [your] height will
be that of the whole of the height and depth to the diameter.
Therefore let the depth of the well be ab, the diameter ac,
the height [of the measurer] cd, the reed ce, the remaining
value cf. cd is quadruple ce. Therefore df is quadruple
ac. Having subtracted dc, cf remains. Therefore the depth
of the well, namely ab, is known.

[Fig. 50]

[Fig. 51]

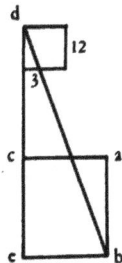

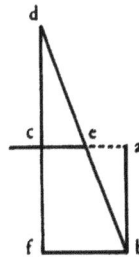

---

[70]"Idem scies . . . scil. ab" (II, 34.17-26). This method is similar to the preceding one, except that the smaller
triangle is made up of the stick and the measurer's height.
See fig. 51. Geom. incert., ed. Bubnov, 1899: 327.3-19
seems to be the source of this proposition; the text is
very close to that of Bubnov's group E, especially MS Chartres 214. Cf. also Geom. incert., ed. Bubnov, 1899: 331.6-
16, Hugh, ed. Baron, 1966: 45.571-578 and Adelard, ed. Willner, 1903: 31.8-23.

[II, 35]   PUTEI CAPACITATEM INVENIRE.

Nota est altitudo putei per precedentem. Nota est
etiam putei area per primum librum, scil. per semi-
dyametrum et semicircumferentiam. Illa area multiplicetur in <al>titudinem. Productum dabit capacitatem pedum   5
crassorum, ita quod regularis sit circulatio putei.

---

2 altitudo putei tr.H / per precedentem om.H
2-3 nota (2) . . . putei om.H
3 post per (2) add. diametrum H
4-5 multiplicetur: feriatur H
5 <al>titudinem: latitudinem MEH
6 circulatio putei tr.H

II, 35. TO FIND THE CAPACITY OF A WELL.[71]

The height of the well is known by means of the preceding proposition. The area of [the surface of] the well is
also known, by means of the first book, namely by means of
the radius and semicircumference. Let that area be multiplied by the height. The product will give the capacity in
cubic feet, given that the circle of the well is regular.

---

[71]"Putei capacitatem . . . putei" (II, 35.1-6). This
proposition which computes the capacity of a well more correctly belongs to Book III, which deals with volumes. The
area of a circular surface was taught in I, 22. Cf. Geom.
incert., ed. Bubnov, 1899: 358.18-21 and Trac. quadrantis,
ed. Britt, 1972: p. 201.

[II, 36] AD ALTITUDINEM METIENDAM QUADRATUM IN POSTICA
WALZAGORE CONSTITUERE.

Dividatur postica in 4 quadras per lineam orientalem
a et meridionalem b. Spatium autem inter occidentalem
lineam et septentrionalem in 2 equa divide per punctum   5
c, aliam quartam inter occidentalem et australem lineam in
2 equa per punctum d, et aliam inter septemtrionalem
et orientalem in 2 similiter per punctum e. Positaque
regula in punctis c, d, lineam rectam fac. Inter e et
c ponens regulam, aliam lineam rectam fac. Et habebis   10
quadratum orthogonium. Et data latitudine ad libitum
divide in sena spatia, et in 12 utrumque latus. Illi
erunt 12 digiti katheti per quos scitur umbra, et
altitudo cum allidada f.

---

1 altitudinem metiendam: altitudines metiendas E altum
   metiendum H
3 lineam: l et a H / orientalem: orientale E
4 meridionalem: meridionale E
5 lineam om.H / post septemtrionalem add. lineas H
6 post c add. posita que regula in punctis c, d M /
   lineam om.H
7 equa om.H / aliam: alium H
9 punctis: puncti E / c, d: e, d H / lineam rectam: linea H
10 ponens: pones H
11 orthogonium: orthogonii H
12 in (2) om.E
13 12: 13 H
14 post altitudo add. est H / f om.E

II, 36. TO SET UP A SQUARE TO MEASURE HEIGHT ON THE BACK
OF THE ASTROLABE.[72]

Let the back be divided into four quarters by an eastern
line a and a southern line b. Divide the space between the
western and northern line into two equals with point c, and
the other quarter between the western and southern line into
two equals with point d, and the other between the northern
and eastern similarly into 2 with point e. Then having set
a ruler on points c [and] d, make a straight line. Putting
the ruler between points e and c; make another straight line.
And you will have a right square. And having been given any
width you please, divide [it] into six parts [and use it to
divide] both sides into 12. And those will be the 12 digits
of the vertical by means of which the shadow and height are
known with the allidad f.

[Fig. 52]

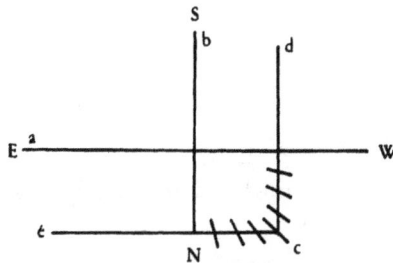

[72]"Ad altitudinem . . . allidada f" (II, 36.1-14). Constructing the shadow square on the back of an astrolabe,
fig. 52, is fairly straightforward, once one remembers that
the directions in an astrolabe are reversed. I assume the
arbitrary width divided into six parts is an auxiliary for
the division of the given side of the square into six parts,
each of which must then be halved. Convenience in use would
suggest that we double the square given here, as in fig. 2,
in I, 1, above. Hugh of St. Victor describes the construction of a triple shadow square, ed. Baron, 1966: 25.55-26.74.
Cf. Liber de astrolabio in Bubnov, 1899: 132.15-21.

[II, 37]   ADDITUM ALTUM METIENDUM QUADRATUM IN QUADRANTE
DESCRIBERE.

Sumatur lignum quadratum, et ducantur due linee

equedistantes a b in latere. Positoque pede circini

immobili in angulo c, describatur 4^a pars circuli.   5

Residuum abscindatur. Fiantque due pinnule in capitibus

servate super equedistantes que perforentur per a.

Dividaturque 4^a pars circumferentie in 3 partes equales,

l, m, n, et unaqueque in 6. Positoque perpendiculo o

in c, altitudinem per hoc mensurabis. Ducaturque linea   10

recta dc ad medietatem illius 4^e circumferentie. Et

in termino linee protense a c usque ad mediam sectionem,

fiat ibi quadratum orthogonium. Dataque latitudine ad

---

1 additum altum: datum E ad altum H / quadrante:
   quadrantem H
3 lignum quadratum tr.M / ducantur: sumantur E
5 immobili om.H / describatur: discribatur M
8 4^a om.H
9 l, m, n: l, n, m M l, n E
10 altitudinem ?H / mensurabis: mensurat H
11 dc: ac E ae H / illius om.E
12 in termino: inde H / post linee add. tunc H /
   sectionem: sectione E
13 orthogonium: orthonium M

II, 37. TO DRAW AN ADDED SQUARE FOR MEASURING HEIGHT IN
THE QUADRANT.[73]

Let a wood square be taken and let two lines, a, b, parallel to the side be drawn. Then having placed the immobile
foot of the compass at the angle c, let a quarter of a
circle be drawn. Let the rest be removed. And let two
pinnules be made in the headings saved above the parallels;
let them be pierced by [a line parallel to] a. Let the
quarter of the circumference be divided into three equal
parts, l, m, n and each one into six. And then having put
a plumb line o at c, you will measure height by this. And
let a straight line dc be drawn to the middle of that quarter circumference. And at the end of the line drawn out
from c as far as the midsection let a square be made there.
And given any width you please, let the two sides be divided

---

[73]"Additum . . . mensuratur" (II, 37.1-16). The shadow
square in the quadrant may be used in conjunction with the
angular scale on the quadrant to measure angles and their
shadows. This proposition explains the construction of both
of these scales; see fig. 53. The division of the quarter-
circle into three and then 18 parts provides angular mark-
ings every five degrees. To use a quadrant one sights
through the pinnules along the edge; the plumb line marks
off the digits on the shadow square. The treatise on the
quadrant gives the construction of a more complete quadrant;
those parts comparable to the text are ed. Britt, 1972:
pp. 106-112.

libitum, dividantur duo latera in sena et in 12 spatia.

Illi notabunt digitos katheti terminati a perpendiculo.

Et per hec finita umbra et altitudo mensuratur.

---

14 dividantur: dividatur E / ante spatia add. duo H
15 digitos om.E / terminati ?ME   om.H
16 finita: funira vel finira ?E   sumitur H / et om.E /
    mensuratur: mensurantur E   om.H

into six and into 12 spaces. These [spaces] mark out the
digits of the altitude defined by the plumb line. And by
means of this, when completed, the shadow and height are
measured.

[Fig. 53]

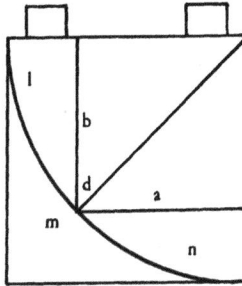

[II, 38] AD UMBRAM ALTITUDINIS HABENDAM COMPONERE
INSTRUMENTUM.

   Preparetur circularis superficies. Et posito pede
circini in centro, fiat circulus cuius semidyameter sit
trium digitorum. Item super idem centrum fiat maior   5
circulus cuius semidyameter duplus sit ad primum. Item
maior cuius semidyameter triplus sit ad primum. Item
maior cuius semidyameter sit quadruplus ad primum. Ita,
quotvis facies circulos quorum distantie erunt equales.
Unamquamque distantiam in 12 divides partes diversi   10
coloris. Fiant circuli super idem centrum. Ad quantitatem primi semidyametri in centro infigatur cathetus qui
est 12 digitorum primi circuli. Unius coloris katheto

---

1 componere: imponere E om.H
5 item: iterum E vel H / fiat: sit M
6 semidyameter: diameter E / sit: tr.ante duplus M fit E /
   item: iterum E vel H
7 item: iterum E vel H
8 sit om.E / post quadruplus add. fit E / ita: item H
10 in 12 om.H / divides: dividas H / ante partes add. in
   3 H / post partes add. et 12 E
11 centrum: e centrum H
11-12 ad quantitatem om.H
12 primi: primum H / infigatur: infingatur E in figura H
13 est 12: 12 est E 12 H / circuli: coloris M / coloris:
   circuli ?M
13-14 katheto . . . colores om.E

II, 38. TO CONSTRUCT AN INSTRUMENT FOR HAVING THE SHADOW OF THE ALTITUDE.[74]

Let a circular surface be prepared. And having put the foot of a compass in the center, let a circle the radius of which is 3 inches be made. Then on the same center make a larger circle the radius of which is double the first, then a larger one the radius of which is triple the first, then a larger one the radius of which is quadruple the first. Thus you will make as many circles as you want, the distances between which will be equal. Divide each distance into 12 parts of different color. Let the circles be made on the same center. Let a vertical the size of the first radius be stuck in at the center. [The vertical] is [the size] of the 12 digits of the first circle. The diverse

---

[74]"Ad umbram . . . per primam" (II, 38.1-17). The device described here, fig. 54, is an unusual one, but its principle is simple. It amounts to no more than a gnomon of a certain length and a series of measuring scales surrounding it. Because the scales are circular, one may read the length of the shadow cast by the sun in any direction. We may probably suppose that the divisions are not each of a different color but that alternating rings are of two different colors, e.g. white and black. This device facilitates the measurement in II, 1 and could be used for II, 3 as well. I know of no other description of a similar instrument.

respondent diversi colores digitis. Hoc instrumento in

plana superficie posito sumatur umbra solis. Et videbis   15

quot erit digitorum. Et per hoc scies altitudinem solis

per primam.

Et hec de altimetria astronomie necessaria in

Almagesti probata sufficiant.

Restat ut in tertia partitione corporum regularium et   20

irregularium capacitates et crassitudines examussim

mensurare doceamus.

---

14 respondent diversi tr.H

15 superficie: superficiei E / videbis: videbitur H

16 et om.H / post solis add. ut M

18-19 astronomie . . . Almagesti om.H

19 post sufficiant add. explicit liber practicorum geometrie
   H

20-22 om.FH

22 doceamus: docemus E / post doceamus add. incipit liber
   tertius E

colors for the digits correspond to the vertical of one
color. When this instrument is put in a plane surface, the
shadow of the sun may be taken. And you will see how many
digits it is. And by means of this you will know the altitude of the sun using the first proposition.

And these things proved in the Almagest suffice concerning the altimetry necessary to astronomy.

It remains for us in the third part to teach how to measure exactly the capacities and volumes of regular and irregular solids.

[Fig. 54]

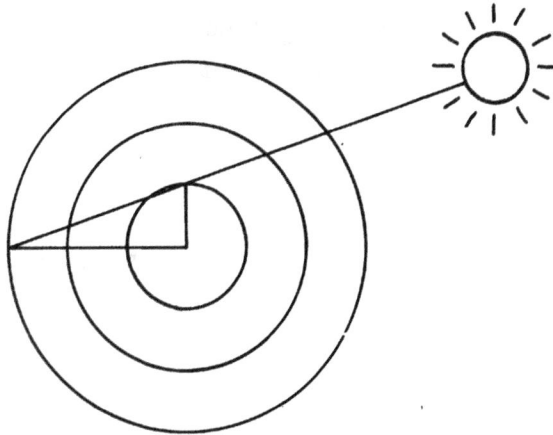

[III, 1] CORPORIS CUBICI CRASSITUDINEM INVENIRE.

Sumatur quantitas lateris que ducta in se faciat
aream eius per primum librum. Illa area numeret ipsum
latus. Productum dabit crassitudinem, et pedes crassos.
Hoc idem fit si cubicetur latus. Et hoc probatur ductis   5
equedistantibus singulis sectionibus per equalitatem
laterum et angulorum et arearum. Ut si 7 pedum est
latus cubici, 343 erit cubus.

---

1 ante hanc enunciationem add. incipit crassimetria H /
   corporis cubici tr.FH / invenire: add. vel investigare E
   investigare FH
3 illa area: aream E / numeret: numeres E feriatur in FH
5 et om.E
7 et arearum: vel archarum F / est om.H
8 cubici: cubicum E / cubus: cubum E cubicus H

III, 1.   TO FIND THE VOLUME OF A CUBIC SOLID.[1]

Take the quantity of a side (fig. 55) which when multiplied by itself produces the area of its [base], [as shown]
by Book I. Let that area be multiplied by that side. The
product gives the volume and the cubic feet. This is the
same if the side be cubed. And this is proved by drawing
parallels to the single sections[2] by means of the equality
of sides and angles and area. So that if the side of the
cube is 7 feet (fig. 55), the cube will be 343 [cubic feet].

Fig. 55

latus cubi 7 pedum

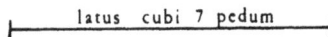

[1]"Corporis . . . latus" (III, 1.1-5). The procedure is
the obvious one for the volume of a cube. The author considers the solid a product of a surface, in this case the
square of side s and area A, by a line, here again the side
s. Thus the volume, V, of the cube may be expressed by the
formula $V = s \cdot A = s^3$. Cf. Adelbold's letter to Silvester
II, ed. Bubnov, 1899: 306.4-7.

[2]"Et hoc probatur . . . cubus" (III, 1.5-8). The proof
technique is the three dimensional analogue to that of I,
2.14-31; it amounts to cutting a cube into smaller cubes
of side one, just as I, 2.14-31 divided a rectangle into
small squares.

[III, 2] COLUMPNE CRASSITUDINEM REPERIRE.

Sumatur area ipsius per primum librum, que multiplicetur in longitudinem eius. Productum dabit propositum. Quod probatur sicut et precedens per lineas equedistantes ductas.

---

1 reperire: perscrutare $\underline{M}$

2 sumatur $\underline{mg. \ hab.}$ Hoc opus commune est ad columpnam rotundam et lateratam. $\underline{H}_a$ / que $\underline{om.H}$

3 multiplicetur: feriatur $\underline{FH}$ / eius $\underline{om.FH}$

4 quod: que $\underline{E}$

4-5 per . . . ductas: ductis lineis equidistantibus $\underline{F}$ per ductas equidistantes lineas $\underline{H}$

III, 2. TO OBTAIN THE VOLUME OF A COLUMN.[3]

Let the area of its [base] be taken by Book I and be
multiplied by its length (fig. 56). The product gives
what was proposed. This is proved just as the preceding
by drawing parallel lines.

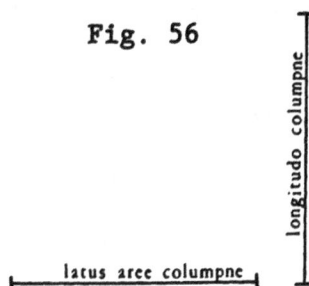

Fig. 56

latus aree columpne

longitudo columpne

---

[3]"Columpne . . . ductas" (III, 3.1-5). This same technique for finding the volume of a cylinder is also given in
II, 35, and III, 14. Cf. Geom. incert., ed. Bubnov, 1899:
358.18-21 and Trac. quadrantis, ed. Britt, 1972: pp. 201-
202. Fig. 56 might lead us to think "columpna" means both
cylinder and rectangular solid; indeed, the marginal note
H_a in H, the same manuscript containing the drawing, indicates that the technique may be applied to both. The proof
technique of III, 1.5-8 will apply only after the circular
base of the column is squared, as in I, 26.

[III, 3]   CIRCULI SPERICAM CRASSITUDINEM PERSCRUTARI.

Nota erit circuli diametros per primum librum,
cuius crassitudinem volumus. Illa cubicetur, habemus
crassitudinem cubi, qui spericam lateribus contingit.
Angulis et lineis ab angulo in angulum procedentibus   5
excedit, quem excessum necesse est rescindere ut spere
circuli quam querimus remaneat soliditas. Quod sic facies.
Summam cubi divide per 21, denominationem ferias vel
multiplices in 10; productum dabit excessum cubi ad speram
circuli, scil. recisiones. Eamdem multiplices in 11 et   10
habes globositatem spere, quam queris. Ut si 7 pedum
est dyameter circuli, illam dyametrum cubica, et surgunt
343 pedes, 21 pars horum est 16 pedum et 3^a pars pedis,

---

3 cuius om.F / cubicetur: cubuetur M
4 post spericam super scr. crassitudinem circuli $\underline{H}_a$ /
   lateribus tr.post contingit M
5 angulum: angulis FH / post procedentibus add. per 20 H
6 necesse est: oportet FH / ante rescindere add. spere
   circuli H
7 quam: quem F / querimus: prius habuimus H / facies:
   facias FH
8 denominationem: denominatione F
8-9 vel multiplices: vel multiplicet E   om.FH
9 10: 11 H
9-10 productum . . . in 11 om.H
10 recisiones: recessiones M / multiplices: ferias F
11 globositatem: globose F
12 illam: illum H
13 post pedes add.sup. quos si dividas per 21 H / ante 21
   pars add. quasi FH / 21 om.F   tr.post horum H /
   pedum: pedes FH / pedis: pedum M

III, 3. TO INVESTIGATE THE VOLUME OF A SPHERE.[4]

The diameters of the circle of which we want the volume
will be known by Book I. Let that [diameter] be cubed; we
have the volume of the cube, which touches the sphere at
the sides (fig. 57). And it is exceeded by the corners
and by the lines proceeding from corner to corner. This
excess [over the volume of the sphere] ought to be taken
from it [i.e., from the diameter cubed], so that the volume
of the sphere which we want remains. This you will do as
follows. Divide the product of the cube by 21, and multiply the quotient by 10; the product gives the excess of the
cube over the sphere of the circle, that is the "cutoffs."
Multiply the same [quotient] by 11, and you have the volume
of the sphere, which you desire. So that if the diameter
of the circle is 7 feet, cube that diameter, and 343 [cubic]
feet result, the 21st part of which, namely the quotient,
is 16 1/3 feet, which multiplied by 10 makes 163 and a

---

[4]"Circuli . . . redditur cubum" (III, 3.1-18). This
method for finding the volume of a sphere amounts to applying the formula $V_o = (11/21) \cdot d^3$. This is equivalent to the
formula of III, 4.8-12 if π is taken as 22/7 since
$(2/3) \cdot A_o \cdot d = (2/3) \cdot (11/14) \cdot d^2 d = (11/21) \cdot d^3 = V_o$. The text
seems to come directly from Adelbold's letter to Silvester,
ed. Bubnov, 1899: 304.17-306.12 and the variant t to 305.11,
a gloss containing phrasing very similar to ours. Cf. also
Geom. incert., ed. Bubnov, 1899: 358.1-3.

scil. denominatio. Que ducta in 10 facit 163 et 3^{am}

pedis, scil. excessum cubi ad speram. Eadem ducta in   15

11 dat soliditatem spere, scil. 179 et duas 3as pedis

quam querebamus. Cui si addatur excessus, redditur

cubus. Quod probatur per precedentes per equidistantiam

laterum in singulis sectionibus.

---

14 163: 162 <u>EM</u>

16 11: 12 <u>F</u> / soliditatem: soliditate <u>?M</u> / <u>post</u> scil. <u>add.</u>
   pedes <u>H</u>

17 quam querebamus <u>om.H</u> quam duc et alius <u>E</u> / addatur
   excessus: addas excessus <u>F</u> denominationem addas ductam
   in 10, que est excessus <u>H</u> / redditur: reddetur <u>F</u>

third feet, which is the excess of the cube over the sphere.
The same [quotient] multiplied by 11 gives the volume of
the sphere we were seeking, namely 179 and two-thirds feet.
If the excess were added to it, the cube would be restored.
This is proved by the preceding propositions[5] by a parallel
through [each of the] single sections of the sides.

Fig. 57

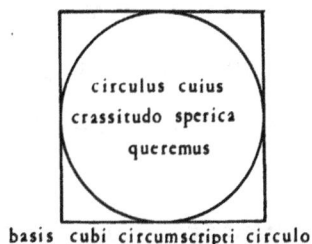

circulus cuius
crassitudo sperica
queremus

basis cubi circumscripti circulo

---

[5]"Quod probatur . . . sectionibus" (III, 3.18-19). The
appeal to the proof method of III, 1.5-8 can show both the
volume of the cube and the volume of the rectangular solid
having as its base a square of area equal to A_o and as its
height $(2/3)\cdot d$, where d is the height of the column. The
technique does not, of course, prove anything about the
sphere itself.

[III, 4] TERRE, SOLIS, ET LUNE, QUANTITATES SPERICAS
INVESTIGARE.

Hec probatur per precedentem. Cum enim notus sit
dyameter terre, ut probatur in Almagesti qui est 120
graduum ipsius cubicati 21 pars sumpta, ducta per 11   5
reddet spericam eius crassitudinem. Continet autem
gradus 60 miliaria.

Similiter metietur quantitatem solis per dyametrum
eius qui notus est quia continet dyametrum terre quinquies
et eius medietatem ut probatur in Almagesti. Cuius   10
cubicati 21 pars percussa seu multiplicata per 11 reddet
eius quantitatem.

Similiter nota erit quantitas lune, cum nota sit

1-18 In codice E haec propositio illi III, 5 successit.
1 et om.E
2 investigare: invenire FH
3 hec: hoc E
4 est: sit H
5 graduum: gradus E / 21 pars: vicesima pars F vicesima pars
     et una H
6 reddet: reddit FH / continet: sunt FH
7 60 miliaria: 60000 F 60000 miliaria M 905142 et pars
     gradus et 21^a H
8 metietur: metiantur M metietatur F metiemur H
9 eius: ipsius F illius H / quia: qui M
10 medietatem mg. hab. scil. novem 17as H$_a$ / post medietatem
     add. fere H / ut . . . Almagesti om.E
11 21: 20 F / seu multiplicata: vel multiplicata M om.FH /
     per: in E / reddet: reddit H
13 erit om.M

III, 4. TO INVESTIGATE THE SPHERICAL VOLUMES OF THE
EARTH, SUN AND MOON.[6]

This is proved by the preceding proposition. Since indeed the diameter of the earth is known and is 120 degrees,
as is proved in the Almagest, the twenty-first part of its
cube, multiplied by 11 will give its spherical volume.
Moreover the degree contains 60 miles.

Similarly the volume of the sun will be measured by its
diameter (fig. 58),[7] which is known because it contains the
diameter of the Earth five and a half times as is proved in
the Almagest. The twenty-first part of [the diameter] cubed
when multiplied by 11 will give its volume. Similarly the

---

[6]"Terre . . . miliaria" (III, 4.1-7). Ptolemy in his
table of chords uses 60 parts as the radius of the unit
circle (Almagest, I, 10); this gives a diameter of 120
parts. Based on the 360 degrees of the circle, the diameter
would be 114 and 6/11 degrees, if π is taken as 22/7. The
technique for finding the volume of the sphere is that of
III, 3. I do not know the source of the author's value of
60 miles per degree; Hugh of St. Victor gives 87½ miles or
70 stadia, ed. Baron, 1966: 49.22-27, a value taken from
Macrobius. Cf. Geom. incert., ed. Bubnov, 1899: 362.5-
363.5.

[7]"Similiter . . . Almagesti" (III, 4.8-18). The ratios
of the diameters of the Sun and Moon to that of the Earth
are from the Almagest, V, 16. The Almagest gives ratios of
volumes, not actual measurements.

eius dyametros, que nota est quia dyameter terre

continet dyametrum lune ter et insuper eius duas 5^{tas},   15

ut probatur in Almagesti. Ex qua colligitur eius

quantitas sicut quantitas solis et terre per precedentia

potest probari. Aliter etiam hec probantur in Almagesti.

---

14 nota est tr.FH
15 eius duas: tr.?F
16 ex qua: et quia FH / eius om.E
17 quantitas om.E / terre: lune H / precedentia: precedentem FH
18 potest probari om.E / etiam: ?F vero H / hec: eorum
   quantitates ?F quantitates eorum H / post Almagesti
   add. quod Almagestantibus relinquimus FH

size of the moon will be known since [the size] of its diameters is known. That is known because the diameter of the
earth contains the diameter of the moon three and two-
fifths times, as is proved in the Almagest. From this its
volume is inferred, just as the volume of the sun and the
earth can be shown by the preceding propositions. These
are also proved in another way in the Almagest.

Fig. 58

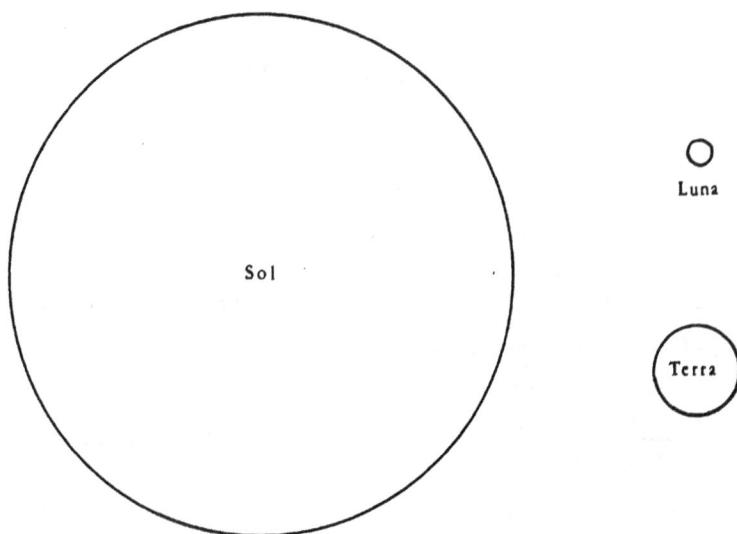

Sol

Luna

Terra

[III, 5] QUANTITATUM SOLIS ET LUNE ET TERRE PROPORTIONES
INVENIRE.

Hec probatur per precedentem. Noti enim sunt

dyametri ipsorum. Sed que est proportio dyametri ad

dyametrum eadem triplicata est spere ad speram, et cubi   5

ad cubum, ut probatur in Euclide.

Ponamus dyametrum terre partem unam. Cum cubum

unitatis non sit nisi unum, et cum dyameter solis sit

5 et dimidium respectu illius, ut ostensum est, cubum

vero solis, 5 et dimidii, est 166 et 4^a et 8^a. Patet   10

igitur quod sol maior est quam terre 166^{es} et 4^a et 8^a

eius.

Item quod dyameter terre continet dyametrum lune

1-22 In codice E haec propositio illam III, 4 antecessit.

1-2 quantitatum . . . invenire: terre, solis et lune quan-
    titates spericas investigare E

1 quantitatum: quantitatem H / et (1) om.FH

7 ante ponamus add. igitur si H / ante dyametrum add. per
    M / cubum: cubus EH

9 respectu illius: cubum lune erit unum M cubum lune erit
    ?F fere erit H / cubum: cubus E

10 vero om.H / solis om.E / 5 et dimidii est om.FH /
    post 8^a add. eius M

10-11 patet igitur: et igitur patet E

11 est: sit E om.H

13 item quod: et quia FH

III, 5. TO FIND THE RATIO OF THE VOLUMES OF THE SUN, MOON
AND EARTH.[8]

This is proved by the preceding proposition, for their
diameters are known. But whatever is the ratio of diameter
to diameter, the same ratio tripled is that of sphere to
sphere, and of cube to cube, as is proved in Euclid.

Let us suppose the diameter of the earth [to be] one
part.[9] Since the cube of unity is nothing unless one, and
since the diameter of the sun is 5 and a half with respect
to it, the cube of the sun will be 166 and a fourth and an
eighth as has been shown. It is therefore clear that the
sun is larger than the earth 166 and a fourth and an eighth
times.

Likewise because the diameter of the earth contains the
diameter of the moon three and two-fifths times, if we

---

[8]"Quantitatum . . . Euclide" (III, 5.1-6). Proposition
III, 3 gives the procedure for finding the volume of a
sphere, given its diameter. Euclid XII, 18 gives the ratio
of volumes of spheres in terms of their diameters.

[9]"Ponamus . . . 6644" (III, 5.7-22). Ptolemy gives two
sets of values for the radii, and hence diameters, of the
moon, earth and sun in Almagest V, 16. With respect to the
earth's radius as one part, written 1P, the sun's radius is
5½P and the moon's 0P 17' 33". Taking the moon's radius as
1P, he gives 3 and 2/5 parts for the earth's radius and 18
and 4/5 parts for the sun's. In cubing these ratios, the
author is more painstaking than Ptolemy (Almagest, V, 16).
The ratio of the volume of the sun with respect to that of
the earth is nearly 170, according to Ptolemy; the author
gives 166 + (1/4) + (1/8), which is precisely the cube of
5½. For the ratio of the sun's volume to that of the
earth, the author uses a value of 6644, a figure which
differs from the cube of 18 and 4/5 by .672. The author's
value for the ratio of the volume of the earth to that of
the moon is 39 + (1/4) + (1/20), which is closer to 39.304,
the correct value for the cube of 3 and 2/5, than is Ptolemy's
39¼.

ter et eius duas quintas, si cubum ex dyametro lune
ponamus partem unam. Cum cubum ex dyametro terre sit   15
39 et 4^a et 20^a, patet quod quantitas terre continet
quantitatem lune 39 et 4^{am} et 20^{am}.

Item cum dyameter solis contineat dyametrum lune
18 et 4 5as eius si dyametrum lune ponamus partem unam,
cuius cubum est unum. Cum cubum 18 et 4 quintarum sit   20
6644, ut probatur in precedentibus, patet quod sol est
maior quam luna 6644.

---

14 quintas: quantitas F
15 cubum cubus E / terre: eius F
18 contineat: continet H / lune om.F
19 18 ?F / post eius add.sup. fere H
20 cuius tr.post cubum F / cubum (1): cubus E / cubum (2):
   cubus EH / 18: centum MF octodecim (corr.ex centum) H /
   sit: fit H
21 post 6644 add.sup. et medietas fere H / precedentibus:
   sequentibus FH
22 post 6644 add. et medietate fere H add. quod Almagestantibus patet FH

suppose the cube [made] from the diameter of the moon [to be] one part. Since the cube [made] from the diameter of the earth is 39 and a fourth and a twentieth, it is clear that the volume of the earth contains the volume of the moon 39 and a fourth and a twentieth times.

Likewise, since the diameter of the sun contains the diameter of the moon 18 and four fifths times; if we suppose the diameter of the moon [to be] one part, its cube is one. Since the cube of eighteen and four fifths is 6644 as is proved in the preceding, it is clear that the sun is larger than the moon 6644 times.

[III, 5 bis]   SOLIS ET LUNE ET TERRE MAGNITUDINES ET EORUM
PROPORTIONES INVENIRE.

In primo libro docuimus per dyametrum circuli

invenire eius circumferentiam, et per semicircumferentiam

invenire eius aream. Cum autem notus sit dyameter solis   5

et lune ad positam lineam rationalem sicut in Almagesti

probatur, et cum circuli ductus in aream eius faciat

columpnam notam que sexquialtera est ad speram circuli

propositam, ergo soliditas spere nota est.

Ergo cum nota sit dyametros tam solis quam lune   10

quam terre, ut probatur in Almagesti, soliditates erunt

note. Item probatur ibi quod dyameter solis continet

dyametrum terre quinquies et eius medietatem. Dyameter

vero ad dyametrum est proportio que spere ad speram

1-30 In codices M et E haec propositio illam III, 1 ante-
    cessit. / solis . . . probatur om.FH
4 circumferentiam: circumferentias M
8 sexquialtera: sex quartas ?E

III, 5 bis. TO FIND THE SIZES OF THE SUN, MOON AND EARTH
AND THEIR PROPORTIONS.[10]

In the first book we taught how to find the circumference of a circle from its diameter and [to find] its area
from the semi-circumference. Since however [the ratio of]
the diameter of the sun and moon to a fixed rational line
is known, as is proved in the Almagest, and since multiplying [the diameter of] a circle by its area makes a column
[of] known [volume], which is in sesquialter [proportion]
to the proposed sphere of the circle, therefore the volume
of the sphere is known.

Therefore since the diameters of the sun as well as of
the moon and earth are known, as is proved in the Almagest,
the volumes will be known. It is also proved there that
the diameter of the sun contains the diameter of the earth
five and one half times. Indeed the ratio of sphere to

---

[10]"Solis . . . probatur" (III, 5[bis].1-30). Because this
proposition is so similar to III, 5 and because it is not
found in F and H, it is likely to be an addition not present
in the archetype. I have therefore labeled it III, 5[bis],
even though it precedes III, 1 in those manuscripts in which
it occurs.

The circumference and area of a circle were shown in I,
21 and I, 22. The relation between the volume of a sphere
and the volume of its circumscribed cylinder is discussed
in Bubnov, 1899: pp. 306.13-308.9 ("Epistola Adelboldi ad
Silvestrum II papam"). The author of the extra proposition
is, in effect, giving us a means of finding the volume
of the sphere, V_s, in terms of the volume of the circumscribed column, V_c, or in terms of the area, A_o, and
diameter, d, of a circle. The method is equivalent to the
formulas $V_s = (2/3) \cdot V_c = (2/3) \cdot A_o \cdot d$. Nonetheless, the author does not use the formula; all he needs here is the
fact that spheres are to one another as the cubes of their
diameters, which he notes in lines 16-18.

triplicata, que etiam cubi ad cubum. Sed si dyametrum   15

terre ponamus partem unam cum cubus unitatis non sit

nisi unitas, cubus vero 5 et dimidii est 166 et 4^a et 8^a

unius, manifestum est quod magnitudo solis continet

magnitudinem terre 166^{es} et insuper eius 4^m et 8^{am}.

Rursum probatur ibi quod dyameter solis continet   20

dyametrum lune 18^{es} et 4 quintas eius. Si dyametrum

lune ponamus partem unam, cuius cubum est unum, cubus

vero 18 et 4 quintarum, 6644 et dimidium fere, patet

omnibus quod magnitudo solis continet magnitudinem 66^{es}

44ter et insuper eius medietatem.   25

Item probatur ibi quod dyameter terre continet

dyametrum lune ter et insuper eius 2as 5as. Si cubum

est ex dyametro terre, surgens est 39 et 4^a et 20^a.

Ergo notum est quod magnitudo terre continet magnitudinem

lune 39^{es} et 4^{am} et 20^{am} eius. Hec in Almagesti probatur. 30

---

15 <u>post</u> etiam <u>add.</u> est <u>E</u> / sed <u>om.</u> <u>E</u>

17 nisi <u>add.sup.</u> <u>E</u> <u>om.</u> <u>M</u> / <u>post</u> vero <u>add.</u> huius <u>E</u>

19 8^{am}: 8 <u>M</u>

27 2as 5as: 2 5 <u>M</u>

30 probatur: probantur <u>M</u>

sphere is triple that of diameter to diameter, which is also
that of cube to cube. But if we take the diameter of the
earth [to be] one part, since the cube of unity is nothing
but unity, and [since] the cube of 5 and a half is 166 and
a quarter and an eighth, it is manifest that the size of
the sun contains the size of the earth 166 plus a quarter
and an eighth times.

In turn it is proved there that the diameter of the sun
contains the diameter of the moon 18 and four-fifths times.
If we take the diameter of the moon [to be] one part, of
which the cube is one, and [since] indeed the cube of 18
and four-fifths is about 6644 and a half, it is clear to
everyone the size of the sun contains the size of the moon
6644 and one-half times.

Then it is proved there that the diameter of the earth
contains the diameter of the moon 3 and two-fifths times.
If the cube is [taken] of the diameter of the earth, 39
and a quarter and a twentieth arise. Therefore it is known
that the size of the earth contains the size of the moon
39 plus a quarter and a twentieth times. This is proved
in the Almagest.

[III, 6]   CORPORIS COLUMPNARIS PYRAMIDEM INVENIRE.

Sumatur quantitas columpne per 3^{am} huius 3^{ii} libri.
Illa erit <tripla> ad suam pyramidem, ut probatur in
12 Euclidis.

---

1 columpnaris: columpnarum M
2 3^{am}: 2 M   secundam FH / 3^{Ii} om.FH
3 tripla (ed.): sexquialtera ME   sex quarta F   tripla
   (corr.ex sex quarta) H / ut: hec H

III, 6. TO FIND THE PYRAMID OF A COLUMNAR SOLID.[11]

Let the volume of the column (fig. 59) be taken by the third proposition of this third book. That [volume] will be [triple] its pyramid, as is proved in Euclid, Book 12.

Fig. 59

basis columpne
et sue
piramidis

[11]"Corporis . . . Euclidis" (III, 6.1-4). Euclid XII, 10 tells us that the cone is one third the cylinder with the same base and height. The author seems to have erred; the revision in H corrected the error. The faulty value along with the correct reference to Euclid may indicate that the author used an intermediate source, not the Elements themselves.

[III, 7] MODII CAPACITATEM PERSCRUTARI.

Sumatur dyameter fundi et dyameter superior quem
attingerit vinum et equentur medietate excessus maioris
addita minori. Inveniatur quoque area fundi ut de circulo docuimus per 2^{am}. Et videatur quot digitorum sit   5
vel quot palmorum ut docuimus. Illa area feriatur in
altitudinem modii, et habes quantitatem modii quot
palmorum erit vel quot pedum.

Eadem doctrina invenitur quantitas cuiuslibet vasis
columpnaris parvi vel magni, trianguli vel quadranguli.   10
Item, quantitas modii scil. numerus palmorum vel pedum
dividatur per quantitatem minoris. Denominatio dabit

---

1 modii: dolii $\underline{F}$

2 dyameter (1) $\underline{om.}\underline{H}$ / et $\underline{om.FH}$ / dyameter (2) $\underline{om.F}$
   Hoc quod [?] nunc [?] commune est opus sive modius sit
   columpnaris sive ut curta piramis. $\underline{H}_a$ / superior:
   superioris scil. et inferioris $\underline{H}$

2-3 quem . . . vinum $\underline{om.H}$

3 attingerit: attigerit $\underline{M}$ attingit $\underline{F}$ / vinum $\underline{om.E}$ /
   equentur: equetur $\underline{FH}$ / medietate $\underline{om.F}$

4 quoque: que $\underline{E}$ similiter $\underline{FH}$

5 2^{am}: secundum librum $\underline{E}$ / sit $\underline{om.H}$

6 palmorum: palmarum $\underline{E}$ / area $\underline{om.E}$

8 palmorum: palmarum $\underline{E}$ / vel quot pedum $\underline{om.FH}$

9 invenitur: inveniatur $\underline{FH}$ / cuiuslibet: cuiusvis $\underline{FH}$

10 columpnaris: columpnarum $\underline{M}$ / vel . . . quadranguli $\underline{om.FH}$ /
   $\underline{post}$ magni $\underline{add.}$ et $\underline{E}$

11 item $\underline{om.E}$ / $\underline{post}$ modii $\underline{add.}$ dividatur $\underline{FH}$ / palmorum:
   palmarum $\underline{E}$

11-12 vel pedum dividatur $\underline{om.FH}$

12 quantitatem: quantitates $\underline{FH}$

III, 7. TO INVESTIGATE THE CAPACITY OF A BARREL.[12]

Let the diameter of the lower base and the upper diameter that the wine might reach (fig. 60) be taken and let
them be evened by adding half the excess of the greater to
the smaller. Also let the area of the base be found as we
have taught concerning the circle in the second proposition.
And let it be seen how many [square] inches or palms it is
in the way which we have taught. Let that area be multiplied by the altitude of the barrel, and you will have the
volume of the barrel, how many [cubic] palms or feet it will
be.

By the same doctrine[13] is found the volume of any small
or large cylindrical vessel or of a [vessel with a] triangular or quadrangular base. Likewise let the volume of the
barrel, that is the number of [cubic] palms or feet, be
divided by the volume of a smaller [vessel].[14] The quo-

---

[12]"Modii . . . quot pedum" (III, 7.1-8). The method for
finding the volume of a barrel, V_b, may be represented by
the formula $V_b = A_e \cdot h$, where A_e is the area of the "evened"
base and h is the height. The author tells us to find the
area of the "evened" base by using what amounts to the average diameter. A modius, which we have translated as "barrel," seems to have sloping sides, like those of a bucket.
In all then the procedure amounts to
$V_b = (11/14) \cdot ((D+d)/2)^2 \cdot h$, where D is the larger diameter
and d the smaller one, and where the value of π is the author's normal one of 22/7. The area of a circle is not
taught in any Proposition 2 of this work but in I, 22. Cf.
Trac. quadrantis, ed. Britt, 1972: pp. 202-203, which is
very close to our text.

[13]"Eadem doctrina . . . quadranguli" (III, 7.9-10). That
is, multiply the area of the base by the height. Where the
sides are sloped, we may assume the author means for us to
use the "evened" base as above.

[14]"Item, quantitas . . . numeri" (III, 7.11-15). The

quotiens minus vas contineatur in modio. Et sic de

quantolibet vase parvo regulari. Quod probatur ratione

numeri.   15

tient will give how many times the smaller vessel is
contained in the barrel, and similarly for every small
regular vessel. This is proved by argument from number.

Fig. 60

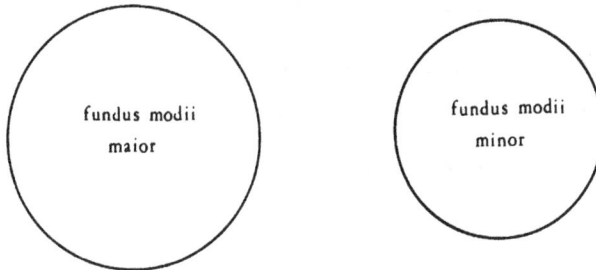

fundus modii
maior

fundus modii
minor

author uses an appeal to numerical example as "proof"
that dividing a larger quantity by a smaller one yields
the number of times the smaller is contained in the larger.
The text is similar to that of Trac. quadrantis, ed. Britt,
1972: pp. 203-204. "Argument from number" is a favorite
proof technique of the author. He usually applies it, as
here, to justify a numerical result of a multiplication
or division.

[III, 8] DOLII CAPACITATEM INVENIRE.

Inveniatur quantitas modii per precedentem.
Inveniaturque dolii area per dyametrum fundi ut docuimus
in 2ᵃ. Sumatur que longitudo dolii secundum vini
continentiam, et feriatur in aream. Denominatio dat   5
quantitatem dolii. Que dividatur per quantitatem modii,
denominatio dabit quotiens modius contineatur in dolio.
Et tunc notum erit quotiens sextarius in modio contineatur,
et quotiens numerata in sextario, notum est quotiens
numerata erit in dolio. Quod etiam ratione numeri   10
probatur.

---

1 post capacitatem add. modiorum E / invenire: inquirere M
3 que om.F / dolii area tr.FH / ante per add. per precedentem, scil. H
4 que om.FH / secundum: scil. E
8 tunc: cum F cum sit H / erit: sit F om.H
8-9 sextarius . . . quotiens om.E
10 erit om.FH / etiam: probatur M
11 probatur om.M

III, 8. TO FIND THE CAPACITY OF A TUN.[15]

Let the quantity of a barrel be found by the preceding
proposition. And let the [cross-sectional] area of the tun
(fig. 61) be found using the diameter of the base as we
taught in the second proposition. Let the length of the
tun be taken in compliance with its contents of wine, and
let it be multiplied by its area. The result gives the
volume of the tun. Let it be divided by the quantity of
a barrel, the result will give how many times the barrel
is contained in the tun. And then [since] how many times
a sextarius is contained in the barrel[16] is known, and
how many times a numerata [is contained] in a sextarius is
known, how many numerata in the tun will be [known]. This
is likewise proved by argument from number.

Fig. 61

area fundi
dolii

---

[15]"Dolii . . . in dolio" (III, 8.1-7). The same basic
technique of multiplying the area of the base by the height
is used to find the volume of a tun. A tun seems to have
straight, or nearly straight sides, since the author does
not tell us to "even" the base. The capacity in terms of
barrels is found by dividing the volume of the tun by the
volume of the barrel. As in n. 12 above, I, 22 gives the
area of the circular base; no Proposition 2 does so. Cf.
Trac. quadrantis, ed. Britt, 1972: pp. 204-206, which is
close to our text.

[16]"Et tunc . . . probatur" (III, 8.8-11). From I, 32
we know that eight sextarii make one modius or barrel; cf.
also IV, 1. The author has not told us how many numerata
make a sextarius. Again appeal is made to numerical example as a proof technique.

[III, 9] DOLIORUM EIUSDEM LONGITUDINIS DYAMETRUM UNIUS DUPLUM AD DYAMETRUM ALTERIUS, CAPACITATEM QUADRUPLAM GENERARE.

Hec probatur per primam, ubi ostensum est, quia ex duplicitate dyametrorum procreatur quadruplicitas arearum. 5 Longitudo autem earum metiens areas non impedit, quod si unus numerus est quadruplus ad alium, et unus numerus eos metiatur, productum maioris erit quadruplum ad productum minoris.

---

1 doliorum eiusdem longitudinis om.E / post unius add. dolii E
2 post alterius add. equalium longitudinum E
4 hec: hoc H / post probatur add. quod F / ubi: non ?F ut H / quod: quia H
6 earum om.FH / areas tr.post impedit FH / si om.E
8 eos tr.ante unus FH

III, 9. TO PRODUCE A QUADRUPLE CAPACITY [IN ONE] OF [TWO]
TUNS OF THE SAME LENGTH, [LET] THE DIAMETER OF ONE [BE]
DOUBLE THE DIAMETER OF THE OTHER.[17]

This is proved by the first [book], where it is shown,
since from a duplicity of diameters is produced a quadru-
plicity of areas (fig. 62). However, multiplying the areas
by their length does not alter this fact since, if one number is quadruple another, and they are multiplied by one
number, the product of the greater will be quadruple the
product of the smaller.

Fig. 62

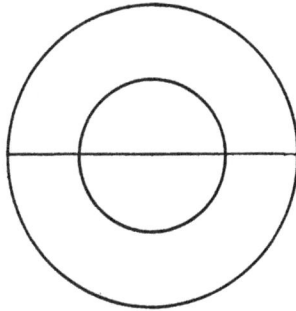

---

[17]"Doliorum . . . minoris" (III, 9.1-9). That doubling
the diameter produces a circle of four times the area is
shown in I, 24. It is therefore clear that "primum" is intended for the feminine "primam" in III, 9.4; MEFH agree
to the latter reading. The justification of the whole procedure is made by appeal to the rule of multiplication in
Euclid VII, 17.

[III, 10] DYAMETRUM UNIUS DOLII DUPLUM AD DYAMETRUM
ALTERIUS DUPLI IN LONGITUDINE, CAPACITATEM DUPLAM GIGNERE.

Hec probatur per precedentem, cum enim area maioris
sit quadrupla ad aream minoris. Si maiorem metiatur
unus numerus, et minorem aream numerus duplus ad illum   5
metiatur, productum maioris aree erit duplum ad productum
minoris, quod probatur ratione numeri.

---

2 longitudine: longitudinem E / duplam: octuplam (corr. ex
    duplam) E alterius M
3 post enim add. erit F
6 metiatur: metietur M / duplum: octuplam (corr. ex duplam) E

III, 10.   TO CREATE A DOUBLE CAPACITY, [LET] THE DIAMETER
OF ONE TUN [BE] DOUBLE THE DIAMETER OF ANOTHER [WHICH IS]
DOUBLE IN LENGTH.[18]

This is proved by the preceding proposition since indeed
the area of the greater is quadruple the area of the smaller (fig. 63). If the larger [area] is multiplied by a number, and the smaller area by a number double that one, the
product of the larger area will be double the product of
the smaller, which is proved by argument from number.

Fig. 63

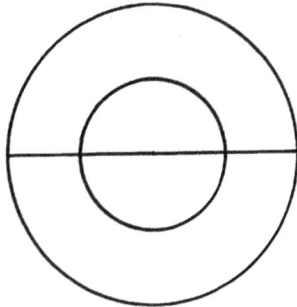

---

[18]"Dyametrum . . . numeri" (III, 10.1-7). This proposition is an extension of III, 9 in that even though the base
of the larger tun is four times that of the small one, its
height is only half as much, effectively dividing it in
half. Proof by numerical example seems easier than an appeal to a rule of multiplication as in the preceding proposition.

[III, 11]   DYAMETRUM DOLII UNIUS DUPLUM AD DYAMETRUM
ALTERIUS DOLII IN LONGITUDINE QUADRUPLI CAPACITATEM UNAM
COEQUARE.

Hec probatur per precedentes, et ratione numeri.

Quia si numerus unius aree est quadruplus ad numerum   5

alterius, si maiorem, aliquis numerus metiatur.   Et

numerus quadruplus ad illum metiatur numerum minoris

aree, idem numerus surget ex tali mensura.

---

1 unius <u>om.</u>FH
2 dolii <u>om.</u>E / <u>post</u> dolii <u>add.</u> quadrupli <u>FH</u> / longitudine:
   longitudinem <u>H</u> / quadrupli <u>om.</u>FH
4 precedentes: precedentem <u>H</u>
5 numerus <u>tr.post</u> aree <u>H</u>
7 minoris: minorem <u>H</u> / aree <u>om.</u>FH
8 surget: surgit <u>H</u>

III, 11.   TO MAKE A SINGLE EQUAL CAPACITY, [LET] THE
DIAMETER OF ONE TUN [BE] DOUBLE THE DIAMETER OF ANOTHER
TUN OF QUADRUPLE LENGTH.[19]

This is proved by the preceding propositions and by argument from number.   If the number of one area is quadruple
the number of the other (fig. 64), let the larger [area]
be multiplied by some number, and let the smaller area be
multiplied by a number quadruple that one, [then] the same
number will arise from that multiplication.

Fig. 64

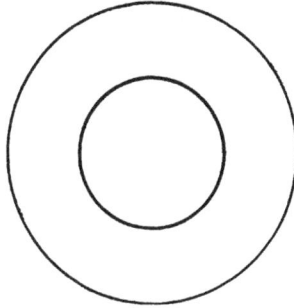

---

[19]"Dyametrum . . . mensura" (III, 11.1-8).   What the
author is saying, in effect, is that if A = 4a, where A and
a are the areas of the larger and smaller bases, then
$\bar{h}\cdot A = H\cdot a$, where H = 4h.

[III, 12] VASIS COLUMPNARIS VEL QUADRANGULI CAPACITATEM
INVENIRE.

Sumatur area vasis per primum quam metiatur altitudo

eius. Productum dat capacitatem, ut probatum est.

---

1 columpnaris: columpnaris vel circularis M circularis FH
3 primum: primam F
4 dat: da M dabit H

III, 12.   TO FIND THE CAPACITY OF A CYLINDRICAL OR
QUADRANGULAR VESSEL.[20]

Let the area of [the base of] the vessel (fig. 65) be

found by the first book and let it be multiplied by its

altitude.   The product gives the capacity as has been

proved.

Fig. 65

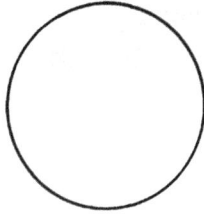

---

[20]"Vasis . . . probatum est" (III, 12.1-4). This proposition extends the method for finding the volume of a cylinder to apply to any right prism. It was already assumed
in the proof of III, 2, if my understanding of the proof is
correct; see n. 5 above. Cf. III, 7.9-10, where a more
general statement of this proposition is given. Trac. quadrantis, ed. Britt, 1972: p. 206 is similar.

[III, 13]   ARCHE CAPACITATEM COLLIGERE.

Metiatur aream altitudo, productum dividatur per
quantitatem vasis circularis vel quadranguli inventam
per precedentem. Denominatio dabit quotiens vas continetur in archa, quod probat ratio numeri.   5

---

1 colligere: eligere M̲
4 per precedentem: prius F̲H̲ / vas o̲m̲.FH
4-5 continetur: contineatur F̲H̲
5 probat ratio: probatur ratione M̲ / h̲i̲c̲ ̲d̲e̲s̲i̲i̲t̲ M, p̲o̲s̲t̲h̲a̲c̲
   c̲o̲e̲p̲i̲t̲ ̲a̲d̲ ̲I̲V̲,̲ ̲1̲.̲3̲

III, 13. TO INFER THE CAPACITY OF A CHEST.[21]

Let the area (fig. 66) be multiplied by the altitude,
and let the product be divided by the volume of a cylindrical or quadrangular vessel found by means of the preceding
proposition. The quotient will give how many times it is
contained in the chest, which argument from number proves.

Fig. 66

---

[21]"Arche . . . numeri" (III, 13.1-5). This is another
proposition telling us how to figure out how many times the
contents of a small vessel will fit into a larger one. Cf.
III, 7.11-15 and III, 8.6-11.

[III, 14]   ARCHARUM QUADRATARUM LATUS UNIUS DUPLUM AD
LATUS ALTERIUS CAPACITATEM OCTUPLAM GENERARE.

Hec probatur per primum librum. Cum enim probatum

sit ibi quod ex latere quadrati duplo ad aliud proveniat

area quadrupla ad aliam aream. Patet quod ex altitudine   5

arche dupla ad aliam altitudinem, altera provenit

multiplicitas. Ex quibus duabus surgit octuplicitas.

Assignatasque varietates doliorum et dyametrorum

variatione et longitudinum et arearum simili diversitate

solers crassimetra de facili poterit invenire.   10

---

4 aliud ?F
5 aream tr.ante aliam H   om.E
5-7 patet . . . multiplicitas: ad hoc alterum latus, scil.
    altitudinis, duplum ad alium, generabit aliam quadru-
    plicitatem E
7 duabus om.F
8 que om.F / varietates: variationesque F   variationes H /
    et: in E
9 longitudinum: longitudinem B / ante et arearum add.
    diversi ?B / et arearum: scil. archarum E
10 solers: solet ?E / de . . . invenire: inveniret ?E

III, 14. TO PRODUCE AN OCTUPLE CAPACITY [IN ONE] OF [TWO] SQUARE CHESTS; [LET] THE SIDE OF ONE [BE] DOUBLE THE SIDE OF THE OTHER.[22]

This is proved by the first book. Since indeed it has been proved there that from the side of a square double another, an area quadruple the other area is produced (fig. 67), it is clear that from the height of a chest double the height of another, the other multiplicity is produced, and from the two of these arises the octuplicity.

And the skillful volume measurer will easily be able to find [in chests], by a similar diversity,[23] the varieties attributed to tuns by the variation of diameters, lengths, and areas.

Fig. 67

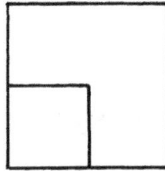

---

[22]"Archarum . . . octuplicitas" (III, 14.1-7). Book I has no explicit proposition discussing the effect of doubling the side of a square. The area of a square is given in I, 10.

[23]"Assignatasque . . . invenire" (III, 14.8-10). In other words, it is left as an exercise to the reader to find for chests the properties given for tuns in III, 9, III, 10 and III, 11.

[III, 15]   DOLII CAPACITATI ARCHAM EQUALEM FABRICARE.

Hec probatur per primum librum cum enim circuli

quadraturam invenire docuimus.   Aree dolii secundum illam

propositionem sumatur radix.   Ad cuius radicis quanti-

tatem fabricetur quorum latus arche longitudinis et   5

latitudinis.   Fiatque quodlibet latus altitudinis equale

longitudini dolii.   Eruntque illa eiusdem capacitatis.

Quod probatur quia aree sunt equales per primum librum

et illas metitur idem numerus.   Ergo producta sunt equalia.

---

1 archam: archam vel aream B   aream F / equalem: equaleam ?H

2 primum: primam F / librum om.FH

2-3 cum . . . invenire: ubi quadrare circulum H

3-9 aree . . . equalia: pro hiis habuit H versus III, 16.3-
    8, docuimus . . . equalia et vice versa; in illis locis
    variantes posuimus. Hic ergo positae sunt lectiones
    quae post docuimus in III, 16.3 ex H sumptae sunt.

3 aree: area E / illam om.F

5 quorum: quodlibet H om.B / post latus add. quodlibet B /
    arche longitudinis tr.E

6 fiatque: feriatque ?B et fiat FH

7 dolii om.B

8 quod om.H / probatur: probat E om.H / primum librum:
    primam FH

9 illas om.F / metitur: metietur F

III, 15. TO MAKE A CHEST EQUAL TO THE CAPACITY OF A TUN.[24]

This is proved by the first book, since indeed we taught how to square the circle. Following that proposition, let the root of the area [of the base] of the tun (fig. 68) be taken. Make the length and width of the chest equal to the size of that root. Then let some side be equal in height to the length of the tun. Then these [i.e., the tun and the chest] will be of the same capacity. This is proved because the areas are equal, from the first book, and they are multiplied by the same number. Therefore the products are equal.

Fig. 68

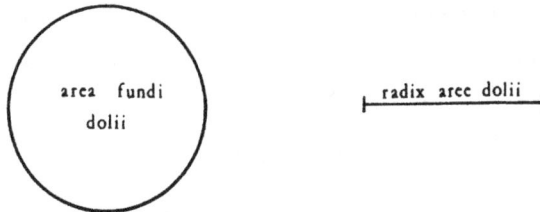

area fundi dolii

radix arce dolii

---

[24]"Dolii . . . equalia" (III, 15.1-9). I, 26 shows how to find a square with the same area as a circle. With the same height and equal areas as bases, the two solids will have the same capacity.

[III, 16] ARCHE QUADRATE DOLIUM EQUALE FABRICARE.

Hec probatur per primum librum ubi quadrati

circulationem docuimus. Et secundum illam propositionem

inventa area arche inveniatur dyameter dolii habentis

consimilem aream. Fabricetur longitudo dolii ad equali-   5

tatem altitudinis arche. Et habetur propositum, quia

ex equalibus areis percussis in eumdem numerum non

proveniunt nisi equalia.

---

2 primum: primam F̲ om.H / quadrati: ?F̲ quadrare H̲

3 circulationem: circulatione ?F̲ circulum H̲ mg.hab. Hoc
   opus docet fabricare archam quadratam equalem dolio.
   Per idem tamen ut patet ex hoc libro et ex primo, potest
   fabricari dolium equale arche quadrate. Istud etiam
   commentum plus videtur comperetere [si̲c̲, competere] 16
   [i.e., III, 15]. Et illud quod ibi ponitur huic. H̲ₐ

3-8 et . . . equalia: pro hiis habuit H versus III, 15.3-9
   docuimus . . . equalia et vice versa: in illis locis
   variantes posuimus. Hic ergo positae sunt lectiones
   quae post docuimus in III, 15.3 ex H sumptae sunt.

3 et secundum: per H̲

4 dyameter: diametrum H̲

6 habetur: habebitur H̲

7 non om.E̲

8 nisi om.H̲ / post equalia add. iterum B versus III, 16.1-2,
   praeter unam variationem, arche . . . ubi. Post ubi in
   III, 16.2 add. iterum B, cum paucis differentiis, versus
   III, 15.2-9, circuli . . . equalia

III, 16.   TO MAKE A TUN EQUAL TO A SQUARE CHEST.[25]

This is proved by the first book where we taught how to
circle a square. And when the area of the chest has been
found following that proposition, let the diameter of the
tun having the same area (fig. 69) be found. Let the length
of the tun be made equal to the height of the chest. And
what was proposed is had because from equal areas multiplied by the same number, nothing results but equals.

Fig. 69

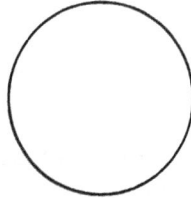

---

[25]"Arche . . . nisi equalia" (III, 16.1-8). The method
for finding a circle equal to a given square is given in
I, 27. Solids of the same height with the given square and
derived circle as bases will have equal volumes.

[III, 17]   PUTEI URNARUM CAPACITATEM COLLIGERE.

Feriatur altitudo eius in aream suam per 2^{am} huius
libri. Productum dividatur per quantitatem urne.
Denominatio dabit numerum urnarum.

Omnes autem regule capacitatum inventarum deserviunt   5
crassitudinibus illorum corporum inveniendis.

Et hec de capacitate corporum regularium et
crassitudine dicta sufficiant. Sed antequam de pyramidem
mensura agamus, de irregularium capacitate et crassitudine
perstringamus.   10

---

1 urnarum tr.post capacitatem FH / colligere: reperire H
   om.F
4 numerum: quantitatem B / post urnarum add. et numerum B
6 corporum super scr. regularium, scil. H$_a$
7 hec: hoc E / et (2): scil. B
8 crassitudine: crassitudinem E / dicta sufficiant. Sed
   om.FH / antequam: ante autem quam H / pyramidem:
   piramide F piramidum H
9 mensura agamus tr.FH
10 perstringamus: prostringamus E

III, 17. TO INFER THE CAPACITY OF A WELL IN URNS.[26]

Let its height be multiplied by its area (fig. 70) following the second proposition of this book. Let the product be divided by the quantity of the urn. The quotient
will give the number of urns.

All the rules of finding capacities serve to find the
volumes of those solids.[27]

And those things said concerning the capacity of regular
solids and concerning volumes suffice. But before we discuss measure dealing with the pyramid, let us touch lightly
on the capacity and volume of irregular [solids].

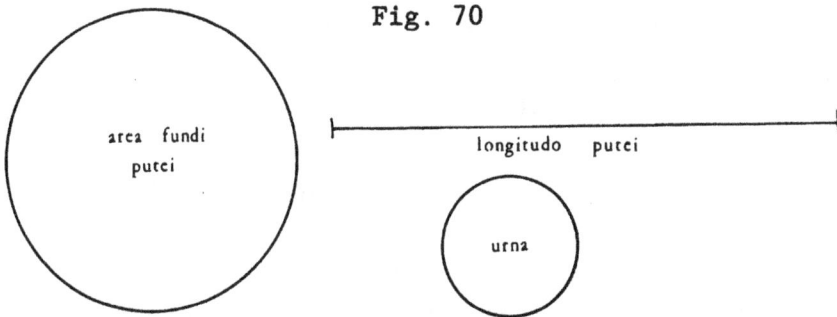

Fig. 70

area fundi
putei

longitudo putei

urna

---

[26]"Putei . . . urnarum" (III, 17.1-4). The volume of a
well was discussed in II, 35. Proposition III, 2 shows how
to find the volume of a columnar solid.

[27]"Omnes . . . inveniendis" (III, 17.5-6). Capacity
seems to mean the number of some measure which a larger vessel contains. Volume, crassitudo, would appear to be a more
abstract measure, perhaps in cubic feet, etc. The author
is thus describing the obvious relation that holds between
the two systems of measure once one knows the cubic contents
of the smaller measure.

[III, 18]   VASIS IRREGULARIS SUMMO, MEDIO ET YMO
CRASSITUDINEM ET CAPACITATEM COLLIGERE.

Esto dyameter vasis inferior 6 pedum, superior

duorum, medium 5, altitudo vasis 12. Sic negociare.

Medium metiatur se; productum metiatur ternarius. Summe   5

illi agregetur productum ex ductu unius dyametri in

reliquum. Eritque totalis summa 87 quam ferimus in 11.

Productum dividamus per 14. Denominationem, scil. 68,

feriamus in 3^{am} altitudinis vasis, 4 scil. Denominatio

dabit eius crassitudinem vel capacitatem quod exercitio   10

geometrie probandum relinquimus. Hec regula necessaria

est ubique consimilis irregularitas summi, medii et ymi,

acciderit, ubique etiam triplex diversitas acciderit.

---

2 crassitudinem et capacitatem: capacitatem et crassitudinem FH

3 inferior: inferius FH / superior: superius FH / post
   superior add. diameter E

4 duorum: ?F

5 productum . . . ternarius: om.H mg.hab. inferius similiter
   latus se, et superius se, et agregentur summe. Deinde
   toti agregate H$_a$

6 agregetur: agregate E / ductu tr.post dyametri B

6-7 unius . . . ferimus om.E

7 87: 47 BF

8 68: 69 BF 64 E

9 feriamus: feriamo ?E / 4 scil.: 8 scil. B om.E

10 exercitio: ex tercio H

11 probandum om.FH

11-12 necessaria est tr.B

12 ubique: ubi F ubique ubi H / consimilis: in simul E

13 acciderit (1): accidit EF / post etiam add. ubi H /
   diversitas acciderit: inciderit diversitas FH

III, 18. TO INFER THE CAPACITY AND THE VOLUME OF AN
IRREGULAR VESSEL BY MEANS OF THE TOP, MIDDLE AND BOTTOM.[28]

Let the lower diameter of the vessel be 6 feet, the upper 2, the middle 5, and the height of the vessel 12 (fig.
71). Proceed as follows: let the middle [diameter] be
multiplied by itself and the product multiplied by three.
And to that result let the product of one diameter by the
other be added. And the whole sum will be 87 which we multiply by 11. We divide the product by 14. Then we multiply
the quotient, namely 68, by one third of the height of the
vessel, namely 4. The product will give its volume or
capacity, which we leave to be proved as an exercise of
geometry. This rule is necessary wherever there is at the
same time an irregularity of the top, middle, and bottom,
indeed, wherever a triple diversity may have occurred.

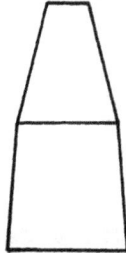

Fig. 71

---

[28]"Vasis . . . acciderit" (III, 18.1-13). The computation in the author's example is correct, accepting that he
rounds off after dividing. Still it is not at all clear
what mathematics are behind his rule for the volume of this
solid, which may be considered as made up of the frusta of
two cones. The procedure may be expressed as the formula
$V = (11/14) \cdot (h/3) \cdot (3b^2+ac)$, where a, b and c are top, middle and bottom diameters, respectively, of the solid of
height h. The author's usual value for $\pi/4$ appears as
11/14. The source of the technique seems to be Geom. incert., ed. Bubnov, 1899: 358.22-359.5 which is very close
to our text, except that for c, Geom. incert. gives 3 where
our text has 6. Leaving the proof as an exercise of geometry may be simply an evasion; I doubt that the author
could have justified the formula.

[III, 19] VASIS IRREGULARIS SUMMO ET YMO CAPACITATEM
VEL CRASSITUDINEM CONCLUDERE.

Esto dyameter inferior 5 pedum, superior 3, altitudo
vasis 9. Sic agendum est. Metiatur dyameter inferior
se, superior se. Agregenturque summe, metiaturque   5
superior inferiorem. Totumque iungatur. Summa, scil.
49, ducatur in 11. Producti 14 pars feriatur in 3^{am}
altitudinis vasis, scil. 3, denominatio dabit propositum.
Hec probatur per probationem precedentis que nemini
morosa esset et difficilis. Illam hic proponeremus.   10

Eadem regula servit in omni consimili irregularitate
summo et ymo tantum considerata. Hiis 2^{lis} regulis
irregularium omni, nisi mira in eis fuerit diversitas,
capacitas colligitur.

---

2 vel: et <u>B</u> / concludere: reperire <u>FH</u>
3 <u>post</u> esto <u>add.</u> eius <u>H</u>
5 <u>ante</u> superior <u>add.</u> diameter <u>H</u>
6 totumque: totum <u>B</u> / scil. <u>ow.E</u>
7 <u>post</u> in 11 <u>mg.hab.</u> deinde operemur ut prius et <u>H</u>_a /
   producti <u>om.B</u>
7-8 producti . . . propositum: productus dabit propositionem <u>H</u>
9 hec: hoc <u>B</u>
9-10 que . . . proponeremus <u>om.E</u>
10 illam hic: hec <u>?F</u>
11 eadem: hec <u>FH</u> / servit: deservit <u>FH</u> / irregularitate:
   irraritate <u>H</u>
12 tantum <u>om.FH</u>
12-14 hiis . . . colligitur <u>om.FH</u>
12 2^{lis}: 3 <u>B</u>

III, 19.   TO DEDUCE THE CAPACITY OR VOLUME OF AN IRREGULAR VESSEL BY MEANS OF ITS TOP AND BOTTOM.[29]

Let its lower diameter be 5 feet, the upper 3, the height of the vessel 9 (fig. 72). This is the way to proceed. Let the lower diameter be multiplied by itself, and the upper by itself. Then let the sum be collected, and let the lower be multiplied by the upper. Let them all be added up. Let the sum, namely 49, be multiplied by 11. Let the fourteenth part of the result be multiplied by a third of the height of the vessel, namely 3. The product will give what was proposed. This is proved by the preceding proof which would be peevish and difficult to nobody. If it were we would have given it.

The same rule applies in every similar [case], so long as an irregular [solid] is considered by top and bottom. Using these rules for all of the irregular [solids], the capacity is inferred, unless there is a surprising diversity in them.

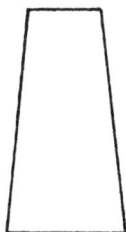

Fig. 72

---

[29]"Vasis . . . colligitur" (III, 19.1-14). The procedure described here may be expressed as $V = (11/14) \cdot (h/3) \cdot (a^2+b^2+ab)$, where a and b are the diameters of the top and bottom, h is the height of the solid and 11/14 is the approximation for $\pi/4$. The proposition seems to have come from Geom. incert., ed. Bubnov, 1899: 359.5-14, which is like our text. Though the formula is the correct one for the frustum, it cannot be derived from the preceding formula. Again, I feel the author is avoiding the proof because he cannot give it.

Sequitur de pyramidum crassitudine metiendarum   15
secundum regularium formarum areas in primo libro
surgentium.

---

15 sequitur de pyramidum: sic pyramide F similiter de
   piramidum H / ante crassitudine add. non ?F /
   metiendarum: metienda ?B / post metiendarum add.
   irregularium E
16 regularium: areas F / post regularium add. crassitudinem
   H / formarum areas om.F / in om.EF / libro: loco H /
   post libro add. positarum BE

Here follows [material] concerning the volume of pyramids to be measured following the areas of regular forms arising in the first book.

[III, 20]　YSOPLEURI PYRAMIDEM INVENIRE.

　　Sumatur area trianguli equilateri per 2^{am} propositionem primi libri.　Illam aream binarius metiatur.
Producto agregetur unum latus trianguli prefati.　Summa
multiplicetur per numerum transcendentem unitate, unum　　　5
latus.　Productum dividatur per 6.　Denominatio dabit
pyramidem ysopleuri.　Et quorum exercitio istud probandum
relinquimus, qui totum Euclidem se scire faceretur.　Hec
regula valet ad omnes figuras rectangulas et equilateras.

---

3 primi ?F / aream ?F
4 trianguli prefati tr.FH
6 productum dividatur tr.F / 6: 60 B
6-7 dabit pyramidem tr.H
7-8 et . . . faceretur om.E
7 quorum: ?F　om.H / istud: istum H
8 faceretur ?B / ante hec add. et FH
9 regula: ?F　figura H / valet: ?F

III, 20. TO FIND [THE VOLUME OF] AN EQUILATERAL PYRAMID.[30]

Let the area of the equilateral triangle (fig. 73) be
found by the second proposition of the first book. Let
that area be multiplied by two. Let one side of the aforementioned triangle be added to the product. Let the sum
be multiplied by a number transcending by one [the quantity
of] one side. Let the product be divided by 6. The quotient will give [the volume of] an equilateral pyramid.
And for an exercise for those would presume to know all of
Euclid, we leave this to be proved. And this rule is useful for all rectangular and equilateral figures.

Fig. 73

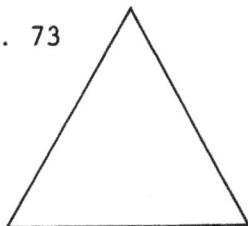

[30]"Ysopleuri . . . equilateras" (III, 20.3-11). The procedure given here amounts to V = (2A+a)·(a+1)·(1/6), where
V is the volume of an equilateral pyramid with a base of
area A and sides of length a. The formula may seem strange,
but it is the correct one for any pyramidal number of side
a, if A is taken to be the polygonal number of side a. Here
A is a triangular number of side a, having as its formula
A = a(a+1)/2. Cf. Cantor, 1875: p. 122. In I, 2 the author
refused to accept the triangular number formula for the area
of a triangle, even though he used the polygonal number formulas for the other polygons in I, 15 to I, 20. Here too
he is using number-sum formulas as if they were surface and
volume formulas. The author seems to have taken this material from Geom. incert., ed. Bubnov, 1899: 348.5-18. Even
those who knew all of Euclid would not find the required
proof there.

[III, 21]   QUADRATI PYRAMIDEM INVENIRE.

Predicta regula deservit theoreumati.   Ut area eius
bis ducta, adiunctoque uno latere, summeque ducte per
numerum unitate transcendentem latus 6 pars constituet
pyramidem.   Quod probatur sicut et precedens.   5

Pentagoni similiter exagoni, et eptagoni, et omni
figurarum equalia latera habentium, quotque angulorum
fiunt pyramides, invenientur per eandem rationem sive
regulam.   Et probabuntur ab Euclidentium subtilitate.
Et probationem huius et illius propositionis que de   10
irregularitate tractat eorum ingenio exercitando ex
industria non ignorantia pretermisimus.

Et hec de crassimetria sufficiant.

---

2 ante predicta add. hec B / predicta regula om.E / ante
   deservit add. huic F
3 bis: eiusdem F / post bis add. iuncta sive B / adiunctoque
   uno latere: unoque latere adiuncto FH
4 transcendentem: transcendente EH / constituet: constitues
   E continet FH
5 et om.FH
6 pentagoni similiter tr.FH / pentagoni: penthagoni BE /
   eptagoni: epthagoni E / omni: omnium F
8 fiunt: ?B / invenientur: inveniatur F inveniantur H /
   eandem: eadem E / rationem sive om.FH
9-12 et . . . pretermisimus om.E
10 huius: eius FH
11 irregularitate: illorum H / eorum: irregularitate H /
   ingenio: ?F
13 et om.EB / crassimetria: cosimetria E

III, 21.   TO FIND [THE VOLUME OF] A SQUARE PYRAMID.[31]

The previous rule applies to the theorem.   So that when
its area (fig. 74) is taken twice and one side is then
added, and the sum is then multiplied by a number transcending the side by one, the sixth part [of the product]
contains the pyramid, which is proved just as the preceding.

Similarly for the pentagon, hexagon, and heptagon, and
all figures having [bases with] equal sides, however many
angles the [bases of] the pyramids contain, they will be
found by the same reasoning or rule.   And they will be
proved by the subtlety of Euclidents.   And we omit the proof
of this and of that proposition which treats of their irregularity, not from ignorance [but] for exercizing cleverness through industry.

And these things suffice for volumetry.

Fig. 74

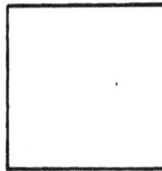

---

[31]"Quadrati . . . pretermisimus" (III, 21.1-12).   As we
remarked in n. 30 to the previous proposition, the formula
$V_n = (2A_n+a) \cdot (a+1) \cdot (1/6)$, where A_n is the polygonal number
of order n and side a, will give the pyramidal number of
order n and edge a.   For the square pyramid, the formula
becomes $V_2 = (2a^2+a) \cdot (a+1) \cdot (1/6)$, which is the formula for
the sum of the first a squares.   Cf. Geom. incert., ed. Bubnov, 1899: 348.5-11 and .19-24 and 349.6-11.   A "Euclident"
seems to be a theoretical geometer or someone who knows all
of Euclid; cf. III, 20.9-10.   The term is comparable to
"Almagestant" in the variants to III, 4.18 and III, 5.22.
I doubt that the author is being honest in claiming to
leave out the proof not because he cannot give it but as
as exercise to test the reader's cleverness and industry.

Sequitur ut de minutiis geometricis et astronomicis
ad predicta necessariis in 4 partitione pertractemus.   15

---

15 in: scil. E / 4: tertia H / post pertractemus add.
    incipit liber 4us E   incipit liber de minutimetria 3us H

Here follows material so that we may deal in the fourth part with geometrical and astronomical fractions necessary for the aforementioned.

[IV, 1] EX CONCURSU INTEGRI SUPER ALIQUOTAM EIUS PARTEM
TOTAM PARTEM PROVENIRE.

Integra geometrie sunt hec: granum, digitus,
palmus, pes, tracubus vel trabicus, pertica, tabula,
sextarius, modius. 4 grana ordei linealiter posita   5
faciunt digitum, 4 digiti palmum, 4 palmi pedem, 4
pedes trabuscum, que omnia secundum longitudinem linee
mensurantur. Quando unum currit super aliud, quod provenit
quadratum accipitur. 4 pertice quadratum includentes
faciunt tabulam, 12 tabule sextarium, 8 sextarii modium. 10

Geometrice autem minutie secundum denominationem
sumuntur aliquarum partium aliquotarum, ut medietas,

---

1 concursu: cursu H
3 integra hic coepit denuo M
4 palmus: palma H / tracubus vel trabicus: tracubus E
   tribucus H
5 ordei linealiter posita: orbiculariter ?H
6 palmum: palmam H / palmi: palme H / 4: 5 E
7 trabuscum: tracubum E tribucum H / linee: prime linee
   posita H
8 post mensurantur add. per omnia M / aliud: ?E
10 12: 13 H / sextarii: sextaria ?H
12 sumuntur: sumantur M tr.post aliquotarum H /
   aliquarum om.H

IV, 1. TO PRODUCE A "WHOLE PART" FROM THE MULTIPLICATION
OF AN INTEGER BY ITS ALIQUOT PART.

These are the integers of geometry:[1] the grain, digit
[or inch], palm, foot, tracubus, perch [or rod], tabula,
sextarius, modius. Four grains of barley set in a line
make an inch, four inches a palm, four palms a foot, four
feet a tracubus, all of which are measured along the length
of a line. When one is multiplied by another what is produced is taken as a square [measure]. Four rods inclosing
a square make a tabula, twelve tabulae a sextarius, eight
sextarii a modius.

However, geometrical fractions are taken following the
denomination of any of their aliquot parts,[2] such as a half,

---

[1]"Ex concursu . . . modium" (IV, 1.3-12). The terms
"whole part" and "aliquot part" will be explained below in
n. 2, where the theorem itself is discussed. Here the author uses "integer" to mean what we might call a unit of
measure. The list of measures is similar to that in I,
32.11-15; cf. n. 92 thereto. Measures are also described
in Gerbert, Geom., ed. Bubnov, 1899: 56.1-63.25, Geom. incert., ed. Bubnov, 1899: 337.1-338.7, "Boethius," Geom. II,
ed. Folkerts, 1970: 145.530-146.550 and 169.930-934.

[2]"Geometrice . . . probare" (IV, 1.13-36). An aliquot
part is a part contained in a whole an exact number of times.
Let us call that number of times a. If we write an integer
as n, then a whole part seems to be n/a. The theorem says
that n·(1/a) = n/a and that such a product is to be considered in square units. The "denomination" of an aliquot part
has a double meaning, for what we call the denominator of a
fraction is in fact what gives it its name or denomination.
The proof of this theorem says, in effect, that linear units
are to be treated as numbers and square units as square
numbers. Cf. John of Seville's Algorism, ed. Boncampagni,
1857: 50.16-20.

3ª, 4ª, 5ª, et sic deinceps. Si ergo currat integrum
super aliquotam eius partem, i.e. si metiatur eam,
provenit tota pars. Quia si currat pes super medietatem   15
pedis, provenit medietas pedis; si super tertiam, tertia;
et sic deinceps. Si duo pedes super tertiam, proveniunt
2 tertie, etc.

Quod sic probatur. Sicut enim ex ductu cuiusvis
numeri in se provenit eius quadratum, ita ex ductu   20
cuiuslibet integri in se provenit eius quadratum. Est
enim in numeris sicut in lineis, optinet ergo pes longus
vel quodvis integrum vicem radicis in numero. Et sicut
ex radice ducta in se surgit quadratus numerus eius,
sic ex pede lineari ducto in se provenit pes quadratus   25
qui planimetrie consecratur. Optinet ergo quaternarius

---

14 si om.H
15 quia: quod M
16 pedis (1): eius M / provenit medietas pedis: proveniunt
   medietates pedum H   pedis M / post tertiam add. et M
17 tertiam: tertia ?M
18 2: ?H
20 eius om.H
20-21 ita . . . quadratum om.M
21 provenit: producitur H
23 quodvis integrum: quivis integraliter H / ante vicem add.
   in ?E
24 ex: in E / post se add. ex E / eius tr.ante quadratus H
25 ex om.E / provenit: proveniet H
26 planimetrie: planimetre M / consecratur: consequatur H /
   optinet: optineat H / quaternarius: quaterius M ?E

third, fourth, fifth, and so on. If, therefore, a whole
is multiplied by its aliquot part, that is, their product
is taken, the whole part is produced. Because if a foot
be multiplied by half a foot, halves of [square] feet are
produced; if by a third, thirds, and so on, if two feet
by a third [of a foot], two thirds are produced, etc.

This is proved as follows. Just as from the multiplication of any number by itself a square [number] is produced,
so from the multiplication of any integer by itself its
square is produced. Since it is for numbers just as for
lines, a linear foot or any unit takes the place of a root
in a number. And just as from a root multiplied by itself,
its square will be produced, so from a linear foot multiplied by itself will be produced a square foot, which is
the domain of planimetry. If therefore four takes the place

vicem pedis linearis, cuius quadratum, i.e. 16 numerus

qui vicem optinet quadrati. Si quaternarius metiatur

medietatem sui, provenit 8^{rius}, scil. medietas 16, scil.

numeri quadrati. Si 4^{am} provenit 4 et sic de quantalibet 30

parte quantum habet radix. Similiter erit in lineis.

Si pes linearis metiatur medietatem sui, provenit medietas

pedis quadrati; si 4^{am}, 4^a; si 3^{am}, 3^a. Et hoc erat

propositum probare.

---

27 <u>ante</u> cuius <u>add.</u> et <u>M</u> / i.e.: est <u>H</u>

27-28 numerus qui <u>om.</u>EM

28 vicem optinet <u>tr.M</u> / <u>post</u> optinet <u>add.</u> pedis <u>H</u> /
    quaternarius: 4 <u>E</u> quaterius <u>M</u>

29 medietatem sui <u>tr.H</u> / provenit: proveniet <u>H</u> /
    8^{rius}: 8^a <u>E</u> 8 <u>H</u> / scil. (1) <u>om.H</u>

29-30 scil. numeri <u>om.E</u> <u>tr.H</u>

30 quantalibet: qualibet <u>E</u>

31 similiter erit: sic <u>H</u>

32 metiatur <u>tr.ante</u> pes <u>H</u> / provenit: proveniet <u>H</u>

33 pedis quadrati <u>tr.M</u>

of a linear foot (fig. 75), then its square, that is, the
number sixteen, takes the place of the square [foot]. If
four be multiplied by its half, 8 which is half of a square
number, namely sixteen, is produced. If by its fourth, a
fourth [part of its square] is produced, and thus for any
part as long as the root has [parts]. It will be similar
for lines. If a linear foot be multiplied by its half,
half of a square foot is produced; if by its fourth, a
fourth [is produced]; if by its third, a third [is produced].
And that is what was proposed to be proved.

Fig. 75

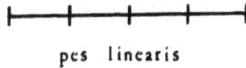

pcs lincaris

[IV, 2]   EX CURSU PARTIS ALIQUOTE SUPER ALIAM ALIQUOTAM
INTEGRI PARTEM, TOTAM PROVENIRE QUOTUS SURGIT NUMERUS
EX MENSURA DENOMINATIONUM.

Hec docet quod si medietas pedis currat super

medietatem, provenit 4^a pars pedis quadrati, sicut ex   5

ductu denominationis metientis se, scil. binarii pro-

venit quaternarius.   Ex cursu 3^e super 3^{am}, provenit

9^a pars pedis, sicut 9 ex ductu ternarii in se.   Ex

cursu 4^e in 4^{am} provenit sextadecima pars, sicut ex

ductu unitatis in se provenit 16^a pars sexdenarii, qui   10

est quadratus quaternarii.   Et hec probat tam ratio

numeri quam fides oculata ductis equidistantibus lineis

ad partes denominatas.

1-2 aliam . . . integri: totam eius E

2 ante totam add. partem E

3 ex: et MH

4 post super add. eius E

6 metientis: medietatis in E / post se add. provenit M

6-7 provenit: surgit E est productus H

9 cursu: ducte H / post cursu add. numero E / provenit om.E /
   sextadecima: 10^a 6^a 8^a ?M / pars om.MH

10 unitatis: quaternarii EH / 16^a pars om.E / sexdenarii:
   16 E

11 post est add. eius E / quadratus om.M / quaternarii om.E /
   tam om.E

12 quam: et E / oculata: occulta H

IV, 2. TO PRODUCE A WHOLE [PART] BY MULTIPLYING AN ALIQUOT
PART BY ANOTHER ALIQUOT PART OF A UNIT, [TAKE] WHATEVER
NUMBER ARISES FROM THE PRODUCT OF THE DENOMINATORS.[3]

This teaches that if half a foot is multiplied by half
[a foot], a fourth part of a square foot is produced, just
as from the multiplication of the measuring denomination by
itself, that is, of two [by itself] four is produced. From
the multiplication of a third of a foot by a third, the
ninth part of a foot is produced, just as 9 from the multiplication of three by itself. From a fourth (fig. 76) multiplied by a fourth a sixteenth part is produced, just as
from the multiplication of one by itself is produced the
sixteenth part of sixteen, which is the square of four.
Argument from number as much as eye witness proves this
by means of parallel lines drawn to the denominated parts.

Fig. 76   [Fig. 77]

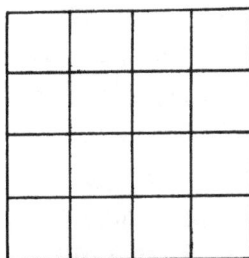

---

[3]"Ex cursu . . . denominatas" (IV, 2.1-13). In effect
the author is saying that $(1/a)\cdot(1/a) = 1/a^2$, where a is an
integer. Cf. Gerbert, Geom., ed. Bubnov, 1899: 57.5-21,
where there is a figure (like fig. 77 here) to which the
proof "by eye witness" will apply; each of the sixteen
squares produced when the sides of the square are quartered
are obviously one sixteenth of the larger square.

[IV, 3]   EX CURSU PARTIS ALIQUOTE SUPER ALIAM TOTAM EIUS
PARTEM, PARTEM TOTAM PROVENIRE QUOTUS SURGIT NUMERUS EX
MENSURA DENOMINATIONUM.

Hec docet quod si medietas pedis currat super eius

3^{am} provenit 6^a, sicut ex mensura binarii in ternarium   5

provenit senarius. Si medietas currat super 4^{am}, pro-

venit 8^a; si super 5^{am}, 10^a etc. ut patet in numeris; si

3^a super 4^{am} provenit 12^a; si 4^a super 4^{am}, 16^a, et sic

in infinitum ut in numeris. Quod probatur ratione

numeri et linearum et protractione ad partes denominatas. 10

Et hec de vulgarium minutiarum sufficiant multi-

plicatione.

Sequitur de astronomicis minutiis in quibus maior latet

difficultas et utilitas, in tractatu quarum subtiliori

quod de hiis deservit suplebimus, de multiplicatione   15

et divisione et radicum extractione deinceps prosequentes.

---

1 aliam om.H / totam super scr. scil. quota ipsam est H_a

1-2 totam eius partem: aliquotam integritatem E

2 partem totam: totam partem E   totam H

4 pedis: eius pedis M

5 post 3^{am} add. partem H

6 currat: provenit E

6-7 provenit 8^a: proveniunt 8 H

7 10^a etc. om.H / patet om.E

8 provenit om.H / 16^a: 61^a ?H

9 ut in numeris: super numeris M   om.H

10 et (2) om.M

11 minutiarum om.H

11-12 sufficiant multiplicatione tr.H

13 minutiis om.H / maior latet om.H

14 post difficultas add. magis latet H / subtiliori:
   subtiliora H

15 quod: que ?H / deservit: fuerint H

IV, 3. TO PRODUCE A WHOLE PART BY MULTIPLYING AN ALIQUOT
PART [OF A UNIT] BY ANOTHER WHOLE PART OF IT, [TAKE] WHAT-
EVER NUMBER ARISES FROM THE PRODUCT OF THE DENOMINATORS.[4]

This teaches that if half of a foot is multiplied by a
third part of it [of the foot], a sixth is produced, just
as from multiplying two by three, six is produced. If a
half be multiplied by a fourth (fig. 78), an eighth is produced, if by a fifth, tenths, etc. as is clear in numbers;
if a third by a fourth, a twelfth; if a fourth by a fourth,
a sixteenth, and thus _in infinitum_ as in numbers. This is
proved by argument from number and lines and by the extension [of parallel lines] to the denominated parts.

And these propositions suffice for the multiplication of
common fractions.

Fig. 78

There follows material concerning astronomical fractions
in which lie hidden greater difficulty and utility; we will
supply the more subtle things concerning these fractions in
the treatise, taking up in turn multiplication and division
and the extraction of roots.

---

[4]"Ex cursu . . . multiplicatione" (IV, 3.1-12). This
proposition is a more generalized form of IV, 2; it may be
expressed as the rule $(1/a)(1/b) = 1/ab$, where a and b are
integers. Cf. J. of Seville, ed. Boncompagni, 1857: 50.20-
22. "Common fractions" is the usual name for what the author
has previously called a "geometrical fraction."

[IV, 4]   EX DUCTU PRIMARUM MINUTIARUM IN SE, SECUNDA
PROVENIRE: ET SECUNDARUM IN SE, QUARTAS.

Hec docet quod si feriantur minuta in minuta

proveniunt secunda, i.e. numerus secundorum, qui in ea

proportione se habent ad unum minutum qua se habet talis   5

numerus minutorum ad unum gradum.   Ut si feriantur 4

minuta in 5 minuta, provenient 20 secunda que sunt 3^a

pars unius minuti, sicut 20 minuta unius gradus.

Similiter ex ductu 2^{orum} in 2^a provenient 4^a.   Ex ductu

3^{orum} in 3^a, provenient 6^a secundum eiusdem proportionis   10

normam.

Si enim habeamus 2 numeros proportionales ad 3^m,

et volumus habere 4^{am} proportionalem 3°, feriamus primum

in 2^m, et productum dividamus per 3^m.   Denominatio dabit

4^m proportionalem 3°.   Hoc est quod docet astronomia: 15

---

1 secunda: secundas E

2 quartas: quarta H

3 post minuta (2) add. inventa H

5 habent: habebit H / qua se: quale M / habet: habebit H

7 minuta (2): om.H / sunt tr.post pars H

8 pars om.E / sicut: fuit M similiter H / 20: 30 H

9 provenient: provenit M proveniunt H

10 provenient: provenit M proveniunt H

12 si enim mg. hab. Hoc sane intellig<atur>, scil. secundum
    19^{am} propositionem 7^i Euclidis. H_a / habeamus: habes H

13 et: etiam H / volumus: voluimus M / post habere add.
    ad M / proportionalem: proportionale M / 3°: 2° M om.H

IV, 4. TO PRODUCE SECONDS BY MULTIPLYING MINUTES BY THEM-
SELVES: FOURTHS BY MULTIPLYING SECONDS BY THEMSELVES.[5]

This teaches that if minutes are multiplied by minutes,
seconds are produced, that is, the number of seconds which
have that proportion to one minute that such a number of
minutes has to one degree. So that if 4 minutes are multiplied by 5 minutes (fig. 79), 20 seconds will be produced,
which is the third part of one minute, just as 20 minutes
[is one third] of a degree. Similarly by multiplying seconds by seconds, fourths will be produced. By multiplying
thirds by thirds, sixths will be produced, according to a
rule of same proportion.

Since if we may have two numbers proportional to a third
and we want to have the fourth proportional to the third,[6]
let us multiply the first by the second, and divide the product by the third; the quotient gives the fourth proportional to the third. This is what astronomy teaches: to mul-

---

[5]"Ex ductu . . . normam" (IV, 4.1-11). Modern usage requires the terms "minutes, seconds, thirds" etc. even though
the author sometimes uses "first minutes, second minutes,
third minutes" etc. This proposition is equivalent to
$(a/60^r) \cdot (b/60^r) = ab/60^{2r}$, where a and b are numbers of
minutes and r equals 1, or numbers of seconds and r is 2,
or numbers of thirds and r is three. We are to read $a/60$
as a minutes, $a/60^2$ as a seconds, etc. The proposition
would be valid for any integer r, but as the author tells
us below in IV, 6.7-8, the astronomer does not have to go
beyond seconds or thirds.

[6]"Si enim . . . integro" (IV, 4.12-17). The author also
shows us how to reduce a given number of seconds to an equivalent number of minutes by applying the relation $c/60^2$:
$1/60 :: c/60 : 1$, where $c/60^2$ is to be read as c seconds,
$c/60$ as c minutes. The product ab in n. 5 above may be
substituted for c.

ferire minuta in minuta, et dividere per 60 qui semper
est 3us. Pars data dat minuta proportionalia integro.

---

16 in minuta om.E
17 est: erit M / proportionalia: comproportionalia ?H

tiply [one number of] minutes by [the other number of] minutes and to divide by 60 which is always the third [propor-
tional].   The given part [i.e. the product of the numbers
of minutes] gives the [number of] minutes proportional to
the integer.

Fig. 79

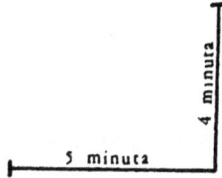

[IV, 5] EX PERCUSSIONE INTEGRI IN MINUTIAS, EIUSDEM
GENERIS MINUTIAS PROCEARI.

Astronomice minutie non sumuntur secundum denominationem partium sicut vulgares, sed secundum denominationem graduum in 60 partes, que vocantur minuta sive   5
prime minutie, minuti in 60 partes que vocantur secunda
vel secunde minutie, secunde in 60 partes que vocantur
3^a vel tertie minutie, et sic deinceps secundum sexagen-
ariam divisionem.

Ferire ergo gradus vel integra in minutias nihil   10
aliud est quam invenire numerum qui in eadem proportione
se habeat ad illos gradus in qua ille minutie se habent
ad 60. Ut ferire 2 gradus in 10 minuta est invenire
vicenarium minutorum, que se habent ad 2 gradus in
eadem proportione in qua se habet 10 ad 60, quod est   15
tantum summe de utroque gradu quantum 10us capit de
60 scil. 6^{am} partem. Unde ex ductu gradus in minuta,
non proveniunt nisi minuta.

1-30 In codice H haec propositio antecessit illam IV, 4.
3 astronomice: astronomie E
4-5 denominationem: divisionem E
5 graduum: gradus E / post graduum add. qui dividuntur H /
    in om.E / vocantur: nominantur H
6 minuti: minuta H / vocantur: denominantur H
7 vel secunde minutie om.H / secunde: secundi E secunda H /
    vocantur: denominantur H
12 qua: quos M
14 que se habent: qui habet se H / 2: secundos E
15 eadem: ea E / habet: habent E
16 10us capit om.H / post 60 add. denarius capit H
17 ductu gradus: graduum ductu H

IV, 5.   FROM THE PRODUCT OF AN INTEGER BY MINUTES, MINUTES
OF THE SAME SORT [ARE] CREATED.[7]

Astronomical minutes are not combined following the denomination of the parts as [are] common [fractions], but
following the denomination of degrees [divided] into 60
parts, which are called minutes or first minutes; minutes
[are divided] into 60 parts which are called seconds or
second minutes; seconds into 60 parts which are called
thirds or third minutes, and so on following sexagesimal
division.

Therefore to multiply degrees or integers by minutes[8] is
nothing other than to find the number which would have the
same proportion to those degrees that the minutes have to
60.   Thus to multiply 2 degrees by 10 minutes (fig. 80) is
to find twenty minutes, which is to two degrees in the same
proportion as 10 is to 60, which is [to say that it repre-
sents] as much of the sum of both of the degrees as 10 does
of 60 namely, a sixth part.   Thus from the product of degrees by minutes nothing is produced but minutes.

---

[7]"Ex percussione . . . divisionem" (IV, 5.1-9).   As remarked in n. 5 above, we shall use terms like "seconds"
rather than "second minutes."   Degrees and integers are both
wholes with respect to minutes, seconds, etc.; the procedures described will apply to either.

[8]"Ferire . . . minuta" (IV, 5.10-18).   The algorithm for
multiplying a degrees by b minutes amounts to finding a number c of minutes, such that $c/60 : a :: b/60 : 60$, where
$c/60$ is again read as c minutes.   Of course c is numerically
equal to ab.   In the author's example, a is 2, b is 10 and
their product c is 20.   Cf. J. of Seville, ed. Boncompagni,
1857: 50.16-23.

Ex ductu gradus in 2^a, non proveniunt nisi
secunda. Si enim percusseris 2 gradus in 10 2^a, pro-   20
veniunt 20 secunda, quod est in eadem proportione summe
de duobus gradibus in qua proportione sumunt 10 2^a de
<60> minutis, scil. <6^{am}> partem unius minuti. Unde ex
ductu gradus in 2^a non proveniunt nisi 2^a. Unde tot
sumantur de quolibet gradu quot fuerunt secunda pro-   25
portionalia in que fit percussio graduum. Similiter ex
ductu gradus in 3^a, proveniunt tertia. Et tot sunt
sumenda de quolibet gradu quot sunt 3^a proportionalia,
que per 60 facta divisione sunt reducenda in 2^a, secunda
similiter in minuta, minuta in gradus.   30

---

19 gradus: graduum <u>H</u>
20-21 proveniunt: provenient <u>H</u>
21 secunda: 3^a <u>M</u> / eadem: ei <u>H</u>
22 in qua proportione: ut 20 secunda <u>H</u> / 10 2^a <u>om.H</u>
23 60 (<u>ed.</u>): 20 <u>ME</u> 2 <u>H</u> / 6^{am} (<u>ed.</u>): 3^{am} <u>MEH</u>
24 unde: ut <u>EH</u>
25 <u>post</u> sumantur <u>add.</u> secunda <u>E</u>
26 que: qua <u>E</u> quo <u>H</u> / fit: fuit <u>H</u>
27 gradus: graduum <u>H</u> / proveniunt tertia <u>om.M</u> / tot sunt:
  sunt tot <u>M</u> tot <u>H</u>
28 de <u>om.M</u> / quot: quos <u>E</u>
29 60 facta divisione: divisione in 60 <u>H</u>
29-30 secunda similiter in <u>om.M</u>
30 similiter <u>om.E</u>

From the product of degrees by seconds nothing is produced but seconds.[9] For if you would multiply 2 degrees by
10 seconds, 20 seconds is produced, which is in the same
proportion to the sum of the 2 degrees as 10 seconds is to
60 minutes, namely the sixth part of one minute, thus from
the product of a degree by seconds nothing is produced but
seconds. Thus whatever part may be taken of any degree,
[there are] as many proportional seconds as there were in
that part by which the degrees were multiplied. Similarly
from the taking of a degree by thirds, thirds are produced.
And however much of any degree is to be taken, there are
that many proportional thirds, which, when a division by
60 has been made, are to be reduced to seconds, similarly
seconds to minutes, minutes to degrees.

Fig. 80

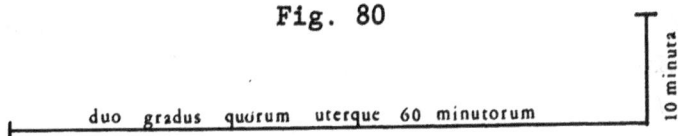

duo gradus quorum uterque 60 minutorum

10 minuta

---

[9]"Ex ductu . . . gradus" (IV, 5.19-30). The method for
finding the product of a degrees by b seconds may be expressed by the relation for finding $c/60^2$, or c seconds,
$c/60^2 : a :: b/60^2 : 60$, where c is the numerical value of
the product ab. In the example, as I have corrected it, a
is 2, b is 10, and their product c is 20.

[IV, 6]   EX DUCTU MINUTORUM IN SECUNDA, 3^a; SECUNDORUM
IN TERTIA, QUINTA; TERTIORUM IN TERTIA, SEXTA PROVENIRE.

Hec per precedentis probatur regulas. Minuta enim
percussa in 2^a generant 3^a. Sed in 3^a generant 4^a et
sic deinceps: secunda in 3^a, 5^a; secunda in 4^a, 6^a; 3^a 5
in 3^a, 6^a. Quod per probationes precedentium ostenditur.
Sed non est ex hiis opus, quia non est necesse astrologo
ultra 2^a vel 3^a procedere. Et ideo hiis supersedendum
in multiplicatione arbitramur.

Sequitur de minutiarum divisione.   10

1-2 secundorum in tertia om.E
2 quinta: 4^a EM
3 precedentis: precedencia E precedentium H / regulas:
   om.E tr.ante probatur H / post probatur add. sive per
   precedentium regulas H
4 sed: minuta H
6 per om.E / probationes: proportiones H / precedentium:
   sequentium H / ostenditur: ostendetur H
7 ex om.H / est (2) om.H
8-9 et . . . arbitramur om.E / hiis om.H
10 sequitur . . . divisione om.E / minutiarum: minutorum M /
   divisione hic desiit M

IV, 6. TO PRODUCE THIRDS BY MULTIPLYING MINUTES BY SECONDS;
FIFTHS, SECONDS BY THIRDS; SIXTHS, THIRDS BY THIRDS.[10]

This is proved by the preceding rules, for minutes taken
times seconds engender thirds; but times thirds, they engender fourths and so on; seconds times thirds, fifths;
seconds times fourths, sixths; thirds times thirds, sixths.
This is shown by the proofs of the preceding. But these
are not needed because it is not necessary for the astronomer to proceed beyond seconds or thirds. And therefore
we decide to omit these in multiplication.

There follows material about the division of minutes.

---

[10]"Ex ductu . . . arbitramur" (IV, 6.1-9). The author
is telling us, in effect, that $(a/60^r) \cdot (b/60^s) = ab/60^{r+s}$.
J. of Seville, ed. Boncompagni, 1857: 50.24-51.4 gives
similar rules. The same general approach of the preceding
propositions would demonstrate this proposition. By using
the "doubling rule" of IV, 4 and the technique for reducing
sexagesimal fractions to sexagesimals of lower order, also
in IV, 4, one could derive all of the given relations.

[IV, 7] SI DIVIDENS NUMERUS HABEAT MINUTIAS, DIVIDENDUM
IN ULTIMAS MINUTIAS DIVIDENTIS RESOLVERE.

    Hec docet quod si maiorem numerum per minorem

volumus dividere, et minor habeat minuta vel 2^a vel 3^a,

utrumque debemus resolvere in ultimas minutias, et   5

post debemus dividere maiorem per minorem. Quod

probatione non indiget.

---

3 hec: hoc E
4 post volumus add. resolvere sive B
6-7 quod . . . indiget om.E

IV, 7. TO RESOLVE THE DIVIDEND INTO LAST MINUTES OF THE
DIVISOR IF THE DIVISOR HAS MINUTES.[11]

This teaches that if we want to divide a greater number

by a smaller, and the smaller has minutes or seconds or

thirds, we should resolve them all into last minutes, and

afterwards we should divide the greater by the smaller.

This does not need a proof.

---

[11]"Si dividens . . . indiget" (IV, 7.1-7). "Last minutes"
must be understood as sexagesimal fractions of the smallest
type or those having the highest power of 60 in their denominators, when expressed in the form $a/60^r$. The proposition means that we must reduce sexagesimal fractions to the
lowest common denominator before dividing. For fractions
of this sort, the reduction amounts to successive multiplication by 60. Cf. J. of Seville, ed. Boncompagni, 1857:
53.7-22.

[IV, 8] EX DIVISIONE TERTIORUM IN 3^a, SECUNDORUM IN
SECUNDA MINUTA, MINUTORUM IN MINUTA, GRADUS IN PARTE
PROVENIRE.

Quid hec intendit, patet, quod potest probari per

percussionem graduum in minuta, minutorum in 2^a, 2^a in   5

3^a; et restaurabitur primus numerus.

Sequitur de radicum extractione. Sed ut pateat in

minutiis, primo de extractione earum in integris erit

agendum. Et primo de formatione quadratorum earum

secundum trigonorum et tetragonorum et ceterorum formam. 10

---

2 in minuta om.E
4 quid: quod H / quod: quia H / potest probari tr.H
4-5 per percussionem: percussione H
5 post minuta add. vel H
8 in om.E
9 earum: eorum H
10 ceterorum: ceterarum B

IV, 8.   TO PRODUCE PARTIAL DEGREES FROM THE DIVISION OF [12]
THIRDS BY THIRDS, SECONDS BY SECONDS, MINUTES BY MINUTES.

What this asserts is plain; it can be proved by multiplying degrees by minutes, minutes by seconds, seconds by
thirds; and the first number will be restored.

There follows material about the extraction of roots.
But so that it may be plain for minutes, first their extraction in integers will have to be treated.   And first the
formation of their squares following the form of triangles
and quadrangles, etc.

---

[12]"Ex divisione . . . numerus" (IV, 8.1-6).   Division
of sexagesimal fractions of the same order may be expressed
as $(a/60^r)/(b/60^r) = a/b$.   If the quotient a/b is not integral it would have to be reduced to sexagesimal fractions
by using an algorithm we might express as $a/b = (60a/b)/60$.
The same rule might have to be applied several times.   The
proof of the division should consist in multiplying the
quotient by the divisor.   It is not clear what the author
intends as proof.

[IV, 9]   EX TRIGONO DUCTO OCTIES ADDITA UNITATE, QUADRATUM
PROCREARI, ET EX EO RADICEM UNITATE DEMPTA DUCTA PARTE
BIS LATUS TRIGONI CONSTITUERE.

Quadratus est numerus qui fit ex ductu radicis in
se, qui fit ex trigonis secundum datam regulam.   Ut   5
primus numerus trigonus est ternarius qui ductus octies
addita unitate facit 25.   Cuius radix, scil. 5, unitate
dempta medietas, scil. binarius, facit trigoni latus.
Similiter in 6^{rio} et in aliis trigonis probatur regula.

---

2 ex eo: eius H / radicem tr.post dempta B / ducta:
   ductaque H
2-3 parte bis super scr. scil. accepta medietate que
   dicitur pars bis H$_a$
3 trigoni: quadrati EB
4 est numerus tr.B
4-5 in se om.E
5 secundum om.E
6 numerus om.E / trigonus est tr.H / post est add. secundum E
8 post dempta add. sumpta E
9 in (2) om.B

IV, 9.   TO BEGET A SQUARE BY ADDING ONE TO EIGHT TIMES A
TRIANGLE, AND FROM THE SQUARE TO MAKE THE SIDE OF A TRIANGLE
BY TAKING HALF OF ONE LESS THAN ITS ROOT.[13]

A square is a number which is made by the multiplication

of a root by itself and is made from a triangle following

the given rule. Thus the first triangular number is

three (fig. 81), which when taken eight times and one added

makes 25. Half of one less than the root, namely 5, of

that [square] makes the side, namely 2, of the [original]

triangle. Similarly the rule is proved for six and for

other triangles.

Fig. 81

---

[13]"Ex trigono . . . regula" (IV, 9.1-9). This proposition deals with polygonal numbers. In contrast to I, 15
through I, 20 and III, 20 and 21, the author is treating
polygonal numbers as such here and in IV, 10 and 11. Converting triangular numbers to square numbers and finding
the side of the original number may be accomplished by the
author's technique. If t_s is a triangular number of unknown side s, then $Q = 8t_s+1$ will be a square number, and
$s = ((8t_s+1)^{\frac{1}{2}} - 1)/2$ will be the whole number which is the
side of the original triangle. Cf. Geom. incert., Ep. et
Vit. Ruf. and Varro? Aliud fragmentum geometriae, ed. Bubnov, 1899: 350.18-20, 532.14-533.5, and 504.24-505.2 and
"Boethius," Geom. II, ed. Folkerts, 1970: 149.598-606.
The intent of the procedure is primarily to find the side
of a triangular number; thus its place here is questionable.

[IV, 10]   EX TETRAGONO DUCTO 16, QUADRATUM GENERARI; ET
EIUS RADICEM DUCTA PARTE 4$^{\underline{a}}$, LATUS TETRAGONI CONSTITUERE.

Hec regula patet in omnibus tetragonis ut quaternarius
sic ductus facit 64, cuius radicis, scil. 8, 4$^{\underline{a}}$ pars,
scil. binarius, facit latus tetragoni. Similiter in   5
novenario, etc.

---

1 ducto: ducto in B̲ om.H̲ / generari: fieri H̲
2 ducta s̲u̲p̲e̲r̲ s̲c̲r̲. acepta, scil. H̲$_a$
4 sic ductus s̲u̲p̲e̲r̲ s̲c̲r̲. sedecies, scil. H̲$_a$
5 tetragoni: trigoni B̲

IV, 10. TO MAKE A SQUARE BY TAKING A SQUARE SIXTEEN TIMES
AND TO MAKE THE SIDE OF THE [ORIGINAL] SQUARE BY TAKING THE
FOURTH PART OF ITS ROOT.[14]

This rule is plain for all squares, as 4 (fig. 82) multiplied this way [i.e. times 16] makes 64. The fourth part
of that root, namely 8, makes the side of the [original]
square, namely 2. Similarly for 9, etc.

Fig. 82

•   •

•   •

[14]"Ex tetragono . . . novenario, etc." (IV, 10.1-6). This
proposition gives us a rule for finding the side of a square
number; it is analogous to the rule for the triangular number in IV, 9 in that it first makes the given number into a
larger square number. The larger square Q is derived from
the smaller square q_s of unknown side s by the formula
$Q = 16q_s$. The side will be found by the formula
$s = (16q_s)^{\frac{1}{2}}/4$. Cf. Geom. incert., ed. Bubnov, 1899: 350.20-
351.2.

[IV, 11] PENTAGONUM VICESIES QUATER DUCTUM, ACCEPTA
UNITATE, QUADRATUM FACERE, ET EIUS RADICEM ACCEPTA
UNITATE ET DUCTA PARTE 3^a LATUS PENTAGONI REDDERE.

Hec ex precedentibus patet. Ut ex quinario sic

ducto adiuncta unitate sit 121, et ex radice, scil. 11,   5

addita unitate sumpta 3^a pars facit latus pentagoni.

Similiter in 12, etc. Et exagono et ceteris secundum

debitam et preostensam regulam crescendo, invenientur

quadrata, et eorum latera.

Sequitur de radicis cubi extractione in integris,

ut post pateat in minutiis.

---

3 parte 3^a mg. hab. Videtur quibusdam hec de 6^a. H$_a$

5 adiuncta: iuncta H / 121: 132 B / scil. om.H

6 unitate om.B / sumpta: si inpari H om.B / 3^a: 2^a E /
   pentagoni om.H

7 et (1) om.E

8 debitam: debitum H / regulam: regulam vel rationem E
   rationem H

10-11 sequitur . . . minutiis om.E

10 cubi: cubici H

IV, 11. TO MAKE A SQUARE BY ADDING ONE TO 24 TIMES A
PENTAGON, AND TO GET THE SIDE OF A PENTAGON BY TAKING A
THIRD OF ITS ROOT PLUS ONE.[15]

This is plain from the preceding. As 5 (fig. 83) multiplied this way [i.e. times 24, and] one added, is 121, and
when one has been added to the root, namely 11, the third
part makes the side of the [original] pentagon. Similarly
for 12, etc. And for the hexagon and the others[16] according to the required and already revealed rule by increasing
[the multiplier], the squares and the sides of the figures
will be found.

Fig. 83

There follows material about the extraction of cube roots
for integers so that afterwards, it may be clear for minutes.

---

[15] "Pentagonum . . . in 12, etc." (IV, 11.1-7). The rule
for finding the side of a pentagonal number converts the
given number into a larger square number $Q = 24p_s+1$, where
p_s is the pentagonal number of unknown side s. The formula
$s = ((24p_s+1)^{\frac{1}{2}}+1)/3$ expresses the method for finding the
side. Cf. Geom. incert., Ep. et Vit. Ruf. and Varro?, ed.
Bubnov, 1899: 351.2-4, 535.6-13 and 505.3-5.

[16] "Ex exagono . . . latera" (IV, 11.7-9). There is a
formula for finding the side s of a general polygonal number of order a; it may be derived from the expression given
for the general a-gon in n. 63 to I, 20. Cf. Cantor, 1875:
p. 122. The formula may be expressed as

$$s = \frac{\sqrt{8(a-2)A_a(s) + (a-4)^2} + (a-4)}{2(a-2)}.$$

[IV, 12] CUBI NUMERI RADICEM INVENIRE.

Distinctis figuris per ternarios, sub figura
sequente ultimum ternarium scribatur digitus, cuius
cubus a superiori vel a superioribus possit subtrahi.
Et facta subtractione, eo deleto, 2° loco scribatur eius   5
quadratum, post aliud scribatur eius digitus. Et sub
quadrato scribatur duplum illius digiti. Et sub subduplo
iterum scribatur duplum ipsius digiti primo secundum.
Deinde ante secundum duplum inveniatur digitus qui
ductus in quadratum primi digiti subtrahatur a superiori   10
quadrato quantum poterit. Et idem ductus in primum

1-59 cubi . . . probatur om.E
1 cubi: cubici H
2 ternarios: 3 B
3 ternarium mg. hab. vel sub prima figura ultimi ternarii
   si non sequatur ipsum alia [figura] H$_a$
4 cubus: cubum H / subtrahi: substrahi B
5 et . . . subtractione om.B
6 post aliud: antequem H
7 sub subduplo: subduplo B
8 primo secundum om.H
10 subtrahatur: subtrahat H
11 quadrato ?H / ductus: punctus B

IV, 12. TO FIND THE ROOT OF A CUBE.[17]

When the figures have been separated into [groups of]
threes [starting at the right], under the figure following
the last [group of] three, let a digit be written, the cube
of which could be subtracted from the number or numbers
above. And when the subtraction has been made and [the cube]
deleted from the number above, let the square of the digit
be written [under] the second place. After, let its digit
be written another [time]. And beneath the square, let the
double of that digit be written. And below the subdouble
let the double of this digit be written again before [i.e.,
to the right of] the second. Then before the second double
let the digit be found which when multiplied by the square
of the first digit may be subtracted from what is above the
square as much as possible. And the same [number] having

---

[17]"Cubi . . . duplati" (IV, 12.1-23). This method for
finding the cube root is rather difficult for the modern
reader to understand; however, it will yield a correct result. The only real problem is determining the second digit
of the root, which the author does not adequately explain.
An aid in understanding is the following identity, upon
which all of the "longhand" techniques for extracting cube
roots are based: $(a+b)^3 = a^3 + 3a^2b + 3ab^2 + b^3$. Any two
digit number is conveniently expressed as a sum of the type
$a + b$ where b is the units and a is the tens, e.g., for the
number 23, a is 20 and b is 3. Thus if the number of which
we want the cube root is N and we let $N = (a + b)^3 + r$, since
any number may be expressed as the sum of a cube and a remainder r, then $a + b$ is the largest cube root, if it was
chosen correctly. Rather than explaining every step of the
author in extracting the root of a cube, let me simply indicate the successive subtractions which reduce N to its
smallest remainder. I shall do this in the following schema
where the right-hand column indicates the line number of the
text where the subtraction in that row is carried out:

digitum, et productum multiplicatum per remotius duplum

quod est sub quadrato, subtrahatur a superiori ipsius

quadrati quantum poterit.   Et eodem digito ducto in

propinquius duplum primi digiti, productum multiplicetur   15

per eundem digitum.   Et productum subtrahatur a superiori

propinquioris dupli.   Et quadrato eiusdem digiti ducto

in primum digitum, productum subtrahatur a superiori

primi digiti.   Et eodem quadrato ducto in suam radicem,

---

13 subtrahatur: subtrahat H
14 et: vel B
17 et: vel B
18 a superiori om.B
19 et: vel B

been multiplied by the first digit, and the product multiplied by the more distant double which is under the square,
let the [result] be subtracted from what is above its square
as much as possible. And when the same digit has been multiplied by the nearer double of the first digit, let the product be multiplied by the same digit. And let the product
be subtracted from what is above its nearest double. And
when the square of the same digit has been multiplied by
the first digit, let the product be subtracted from what
is above the first digit. And when the same square has been
multiplied by its root, let the product be subtracted from

$$N = a^3 + 3a^2b + 3ab^2 + b^3 + r$$

$-a^3$	4
$- a^2b$	10
$-2a^2b$	13
$-2ab^2$	16
$- ab^2$	18
$-b^3$	21

After all the subtractions have been made, only r remains.
Since the author is finding the root of a cube, he will have
no remainder, i.e. r = 0.

The actual working out of the root requires a careful
place notation so that the successive subtractions may be
made with concern only for the numerical values of a and b.
Thus for the author, if the root is 23, a is simply 2 and b
is 3, but he keeps those digits in places such that they
function as 20 and 3. The placement of the various digits,
as indicated in lines 2-3 and 5-9, may be shown as follows,
where the six x's in the top line indicate the maximum number of digits the cube of a two digit number may occupy:

```
x   x   x   x   x   x
        a
      a²  a
      2a  2a  b
```

productum subtrahatur a suo superiori. Hoc facto, si   20
nihil remanserit de suprapošito numero, ipse suprapositus
habet radicem, cuius radix est numerus constans ex
digito ultimo invento et subduplo duplati.

  Verbi gratia, radix huius numeri 12167 est hic numerus 23, quod probatur quia si ille numerus cubicetur,   25
habebitur prior numerus cubicus, ut sic scribatur cubus.

  Si vero plures fuerint figure in numero cuius radix
est extrahenda, ita quod oporteat plures quam duos
digitos inveniri, facies consequenter sic. Secundo
digito iam invento anterioretur quadratum eius uno loco   30

---

22 habet ?B
24 12167: corr. ex 13167 H   13167 B
25 23: 232 B
28 oporteat: oportet H

what is above it. This having been done, if nothing remains
of the number placed above, that [number] placed above has
a root, which root is a number arising from the digit
last found and the subdouble of the doubled [number].

For example,[18] the root of this number: 12167, is this
number: 23, which is proved because if that number be cubed,
the former cubic number will be had, and thus may a cube be
written.

If indeed there were several figures in the number of
which the root is to be taken,[19] such as occurs [when] more
than two digits [are] to be found, then proceed as follows.
When the second digit has once been found, let its square

Other treatments of the cube root are found in Petri
Philomeni de Dacia in Algorismum vulgarem Johannis de Sacrobosco commentarius, ed. Curtze, 1897: pp. 17-19 (Sacrobosco) and 84-92 (Peter of Dacia), and Leonardo Pisano,
Liber abbaci, ed. Boncompagni, 1857-1862, v. 1: pp. 381-384.
Even though my practice has been not to cite treatises written after the time of the author, I have included these as
an aid in understanding the rather complex method for finding cube roots.

[18]"Verbi . . . cubus" (IV, 12.24-26). The example given
is a correct one, as 23 cubed is 12167.

[19]"Si vero . . . probatur" (IV, 12.27-59). The rest of
this proposition is an extension of the method above to account for larger numbers. The text is somewhat corrupt; not
all of the cube of a three digit number would be accounted
for. If we write the root of our larger cube as $a + b + c$,
then for its cube we have $(a + b)^3 + 3a^2c + 6abc + 3b^2c + 3ac^2 + 3bc^2 + c^3$. The first term was taken care of by the
procedure in n. 17; the succeeding terms should be eliminated
in this part of the method. Most of them are, but after all
the subtractions indicated in the text are completed, there
still remains $4abc + bc^2$.

versus dextram sequentibus prius inventis per ordinem
continue. Et ante quadratum ponatur sua radix, et sub
quadrato duplum radicis, et iterum sub radice aliud
duplum. Vel sub duplo 2^i digiti posito sub eius quadrato
ponatur duplum primi digiti. Et sub illo duplo ponatur   35
primus digitus. Deinde invenias 3^m digitum antecedens
idem qui ductus in quadratum primi digiti subtrahatur a
superiori ipsius quadrati quantum poterit. Idem ductus
in primum digitum et productum multiplicatum per remotius
duplum primi digiti subtrahatur a superiori primi digiti. 40
Item, idem ductus in 2^m digitum, et productum cum
multiplicatis per propinquius duplum primi digiti
subtrahatur a superiori primi digiti. Item, ductus in
quadratum 2^i digiti subtrahatur a superiori quantum
poterit. Item, idem ductus in 2^m digitum, productum   45
multiplicatum per remotius duplum 2^i digiti subtrahatur
a superiori remotioris dupli 2^i digiti. Item, quadrato

36-37 antecedens idem: ante secundum duplum H
37 subtrahatur: subtrahat H
40 post superiori add. ipsius remotioris dupli H
41-42 cum multiplicatis: multiplicatum H
42 propinquius: remotius H / primi: secundi H
44 quadratum ?H / subtrahatur: subtrahat H / post superiori
   add. primi digiti H

be moved one place towards the right following in continual
order those found previously. And before the square let
its root be placed, and below the square the double of the
root, and again below the root another double. Or let the
double of the first digit be placed under the double of the
second digit placed under its square, and let the first
digit be placed under that double [of the first digit].
Then you may find the third digit before that same [second
double]. Let that [digit] multiplied by the square of the
first digit be subtracted from what is above its square as
much as possible. Let the same [digit], multiplied by the
first digit, and the product multiplied by the more distant
double of the first digit, be subtracted from what is above
the first digit. Then let the same [digit] multiplied by
the second digit, and the product when it has been multiplied by the nearer double of the first digit, be subtracted
from what is above the first digit. Then let [the same
digit] multiplied by the square of the second digit be subtracted from what is above as much as possible. Then let
the same [digit], multiplied by the second digit, the product multiplied by the more distant double of the second
digit, be subtracted from what is above the more distant
double of the second digit. Then when the square of the
third digit has been multiplied by the double of the first

3^i digiti ducto in duplum primi digiti quod est sub

quadrato 2^i digiti, productum subtrahatur a superiori

quadrati 2^i digiti.   Item, eodem quadrato ultimi digiti   50

ducto in primum digitum quod est sub quadrato 2^i digiti,

productum subtrahatur a superiori quadrato 2^i digiti.

Item, ultimo digito ducto in propinquius duplum 2^i digiti.

Productum multiplicatum per ultimum digitum, subtrahatur

a superiori 2^i digiti.   Item, ultimo digito ducto in suum   55

quadratum, productum subtrahatur a suo superiori.

Et secundum predictam regulam per digitos inventos

subduplos duplatorum, habebitur radix cubica.   Quod per

cubicationem eius probatur.

---

56 subtrahatur: protrahatur H

digit which is under the square of the second digit, let
the product be subtracted from what is above the square of
the second digit. Then when the same square of the last
digit has been multiplied by the first digit which is under
the square of the second digit, let the product be subtracted from what is above the square of the second digit.
Then when the last digit has been multiplied by the nearer
double of the second digit, let the product, multiplied by
the last digit, be subtracted from what is above the second
digit. Then when the last digit has been multiplied by its
square, let the product be subtracted from what is above it.

And following the aforesaid rule the cube root will be
had from the digits [which are] subdoubles of those doubled.
This is proved by cubing it.

[IV, 13]   RADICEM CUBICI IN MINUTIIS SECUNDUM DENOMINA-
TIONEM PARTIUM REPERIRE.

   Hec docet quod si volumus radicem numeri cubici per

minutias invenire, oportet sumere geometricas minutias

que super partem aliquotam sumuntur vere.   Quod probatur   5

quia radix inventa per cursum integri in suas partes

secundum primam propositionem huius 4^i libri et secundum

2^{am}, reddet illum cubicum.

---

4 sumere: sumi H
5 vere om.H
6 per cursum: percussim E / in: per H
7 propositionem: proportionem H / 4^i: tertii H /
   libri om.B / secundum om.H

IV, 13. TO FIND THE CUBE ROOT IN FRACTIONS IN COMPLIANCE
WITH THE DENOMINATION OF THE PARTS.[20]

This teaches that if we want to find the root of a cubic

number with fractions we should take geometric fractions

which are really taken over an aliquot part. This is proved

because the root found will give that cubic [number] once

again, when its integer is multiplied by its parts following

the first and second propositions of this fourth book.

---

[20]"Radicem . . . cubicum" (IV, 13.1-8). The author seems
to be saying that a fractional number which is a cube should
be reduced so that it is all written over the same denominator. For example, 3 and 3/8 is a cube; before extracting the
root, one should convert is into the form 27/8. This form
shows clearly its aliquot nature; 3 and 3/8 is the eighth
part of 27. Propositions IV, 1 and 2 deal with the conversion of a fraction into the whole of which the aliquot part
is part.

[IV, 14] SI IN CUBICUS HABUERIT MINUTIAS, RADICEM
MINUTIAS HABERE NECESSE EST.

Quod hec proponit, patet, eiusque conversam patet
esse veram. Sed quod ita sit in re, difficilimum est
probare. Nos ergo probemus illud. Queramus ergo per   5
theoreumata premissa cubicum numerum 5 et dimidii, scil.
5 pedum et semis. Currant ergo 5 pedes et semis super
5 et semis, excrescent 30 pedes et 4^a pedis, scil.
quadratum 5 et dimidii. Quod probatur per primam et
2^{am} huius ultimi libri, quia ex cursu 5 pedum super 5   · 10
pedes, excrescunt 25; ex cursu 5 pedum super dimidium,
excrescunt quinque dimidia; item ex cursu 5 super dimidium
excrescunt 5 alia dimidia. Ecce 10 dimidia ex quibus
fiunt 5 pedes, qui iuncti aliis faciunt 30. Ex cursu

---

1 habuerit minutias tr.H
2 minutias: minutiarum H
3 eiusque: quia eiusdem E
4 veram: umbram E
5 ergo (1): autem H / ergo (2) om.H
7 semis (1): dimidii E / currant . . . semis (2) om.E /
   post super add. pedes ?E   sedes H
8-9 scil. quadratum tr.H
9 et (1) om.E
10-11 super . . . pedum om.E
11 5 om.H
12 excrescunt: surgunt E / dimidia: dimidum ?E   dimidii H /
   item: iterum E   et H
13 alia dimidia: alii dimidum ?E   alii dimidii H / dimidia
   (2): dimidum ?E   dimidii H
14 fiunt: sunt B

IV, 14. IF THE CUBE HAD FRACTIONS, THE ROOT MUST HAVE
FRACTIONS.[21]

What this proposes is clear, and its converse is clearly
true. But that it is so in reality is most difficult to
prove. However, let us prove it. Let us seek, using the
theorems put before, the cube of 5½, that is of 5½ feet.
Therefore let 5½ feet be multiplied by 5½ feet, there will
arise 30¼ feet, that is, the square of 5½. This is proved
by the first and second propositions of this last book, because from the multiplication of 5 feet by 5 feet, 25 arises;
from the multiplication of 5 feet by a half, 5 halves arise;
then from the multiplication of 5 by a half arise another 5
halves. Here are 10 halves, from which are made 5 feet,
which, added to the others makes 30. From the multiplica-

---

[21]"Si in cubicus . . . inveniatur" (IV, 14.1-27). The
proof the author offers us is not particularly satisfying.
He shows us the converse first, using a single numerical
example to demonstrate that the cube of a number with
fractions will have fractions. But if irrational fractions
are included the converse is false. For example the cube
root of 2 is an irrational fraction; its cube, 2, has no
fractions. If we exclude irrationals, then the author's
claim is correct. The author's "proof" of the converse
amounts to a detailed explanation of all the steps in cubing
a whole number plus a fraction. The proposition would have
been better labeled as such. After the example showing the
converse, the author appeals to IV, 12 for the proof of the
theorem. I assume he would have us find the cube root of
166 plus 1/4 plus 1/8 as his demonstration.

dimidii super dimidium provenit 4 pars pedis. Hoc   15

probatur per primas huius ultimi libri. Quadratum ergo

5 et dimidii, scil. 30 et 4^a pedis, feriatur iterum in

5 et dimidium, et cubus generetur. Si 5 pedes currant

super 30 et surgent 150. Item dimidius pes super 30

pedes currat, et surgent 30 dimidii pedes, scil. 15 in-   20

tegri qui iuncti aliis faciunt 165. Item currant 5

pedes super 4^{am} pedis et surgent 5 4^e unde fiet unus

pes, et remanebit 4 pars pedis. Habes igitur 166 pedes

et 4^{am}. Item currat medietas pedis super 4^{am}, proveniet

8 pars pedis. Igitur cubus 5 pedum et dimidii est 166   25

et 4^a et 8^a pars unius pedis. Similiter reciproce per

antepenultimam prefati cubi dicta radix inveniatur.

---

15 dimidii: dimidum ?E / dimidium: dimidum ?E / hoc: hec H

16 primas om.H / libri om.H

17 dimidii: dimidum ?E / scil.: est E / 30: 20^a B /
   pedis om.H / in: scil. E

18 dimidium: dimidum ?E / si: sic H

19 30 (1): 20 B / surgent: surgunt H / item: iterum E et H /
   30 (2): 20 B

20 surgent: surgunt H / 30 dimidii: 30 dimidum ?E
   dimidii 3 H

20-21 integri: integra B

21 faciunt: facient E / item: iterum E et H

22 pedis: partis B / surgent: super H

23 remanebit: remanebis E / habes: habet E

24 4^{am} (1): 4^{am} pedis E / item: iterum E et H / pedis:
   pedum H

25 5 pedum om.H / dimidii: dimidum E

26 pars om.H / pedis om.H

27 antepenultimam: ?E

tion of a half by a half is produced the fourth part of a
foot. This is proved by the first propositions of this last
book. Therefore let the square of 5½, namely 30¼, be multiplied again by 5½, so that the cube is generated. If 30 is
multiplied by 5 feet, then 150 feet will be produced. Then
let 30 be multiplied by a half foot, and 30 half feet will
be produced, that is 15 whole feet, which added to the
others make 165. Then let a quarter foot be multiplied by
5 feet and there will be produced 5 fourths, from which will
be made one foot, and a fourth part of a foot will remain.
Therefore you have 166¼ feet. Then take a half foot times
a quarter, and an eighth part of a foot will be produced.
Thus the cube of 5½ feet is 166 and a quarter and an eighth
part of one foot. Similarly and conversely by the proposition before the last, the said root of the aforementioned
cube may be found.

Et hoc valet ad 5 propositionem crassimetrie, scil.
ad probandam proportionem quantitatis solis ad quantitatem
terre, quia dyameter solis est 5 et dimidium respectu   30
dyametri terre, ut primo ostensum est ibi. Et ideo
necessario sol 166, 4 et 8 parte est maior terra. Similiter solers minutimetra probabit terram esse maiorem luna
39 et 4^a et 20^a, quia dyameter terre respectu dyametri
lune est trium partium et duarum quintarum. Similiter   35
proportio solis ad lunam probatur per cubicationem
dyametri solis respectu dyametri lune, scil. 18 et 4
quintarum.

Et ita quivis studens et inveniendo deficiens scias
quantum laboris in huius rei examinatione et novi libri   40
compositione impenderimus.

Sequitur de radice quadratorum et primo de integris.

---

28 hoc: hec H / crassimetrie, scil.: arismetrie H
29 proportionem quantitatis: quantitatem E / ad: et E
30 est . . . dimidium om.H / dimidium: dimidum ?E
31 post terre add. est 5 et dimidii H / primo: ibi H /
    ostensum: ille preostensum E
32 post necessario add. et ideo H
33 solers: solis ?H / esse om.H
34 20^a: 50^a EB
36 probatur: probabitur H
37 dyametri . . . respectu om.E
39-41 et . . . impenderimus om.E
39 ita: in hiis H / studens: studes H / scias: sciat H
41 impenderimus: impedimus B
42 sequitur . . . integris om.H

And this is useful for the fifth proposition of the book
of crassimetry,[22] namely for showing the proportion of the
size of the sun to the size of the earth, because the diameter of the sun with respect to the diameter of the earth
is 5½, as was shown there first. And therefore the sun is
necessarily 166 and a fourth and an eighth part larger than
the earth. Similarly the clever fractional measurer will
show the earth to be greater than the moon 39 and a fourth
and a twentieth times, because the diameter of the earth
with respect to the diameter of the moon is 3 2/5 parts.
Similarly the ratio of the sun to the moon is proved by cubing the diameter of the sun with respect to the diameter of
the moon, namely 18 4/5.

And thus anyone studying and failing to find [these
things] may know how much work we have expended on the exam-
ination of this thing and in the composition of a new book.

There follows material about the root of squares and
first about integers.

---

[22]"Et hoc valet . . . impenderimus" (IV, 14.28-41). The
numerical example in IV, 14.1-27 is drawn from III, 5, where
the proportions of the sizes of the sun, moon and earth were
found by cubing the ratios of their diameters. The author's
plea for approbation may indicate that cube roots and cubes
were not normally discussed in his time.

[IV, 15]   EX DUCTU NUMERI IN ALIUM EQUALEM SIBI,
QUADRATUM GENERARI.

Quadratus numerus est qui sit ex ductu sue radicis
in se. Equivalet autem radicem metiri se et metiri
alium equalem sibi, unde quadratum generatur.   5

---

1 <u>post</u> alium <u>add.</u> sibi <u>H</u>
3 numerus est <u>tr.</u>E / sue <u>om.</u>H
4 equivalet <u>bis</u> <u>E</u> / radicem: radicis <u>E</u>
5 unde <u>om.</u><u>H</u>

IV, 15.   TO GET A SQUARE FROM THE MULTIPLICATION OF A
NUMBER BY ANOTHER EQUAL TO IT.[23]

A square number is that which is made by the multiplication of its root by itself. However it is equivalent for a
root to be multiplied by itself and to be multiplied by
another equal to it; whence a square is begotten.

---

[23]"Ex ductu . . . generatur" (IV, 15.1-5). This proposition is nothing more than the definition of a square. It
tells us that $a \cdot a = a^2$, where a is any number. Cf. J. of
Seville, ed. Boncompagni, 1857: 72.27-73.2.

[IV, 16] EX DUCTU QUADRATI IN QUADRATUM, QUADRATUM
PROCREARI.

Hec patet quia si numerus habens radicem metiatur
habentem, productum habebit radicem. Ut si 4^{rius}
mensuret 9, surgunt 36, quadratum. Quod probatur quia   5
radix in se ducta eandem summam reddit, scil. 6 ductus
in se.

1-7 In codice H haec propositio illam IV, 15 antecessit.
3 hec: hoc E / quia: quod H
4 post habentem add. radicem ?E / habebit: ?E
5 mensuret om.H
6 summam reddit tr.H
6-7 scil. . . . se om.H

IV, 16. TO GET A SQUARE FROM THE MULTIPLICATION OF A SQUARE BY A SQUARE.[24]

This is clear for if a number having a root is multiplied by a [number] having [a root], the product will have a root. So if 9 is multiplied by 4, 36, which is a square, arises. This is proved because the root taken by itself gives back the same product, namely 6 times itself.

---

[24]"Ex ductu . . . ductus in se" (IV, 16.1-7). We may restate this proposition by means of the formula $a^2 \cdot b^2 = (ab)^2$, where a and b are any numbers. Cf. J. of Seville, ed. Boncompagni, 1857: 73.3-7, which gives the same example, and Adelard?, Algorism, ed. Curtze, 1898: 25.8-9.

[IV, 17]   EX DUCTU QUADRATI IN NON QUADRATUM, NON
QUADRATUM SURGERE.

Hec patet per precedentia quia si ternarius in 9

ducatur non surget quadratum et sic in reliquis.

---

1 ductu: ducti E

1-2 non quadratum (2): numquam E

2 post surgere add. quadratum E

3 precedentia: precedentem H / quia si ternarius: quia,
   scil. 3us E ut ex ductu ternarii H

4 ducatur om.H / surget: procedit H / quadratum: quadratus H

IV, 17.   TO PRODUCE A NON-SQUARE FROM THE MULTIPLICATION
OF A SQUARE BY A NON-SQUARE.[25]

This is shown by the preceding for if 3 is multiplied

by 9, a square does not result, and so for the others.

---

[25]"Ex ductu . . . in reliquis" (IV, 17.1-4). This
proposition tells us that if a is not a square and b^2 is
any square, then ab^2 is not a square.

[IV, 18]   EX DUCTU RADICIS IN SUUM QUADRATUM, CUBICUM
GENERARI.

Hec patet per precedentia quia si ternarius ducatur

in novenarium procedit cubus, scil. 27.   Et sic in

reliquis.   5

1-5 ex ductu . . . reliquis <u>om.</u>H
3 precedentia: precedenti <u>E</u>

IV, 18.   TO GENERATE A CUBE FROM THE MULTIPLICATION OF A
ROOT BY ITS SQUARE.[26]

This is shown by the preceding for if 3 is multiplied
by 9 a cube results, namely 27, and so for the others.

---

[26]"Ex ductu . . . in reliquis" (IV, 18.1-5).   This is
virtually a definition of a cube as $a \cdot a^2 = a^3$.

[IV, 19] SI EX DIVISIONE NUMERI PER NUMERUM PROVENIAT
QUADRATUM, EX DUCTU UNIUS IN RELIQUUM SURGIT QUADRATUM.

   Hec patet et probatur per ductum radicis in se,

quadratum reddentis, ut si 36 dividatur per novenarium

provenit 4, qui est 4^{tus}. Ergo si 36 metiatur novenarium   5

provenit quadratum, scil. 324. Quod probatur, quia 18

ductus in se, eundem numerum reddit.

---

1 per: in H
2 surgit: surget E / quadratum: quadratus H
3 et . . . ductum: ex ductu H
4 si om.H
5 ergo: ?E grossi H
6 provenit: proveniet H / scil. om.H / quod probatur om.B
7 ductus: ductum E / numerum om.H

IV, 19.   IF FROM THE DIVISION OF A NUMBER BY A NUMBER A
SQUARE IS PRODUCED, FROM THE MULTIPLICATION OF ONE BY THE
OTHER A SQUARE ARISES.[27]

This is clear and is shown by restoring the square from

the product of the root by itself. Thus if 36 be divided

by 9, 4 is produced, which is a square. Therefore if 36 be

multiplied by 9, a square, namely 324, is produced. This

is proved because 18 taken by itself gives back the same

number.

---

[27]"Si ex divisione . . . reddit" (IV, 19.1-7). The
author gives the interesting result that if $a/b = c^2$, then
$ab = (a/b) \cdot b^2 = b^2 c^2 = (bc)^2$. Cf. J. of Seville, ed. Boncompagni, 1857: 73.7-15 and Adelard?, ed. Curtze, 1898:
25.9-13. His justification relies on a particular numerical
example.

[IV, 20]   QUADRATI RADICEM INVENIRE.

Si numerus paribus scribatur differentiis, sub
penultima scribendum est.   Si inparibus, sub ultima,
talis inquantum numerus scribendus est qui ductus in se
destruat superiorem vel superiores vel maiorem partem.   5
Deinde duplatus secundetur.   Et ante illum scribatur
numerus qui ductus in duplatum deleat quicquid est supra
vel maiorem partem, et inde in se, deleat quod supra
se est vel maiorem partem.   Si talis non inveniatur
ponatur ibi cifra.   10

Deinde si plures sint figure supra, inventus numerus
dupletur et secundetur.   Et primus duplatus succedat
loco eius.   Et scribatur alius numerus ante secundum
duplum, qui ductus in duplum deleat secundum quod est
supra vel maiorem partem.   Et inde in 2^{m} duplum,   15

1-24 quadrati . . . primus om.E
2 sub: si B
5 destruat: destruit H
14 deleat om.H
14-15 quod . . . duplum: et primum H

IV, 20. TO FIND THE ROOT OF A SQUARE.[28]

If the number is written in an even number [of digits], under the next to last [digit] a number is to be written such that [when] multiplied by itself, it eliminates what is above or most of it. If [the number is written] in an odd [number of digits, write the number] under the last [digit]. Then let the [first digit of the root] doubled be moved to the second position. And before that [i.e., to its right] let there be written the number which multiplied by the double annuls whatever is above it or the greater part, and then [with that number multiplied] by itself [added to the product just found], let it annul what is above it or the greater part. If such a number cannot be found, let a zero be written there.

Then if there are several figures above, let the number found be doubled and moved to the second place. And let the first doubled number succeed to its place. And let another number be written before the second double [i.e., to its right], which multiplied by the double annuls the second which is above or the greater part. And then proceed similarly, [taking its product] with double the second number,

_____

[28]"Quadrati radicem . . . primus" (IV, 20.1-24). This technique for extracting a square root is based on the relation $N = (a+b)^2 + r$, where r is the remainder. If N has a two-digit root, then b is the units and a is the tens, e.g., for the number 84, a is 80 and b is 4. One may simply extend the formula by replacing b by $(b_1 + c)$ to deal with a number having a three digit root, $a + b_1 + c$, and so on for any number of digits. To demonstrate the author's method, let us extract the root of 7200, the example below in IV, 26. The left-hand column gives the line number from IV, 20 of the various steps shown in the other two columns. In the center

similiter faciat, et idem in se similiter agat. Si

non potest poni, ponatur ibi cifra.

  Hac arte agendum est donec perventum sic ad

ultimam. Si aliquid remanserit, non est quadratus

numerus sed surdus; si nichil, quadratus est. Quod   20

probatur, quia si <de>duplentur duplate, habebitur

eius radix, que ducta in se, quadratum reddit. Si quid

remanserit, iungatur producto ex radice, et habebitur

primus.

---

18 perventum sic: perveniatur <u>H</u>
19 remanserit: recensetur <u>B</u>
21 <de>duplentur: duplentur <u>BH</u>

and then [with that number multiplied by itself added to

it] proceed similarly [i.e. annul...]. If [such a number]

cannot be put [there], let a zero be put there.

One should proceed by this art until the end [is] thus

reached. If anything remains, [the number] is not a square

but a surd; if nothing, it is a square. This is proved

because if the doubled [number] be undoubled, its root will

be had, which, when it has been multiplied by itself, re-

stores the square. If anything remains, let it be added to

the product from the root, and the first [i.e., original

number] will be had.

---

column, I have expressed the steps of the process in terms
of the relation $N = (a+b)^2+r = a^2+2ab+b^2+r$. The right-hand
column gives successive configurations of the process.
These configurations suggest that the calculation was carried out on a slate, where parts of the configuration were
successively removed and replaced by new values.

line	explanation	configurations
2-3	N=7200	7200
4	a=80	8
5	$n-a^2$=800	
6-8	2ab=640, a trial product	800
6	2a+b=164, b=4	164
8-10	$N-a^2-2ab-b^2$=r=144	144
11-13	2a+2b=168	168
21	a+b=84	

Similar techniques for finding square roots are also
given in J. of Seville, ed. Boncompagni, 1857: 75.10-79.5
and Adelard?, ed. Curtze, 1898: 25.20-26.9.

[IV, 21] LIMITES INPARIBUS FIGURIS SCRIPTOS, RADICEM
HABERE; PARIBUS EA CARERE.

Hec probatur per 5^{am} antecedentem. Secundum eam,

enim unitas, scil. primus limes, 1 prima differentia

scriptus, est quadratus numerus, quia surgit ex ductu   5

unitatis in sibi equalem. Similiter 3us limes et 5us

quod per eandem probatur. 2us limes et 4us, paribus

scripti figuris, radice carent, quia ex nullo ductu in

se fiunt.

---

3 probatur: probantur E tr.post antecedentem H
4 enim unitas om.H / 1 prima: una figura sive H
6 sibi equalem: se H
7 quod om.H / 2us limes: duobus lime E / post 4us add. et H
8 scripti: scriptus H / ante figuris add. differentiis, i.e.
   H / radice carent tr.EH / ductu: ducto E

IV, 21. DECIMAL POSITIONS WRITTEN AS AN ODD NUMBER OF
FIGURES HAVE A ROOT; AS EVEN, THEY LACK IT.[29]

This is proved by the fifth proposition preceding. For
according to it, unity, namely the first decimal position,
or 1 written in the first place, is a square because it
arises from the multiplication of one by itself. Similarly
the third position and the fifth [are squares], which is
proved by the same proposition. The second position and
the fourth, written as an even number of figures, lack a
root because they are not produced from anything multiplied
by itself.

---

[29]"Limites . . . fiunt" (IV, 21.1-9). Even though <u>limes</u>
generally means position in decimal notation, here it seems
to be taken as a one in that position. Thus the proposition
says that 1, 100, 10000 and so on are squares, but 10, 1000
and so on are not. The fifth proposition preceding would
be IV, 16; it is not relevant, but IV, 15, the definition
of a square, would be. Cf. J. of Seville, ed. Boncompagni,
1857: 73.15-74.6, which is quite similar to our proposition
and Adelard?, ed. Curtze, 1898: 25.13-15.

[IV, 22] IMPARES ASTRONOMICAS MINUTIAS RADICE CARERE.
PARES HABERE NECESSE EST.

Contrarietas est in integris et minutiis, impares
enim limites in integris habent radicem, ut dictum est;
pares carent. In minutiis vero impares carent radice, 5
scil. prime, 3^e et 5^e. Minuta, 3^a et 5^a et impariter
sumpta, quia ex nullo numero ducto in se fiunt. Pares
vero minutie habent, scil. 2^a, 4^a etc., quia ex ductu
alterius in se fiunt.

---

3 minutiis: in minutiis H
4 in integris om.H / ut dictum est: ut dictum E   om.H
5 carent: quarens E / post carent add. ut probat precedens
   H / impares: partes H
6 scil. om.H / post minuta add. est E   scil. H / impariter:
   imparatur ?E
7 ducto tr.post se H
8 scil. om.H / post 2^a add. et E

IV, 22. ASTRONOMICAL FRACTIONS OF ODD ORDER LACK A ROOT;
THOSE OF EVEN ORDER NECESSARILY HAVE ONE.[30]

The case is reversed for integers and for fractions, for
the uneven position for integers have a root, as has been
said; the even lack a root. In [sexagesimal] fractions the
uneven [orders] lack a root namely the first, third and
fifth. Minutes, thirds and fifths and those unevenly taken
[lack a root] because they are not produced from a number
multiplied by itself. The even [orders], namely seconds,
fourths, etc., however, have [roots] because they come from
the multiplication of another by itself.

---

[30]"Impares . . . fiunt" (IV, 22.1-9). It seems that we
should understand this to mean that $1/60^n$ is a square if and
only if n is even. Proposition IV, 4 could be used to show
that 1 second, $1/60^2$, and 1 fourth, $1/60^4$, are squares. Cf.
J. of Seville, ed. Boncompagni, 1857: 74.7-21 and Adelard?,
ed. Curtze, 1898: 25.15-19.

[IV, 23] IMPARES MINUTIAS PARIUM RADICEM ESSE, ET
ETIAM PARES PARIUM.

Hec probatur per 5^{am} huius ultimi libri, que
argumentatur quia ex ductu minutorum in se, veniunt 2^a,
ergo sit radix eorum. Ex ductu secundorum in se, 4^a,   5
ex ductu 3^{orum} in se, 6^a.

---

3 ultimi libri: ultimi B libri ultimi H / que: quod B qui H
4 quia: quod EH
5 sit: sunt E / secundorum: binorum E
6 ex ductu om.H

IV, 23. FRACTIONS OF ODD ORDER MAY BE THE ROOT OF ONE OF
EVEN ORDER, AND ALSO EVEN OF EVEN.[31]

This is proved by the fifth proposition of this last
book, which is adduced as proof since from multiplying
minutes by themselves, seconds are produced, therefore
[minutes are] their root. From multiplying seconds by
themselves, fourths [are produced]; from multiplying thirds
by themselves, sixths.

---

[31]"Impares . . . 3^{orum} in se, 6^{a}" (IV, 23.1-6). If we
express a sexagesimal fraction having a root as $1/60^{2n}$,
then this proposition tells us that n may be even or odd.
It is IV, 4, not IV, 5 which may be used to prove this.

[IV, 24]   GEOMETRICAS MINUTIAS DENOMINATAS A QUADRATO
NUMERO TANTUM RADICEM HABERE.

Hec probatur per 2^{am} et 3^{am} huius ultimi libri.

4 enim pars integri et 9, quia denominatur a quadratis,

radices habent minutias, scil. denominatas a radicibus   5

quadratorum, scil. medietatem que ducta in se efficit

4^{am}, et 3^{am} ex qua surgit 9^{a}. Medietas et 3^{a} et huius-

modi secundum regulam carent ea.

---

1 geometricas: diometrias H
3 post et add. per E
4 enim om.E / integri: integra EH / denominatur ?B
5 post habent add. scil. E / minutias scil. om.E /
   scil.: secundum H
6 efficit om.H
7 et (3) om.EH
8 carent om.H

IV, 24. GEOMETRIC FRACTIONS DENOMINATED BY A SQUARE NUMBER
HAVE A ROOT CORRESPONDING [TO THE SQUARE NUMBER].[32]

This is proved by the second and third propositions of
this last book, for because the fourth and ninth part of
an integer are denominated by squares, [their] roots have
fractions. That is, [the roots are] denominated by roots
of squares, that is, a half, which when multiplied by itself produces a fourth, and a third, from which arises a
ninth. A half and a third and those like them according
to the rule lack a root.

---

[32]"Geometricas . . . carent ea" (IV, 24.1-8). This prop-
osition may be restated as $(n/a^2)^{\frac{1}{2}} = n^{\frac{1}{2}}/a$. The rules for
multiplying common, or geometric, fractions were given in
IV, 2 and IV, 3; they may be used in the proof of this
theorem. Cf. J. of Seville, ed. Boncompagni, 1857: 74.21-
75.4.

[IV, 25]   NUMERUM SURDUM IN GENUS MINUTIARUM RADICEM
HABENS RESOLVERE.

   Hec docet quod si numeri carentis radice volumus
radicem per minutias habere. Vel si cum integro sint
minutie, totum resolvamus in minutias habentes radicem,   5
scil. in 2^a vel 4^a vel 6^a et huiusmodi. Secundum
maiorem resolutionem, maior erit certificatio.

---

4 radicem: eam H
6 et huiusmodi om.E
7 maiorem: minorem B / post certificatio add. predictorum B

IV, 25. TO RESOLVE A SURD NUMBER INTO ONE OF THE KINDS OF
FRACTIONS HAVING A ROOT.[33]

This teaches how, if we want the root of a number lacking it, to have it [approximately] in minutes. For instance
if there are fractions with an integer, let us resolve the
whole into fractions that do have a root, namely into seconds or fourths or sixths and those of that sort. The further the resolution is carried, the more precise will be
[the root].

---

[33]"Numerum surdum . . . certificatio" (IV, 25.1-7). The
steps given here are preliminary to extracting the root of
a fraction or of a mixed number; they amount to putting the
fraction and whole number over a common denomination. It
is not necessarily the least common denominator since the
least common denominator need not have a root. In other
words, the number is resolved into a form $a/60^{2n}$, which has
a root $a^{\frac{1}{2}}/60^n$. The greater is $\underline{n}$, the more digits there are
in $\underline{a}$, and hence in $a^{\frac{1}{2}}$. Cf. J. of Seville, ed. Boncompagni,
1857: 81.15-82.8 and 83.3-5 and Adelard?, ed. Curtze, 1898:
26.10-12.

[IV, 26] MINUTIARUM RADICEM PERQUIRERE.

Facta resolutione integri in minutias radicem
habentes per precedentem. Notatis differentiis paribus
vel imparibus, queratur radix earum secundum datam
regulam de extractione radicis integrorum per unam   5
precedentium. Deinde duplate deduplentur secundum
regulam integrorum, et habebitur radix.

Et si facta fuerit resolutio in 2^a, tantum minuta
erunt radix inventa. Si in 4^a, tantum 2^a erunt radix,
que reducenda sunt in minuta, minuta vero in integra.   10
Et integra cum minutiis erunt radix numeri, quia ducta
in se reddent cum adiuncto eo quod remanserat.

Ut si duorum integrorum velis radicem, verte ea in

---

2 post integri add. in radicem H
4 datam om.E
5 integrorum: integra H
5-6 per unam precedentium om.E
6 deduplentur: duplentur B decuplentur H / post secundum
   add. datam H
7 integrorum om.H / habebitur radix tr.H
8 et om.E / facta fuerit ?B / resolutio: resolutionem B /
   2^a: bina E
9 erunt: erit B / 2^a erunt radix: bina erunt radix E
   radix erit secunda H
10 post que add. sunt H / reducenda: reducende E / sunt om.H /
   post minuta (1) add. et E / vero om.EH
11 et om.H / quia: que B
12 reddent: reddunt H / cum adiuncto: numerum addito H
13 radicem om.H

IV, 26. TO FIND THE ROOT OF FRACTIONS.[34]

Perform the resolution of the integer into fractions having a root by the preceding proposition. Having noted the
even or uneven positions, then let the root be searched for
following the given rule for the extraction of a root of
integers by one of the preceding propositions. Then let the
doubled [number] be undoubled following the given rule and
the root will be had.

And if the resolution was into seconds, the root will be
so many minutes. If into fourths, the root will be so many
seconds, which are to be reduced to minutes, minutes to integers. And the root of the number will be integers with
minutes, which multiplied by themselves restore the number
when what remains has been added to it.

So that if you want the root of 2 integers,[35] turn it into

---

[34]"Minutiarum . . . remanserat" (IV, 26.1-12). This
tells us that once the number of which we want the root has
been put into the form $a/60^{2n}$, we may treat $\underline{a}$ as if it were
an integer. When the root, $a^{\frac{1}{2}}$, has been found, it is considered as $a^{\frac{1}{2}}/60^{n}$, which is then reduced so there are not
more than 60 of any of the orders of sexagesimal fractions.
Cf. J. of Seville, ed. Boncompagni, 1857: 82.9-14 and 82.30-
83.1 and Adelard?, ed. Curtze, 1898: 26.10-15.

[35]"Ut si duorum . . . in se" (IV, 26.13-19). In the
example I used in n. 28 to explain IV, 20, I extracted the
square root of 7200. The author has converted 2 into 7200
seconds and found the root to be 84 minutes, or 84/60 in our
notation. This is equivalent to 1 and 24/60 or 1 and 24 minutes. The same example is used in J. of Seville, ed. Boncompagni, 1857: 83.7-13 and Adelard?, ed. Curtze, 1898: 26.16-
20.

minuta et minuta in 2^a, et erunt 7200 secunda. Sume

radicem illius numeri secundum regulam integrorum. Et   15

facta deduplatione remanent 84 que sunt minuta. Ergo

unum integrum, factum scil. ex 60 minutis, et 24 minuta,

neglecto residuo fit radix duorum. Quod probatur ex

ductu radicis in se.

---

14 secunda <u>om.</u>B / illius ?<u>E</u>
15 integrorum: numerorum <u>H</u>
16 deduplatione: reduplicatione <u>B</u> duplatione <u>H</u>
17 <u>post</u> integrum <u>add.</u> unum <u>H</u>
18 fit: sunt <u>E</u> / ex <u>om.</u>H

minutes and the minutes into seconds, and there will be
7200 seconds. Take the root of that number following the
rule for integers. And when the undoubling has been made,
there remain 84, which are minutes. Therefore one integer,
made from 60 minutes, and 24 minutes, the remainder having
been neglected, is the root of two. This is proved by the
multiplication of the root by itself

[IV, 27]   RADICEM MINUTIARUM EXPRESSIUS PER CIRCULOS
INVENIRE.

   Hec subtilius perscrutari docet quam precedens

propositio, docet enim quod facta resolutione integri

in minutias per precedentem propositionem illis minutiis   5

habentibus radicem, preponantur quotvis circuli, i.e.

cifre pares tantum modo. Et quanto plures ponuntur,

tanto erit subtilius opus. Totalisque numeri radix

queratur. Dedupleturque numerus. Deinde numeretur ab

initio figurarum medietas circulorum positorum. Quod   10

super erit, sumatur, quia illud est radix minutiarum

sine cifrarum appositione. Removeaturque a radice, et

servetur residuum radicis. Feriatur in 60, et a producto

---

4 propositio om.**H**
6 **ante** radicem **add.** scil. **H** / preponantur: proponantur **E**
7 pares: paros **E** / tantum modo: termino ?**H** / ponuntur:
   ponantur **H**
8 erit **tr.post** subtilius **H** / totalisque: talisque **B**
8-9 radix queratur **tr.E**
9 queratur: queretur **B** / dedupleturque: dupliceturque **B**
   decupleturque **H**
10 positorum: prepositorum **E**
11 est: erit **H**
12 sine: sive ?**E** / appositione: opositione **E**
13 radicis: radix **H** / producto: predicto **B**

IV, 27.   TO FIND THE ROOT OF FRACTIONS MORE CLEARLY BY USING CIRCLES.[36]

This teaches how to work more accurately than the preceding proposition, for it teaches that when the resolution
of integers into fractions has been made by the preceding
proposition, for those fractions having a root, then add on
as many circles, or zeros, as you want, but only in pairs.
The more are added on, the more accurate will be the work.
And then let the root of the whole number be obtained, then
let the number be undoubled.   Then let half [the number] of
the placed circles be counted off from the beginning of the
figures.   Let what is left over be taken, because that is
the root of the fractions without the affixing of zeros.
Let it be taken from the root, and let the rest of the root
be put aside.   Let the root be multiplied by 60, and let
half [the number] of circles be counted off from the product.

---

[36]"Radicem . . . erunt 3^a" (IV, 27.1-19).   The method
here is a curious combination of decimal and sexagesimal
methods.   The author suggests adding pairs of zeros, which
he also calls by the quaint and descriptive name of circles,
to the right of the number of which we want the root.   His
technique of counting off from the result half the number
of originally added zeros is equivalent to placing the decimal point in the correct position.   We are then to convert
the part to the right of the decimal point to sexagesimal
fractions by multiplying by 60.   Successive pointing off
and multiplying will convert the number into smaller and
smaller sexagesimal fractions.   The placement of the products indicates that the author writes fractions from top
to bottom, the sexagesimal having the smallest denominator
written first.   Cf. J. of Seville, ed. Boncompagni, 1857:
86.14-87.29 and Adelard?, ed. Curtze, 1898: 26.33-27.19,
both of which add zeros after integers, as in IV, 28, rather
than after a number of sexagesimal fractions.

numeretur medietas circulorum. Sumaturque residuum,

et separatum subscribatur sub eo quod extraximus sub   15

minutiis, quia hec erunt 2^a si facta fuerit resolutio

in 2^a tantum. Si quid remanet de radice illis subtrac-

tis, feriatur in 60, et a producto numeretur medietas

circulorum. Et residuum sumatur et erunt 3^a.

    Verbi gratia, faciemus in duobus. Si duorum vis   20

radicem resolvere in 2^a habes 7200. Illis 2^{is} prepone

4 cifras. Si pares plures preponeres subtilius ageres.

Cuius numeri sume radicem, que deduplata erunt 8480. Ab

illa radice numera medietatem circulorum, scil. 2

differentias. Sume residuum ab illo et dele, et erunt   25

84 minuta, scil. unum integrum et 24 minuta. Residuum,

---

15 et separatum: esse parium $\underline{E}$ / sub (1): cum $\underline{E}$ /
    extraximus: extractimus $\underline{E}$
16 hec erunt: hic erit $\underline{B}$ / fuerit: fuit $\underline{E}$
17 quid: quidem $\underline{E}$ / remanet: remaneat $\underline{B}$
19 et (1) $\underline{om.H}$ / residuum sumatur $\underline{tr.H}$ / erunt: erit $\underline{B}$
20 gratia $?\underline{E}$ / faciemus $\underline{om.EH}$ / duorum: duorum vel 2^{orum} $\underline{B}$
    $\underline{om.E}$
20-21 vis radicem $\underline{tr.H}$
21 resolvere: resolve $\underline{B}$ / $\underline{post}$ 2^a $\underline{add.}$ et $\underline{B}$ / habes: habens
    $\underline{E}$ / $\underline{post}$ prepone $\underline{add.}$ ex $\underline{E}$
22 $\underline{ante}$ si $\underline{add.}$ et $\underline{H}$ / pares $\underline{om.H}$ / preponeres: adderes $\underline{H}$
23 cuius: eius $\underline{E}$ totius $\underline{H}$ / $\underline{post}$ numeri $\underline{add.}$ totius $\underline{B}$ /
    deduplata: decuplata $\underline{B}$ duplata $\underline{H}$ / erunt: erit $\underline{H}$ /
    8480: 8410 $\underline{B}$ 840 $\underline{H}$
24 numera: unam $\underline{H}$
25 $\underline{ante}$ ab $\underline{add.}$ et $\underline{EH}$ / et: loco $\underline{EH}$ / dele $\underline{om.E}$
26 $\underline{post}$ residuum $\underline{add.}$ scil. $\underline{H}$

And let the rest be taken, and let what was separated be written below that which we have removed, [that is,] below the minutes because these will be seconds if the resolution has been made as far as into seconds. If anything remains of the root, when that has been subtracted, let it be multiplied by 60, and let half of the circles be numbered from the product. And let the rest be taken and they will be thirds.

For example, let us proceed for two.[37] If you want the root of two [you should] resolve it into seconds, and you will have 7200. Put 4 zeros before those seconds [i.e. to their right]. And if you would put more pairs before, you would proceed more accurately. Take the root of that number, which undoubled will be 8480. From that root, number [off] half of the circles, namely 2 places. Take the remainder from it and delete [it], and there will be 84 minutes, that is one integer and 24 minutes. Take the rest,

---

[37]"Verbi gratia . . . reddunt" (IV, 27.20-31). The example takes the root of what we would write as 7200.0000 seconds. By my calculation, the root is 84.85 minutes; the author finds 84.80. After changing 84 minutes into 1 and 24 minutes, he multiplies the remaining .80 minutes by 60 to convert it into 48 seconds.

80, percute in 60, et numera iterum duas differentias.
Et sume residuum, scil. 48 2^a, et subscribe aliis. Et
nichil remanet. Unum ergo integrum, 24 minuta, 48 2^a
sunt radix 2^{orum} integrorum, quia in se ducta fere illa   30
reddunt.

---

27 80: 84 $\underline{E}$ / numera: innumera $\underline{H}$ / iterum duas: 3^m et $\underline{B}$
29 ergo $\underline{om.}H$ / $\underline{post}$ minuta $\underline{add.}$ scil. $\underline{E}$

namely 80, times 60, and number [off] once again the two
places.   And take the rest, namely 48 seconds, and write
it below the others.   And nothing remains.   Therefore one
integer, 24 minutes, 48 seconds are the root of two integers, because multiplied by themselves they nearly return
those [two].

[IV, 28]   RADICEM MINUTIARUM SINE MINUTIIS PERSCRUTARI.

Hec subtilissime docet operari, et vicinissima

veritati est que sine resolutione integri in minutias

docet radicem integri surdi per minutias reperire,

sicut numeri. Surdi integri cuius radicem volumus   5

minutiatim scribamus. Et ante eum ponamus circulos

quotlibet pares dumtaxat. Et quanto plures ponemus,

tanto subtilius agemus. Numerique radicem queramus,

que deduplata scribatur. Numeremusque medietatem circu-

lorum ab uno. Residuum separemus. Et erunt minuta que   10

resolvantur in integra. Et seorsum scribantur. Residuum

radicis feriamus in 60, et a producto numeretur medietas

circulorum. Residuum seorsum scribatur sub minutis,

---

3 veritati est tr.H
4 integri surdi tr.H
5 sicut: sunt E sic H / post surdi add. integri H /
    radicem: radices E
6 minutiatim: per mutatim ?B minutatim ?E / ante eum
    tr.post ponamus H / eum: eam B
6-7 circulos quotlibet: quotvis circulos H
7 pares dumtaxat: pares dumtaxas ?E tr.H
8 tanto om.H
9 que om.H / deduplata: duplata E
10 uno: imo ?H / minuta: iiii ?H
11 resolvantur: solvantur E / scribantur: scribatur B
12 in om.E / producto: predicto B
13 seorsum om.B tr.post minutis H
13-16 scribatur . . . residuum om.B

IV, 28. TO FIND A FRACTIONAL ROOT WITHOUT [USING]
FRACTIONS.[38]

   This teaches how to work most accurately, and it is closest to the truth; it teaches how to find the root in fractions of a surd integer just as [found] for numbers. Let
us write the integers of the surd of which we want the root
in fractions. And let us put before it [i.e., to its right]
as many circles, in pairs only, as wanted. And the more we
put, the more accurately do we work. And then let us obtain
the root of the number, which [root] should be written undoubled. Then let us count off from the end, half [the
number] of circles [originally added]. Let us separate
the rest. And they will be minutes which may be resolved
into integers. And let them be written separately. Let us
multiply the rest of the root by 60, and from the product
[that same] half of the circles will be counted off. Let
the rest be written separately under the minutes, and they
will be the seconds. Also let the rest of the root be multiplied by 60, and from the product let [the same] half of
the circles be counted off. Let the rest be written beneath

---

   [38]"Radicem . . . et 6^{a}" (IV, 28.1-17). This proposition
is a more consistent treatment of converting decimal fractions to sexagesimals. The technique of the previous proposition is applied without a prior conversion to sexagesimal
fractions. Again, the multiplication of the fraction to the
right of the decimal point by 60 converts it into a sexagesimal plus a decimal fraction of that sexagesimal. This
remaining decimal part is converted into a smaller sexagesimal plus a decimal fraction of that smaller sexagesimal.
The process may be continued as long as desired. Cf. J. of
Seville, ed. Boncompagni, 1857: 86.14-87.29 and Adelard?,
ed. Curtze, 1898: 26.33-27.19.

et erunt secunda. Item, residuum radicis feriatur in

60, et a producto numeretur medietas curculorum.   15

Residuum scribatur sub 2^{is}, et erunt 3^a. Si quid super-

est, similiter negociandum est usque ad 4^a et 5^a et 6^a.

  Verbi gratia, si duorum radicem invenire velimus,

ante ipsam scribamus 6 circulos sic 2000000. Illiusque

numeri sumatur radix que deduplata erit 1414. Numeretur- 20

que ab initio medietas circulorum. Et transibit 3

differentias, et remanet unum integrum scil., quod dele

et seorsum scribe. Residuo ducto in 60^a, et de preteri-

tis differentiis remanent 24 minuta, que scribe sub

integro. Residuo ducto in 60, et numeratis tribus dif-   25

ferentiis remanent 50 2^a scil. que subscribantur minutis.

---

14 item: iterum E

16 si: sed E et si H

17 post similiter add. erit H / est om.H

18 radicem invenire velimus: radicem volumus E volumus
   radicem invenire H

19 ipsam: ipsum H / illiusque: illiudque ?E

20 numeri ?H / deduplata: duplata BE / 1414: 144 B

21 3: 2 B

23 de om.H

23-24 preteritis: predictis B / post preteritis add.
   tribus H

24 remanent: remanet B

25 tribus: 2 B

26 remanent: remanet B / scil. om.H / subscribantur:
   subscribatur E

26-28 minutis . . . subscribantur om.E

the seconds and they will be the thirds. And if anything
is left over, it is to be dealt with similarly all the way
to 4ths and 5ths and 6ths.

For example, if we wish to find the root of two,[39] let
us write 6 circles before it, thus: 2000000. Let the root
of that number be taken, which undoubled will be 1414. Let
half [the number] of circles be counted off from the beginning. And it will go past 3 places, and one integer remains; delete it and write it separately. The rest having
been multiplied by 60, and when the places are counted off
24 minutes remain, which you write under the integer. When
the rest has been multiplied by 60, and the three places
have been counted off, fifty seconds remain; let them be
written under the minutes. When the rest has been multi-

---

[39]"Verbi gratia . . . opus" (IV, 28.18-32). The root
of 2 found here, 1.414 would satisfy the modern reader in
that form. The author needs it as a sexagesimal fraction,
the form to which he is accustomed. Cf. J. of Seville, ed.
Boncompagni, 1857: 87.29-90.14, which gives the same example.
Having set aside the integer 1, we may express the author's
procedure as follows: .414 x 60 = 24.840, set aside the
24 minutes; .840 x 60 = 50.400, set aside the 50 seconds;
.400 x 60 = 24.000, set aside the 24 thirds; nothing remains.

Residuo ducto in 60, et numeratis 3 differentiis remanent
24 3^a, que subscribantur 2^{is}. Et nichil remanet. Unum
ergo integrum, 24 minuta, 50 2^a, 24 3^a sunt radix duorum.
Quod probatur per 4^{am} et 5^{am} huius libri. Et si plures   30
circuli quam sex prescribantur integro, subtilius esset
opus. Et hec de radice quadratarum minutiarum dicta
sufficiant.

   Explicit practica geometrie.

---

27 remanent: remanet B
31 quam sex tr.ante circuli H / prescribantur:
   perscribantur E om.H / esset: esse E
32 hec: hic B   hoc H / de om.E / quadratarum: quadratorum H /
   minutiarum: minutiatim H / dicta: quesita H
34 practica geometrie: minutimetria H / post geometrie add.
   hic sunt 120 theoreumata E

plied by 60, and the three places have been counted off,
24 thirds remain; let them be written under the seconds.
And nothing remains. Therefore one integer, 24 minutes,
50 seconds, and 24 thirds is the root of two. This is
proved by the 4th and 5th proposition of this book. And
if more than six circles are written before the integer,
the more subtle would have been the work. And what has
been said suffices concerning the root of squares by minutes.

Here ends the practical geometry.

LE PRATIKE DE GEOMETRIE

EDITION, TRANSLATION AND COMMENTARY

[I, 0] Chi commenche dyometrie

Nous commencerons une oevre sor le pratike de

geometrie, lakele nous deviserons en 3 parties. En

la premiere partie ensegnerons nous a trover le mesure

des planeces, en la seconde a trover le mesure des   5

hautesches et des perfondeces et des crasse mesures,

en la tierce a trover les minuces de gyometrie et

d'astronomie covignable as ii parties devant.[1]

---

1 chi . . . dyometrie <u>om.</u>R

3 geometrie: gyometrie <u>R</u> / lakele: laquele <u>R</u>

6 hautesches: hauteces <u>R</u> / perfondeces: perfundeces <u>R</u> /

   <u>post</u> crasse <u>add.</u> bens ?R

---

[1]"Nous commencerons . . . ii parties devant" (I, 0.1-8).
The notes to the <u>Pratike</u> will be like those to <u>ACC</u>. Just as
in the notes to <u>ACC</u>, I, 0.1-8 refers to Book I, proposition
0 (an arbitrary number for the Prologue), lines 1-8. In
these notes such a number is to be understood as a reference
to the <u>Pratike</u> unless <u>ACC</u> is specified. For the source of
the French adaptation cf. <u>ACC</u>, I, 0.24-30.

[I, 1]   Se tu vels trover la mesure d'une ligne droite,
soit la ligne droite en une planete.   Se tu tiens
l'astrelabre en droit tes iex et tu regardes parmi
le partrieus de le dyagonal le fin de le ligne, et
tu vois combien la ligne deseure la li rendra de lonc   5
del livel en l'ortogone et par quele proportion la
ligne deseure l'evera a se basse.   Par autele proportion s'avera le visée de tes ieus a la fin de le
ligne.

   Che pues tu prover en tel maniere: illuec   10
sont 2 triangle, 1 grans de quel le ligne a mesurer
est <la basse> et li liviax <est> la planece et la
ligne deseure <est> la visée ki trespasse parmi le
dyagonal, et 1 petis triangles en l'ortogone kerré
duquel la basse est de <xii> piés et liviaus de <xii>   15
et ausi com la ligne deseure a se basse del petit triangle ausi fera la ligne deseure du grant triangle.   Cist
doi triangle sont establi sor une misme ligne deseure

---

4 partrieus: pertruis R̲ / dyagonal: diagonal R̲
5 rendra: renda R̲
6 quele: kele R̲
8 le: la R̲ / tes ieus: tex iex R̲
10 che: ce R̲
11 triangle: triangele R̲ / de: del R̲ / le: la R̲
12 la basse (ed.) om.RS̲ / est (ed.): et RS̲
13 est (ed.): et RS̲ / ki: qui R̲
14 kerré: querré R̲
15 xii (1,ed.): xi RS̲ / liviaus: liviax R̲ / xii (2,ed.):
    xi RS̲
17 ausi fera: ausera R̲
18 misme: meisme R̲

dont il ont une misme proportion entraus. Et c'est

une rieule en gyometrie. Dont se li liviax du petit   20

triangle est soustrebles a se basse, li estateur de

la planece sera soustreble a se basse, che pues veoir

a l'astrelabre. Dont se tu ses le mesure de le

planece, tu sauras le mesure de le ligne et par

ceste raison saras tu mesurer tel ligne[2] com tu   25

vauras ou le lei d'une ewe ou le lei d'une flueve

ou le hautesche d'un arbre ou d'un clochier.[3] En

ceste maniere sauras tu mesurer le lonc et le lei

de toutes coses par l'astrelabe. Ceste cose misme

porras prover par le quadrangle.[4]   30

---

19 misme: meisme R / entraus: entrals R

22 la: le R / che: ce R

24 sauras: saras R

25 mesurer tel ligne: tel ligne mesurer R / com: que R

26 vauras: wauras R / ou (1): u R / ou (2): u R / d'une:
    d'un R

27 ou: u R / hautesche: hautece R / ou: u R / clochier:
    clocier R

28 sauras: saras R / lei: ley R

29 toutes: totes R / misme: meisne R

---

[2]"Se tu vels . . . tel ligne" (I, 1.1-25). Cf. ACC I,
1.1-24. Note that planimetra, someone who measures plane
surfaces, is translated as planece, a plane surface.

[3]"Com tu vauras . . . d'un clochier" (I, 1.25-27). The
source of this is not obvious.

[4]"En ceste maniere . . . quadrangle" (I, 1.27-30). Cf.
ACC I, 1.41-42.

[I, 2] Se tu veus trover l'aire du triangle equilarere.
Devise le triangle par le livel en 2 triangles ounis. Mais
saches ke li costes du triangle equilatere est graindres
de sen livel le septisme partie de soi. Dont se cascuns
costes du triangle equilatere est de 7 pars, li liviax   5
sera de 6. Keure dont 1 des costes sor la moitié du livel
ou liviax sor le moitié del ouel coste, la somme fera l'aire
del triangle, de 21 piés.[5]

Che pues prover en ii manieres. En la premiere, se
tu fais 1 petit quadrangle duquel li un des costes soit   10
de 3 piés, ounis a le moitié du livel, et li autres soit de
7 piés, ounis au coste du triangle. Keure li 1 coste sor
l'autre; la somme du quadrangle sera de 21 piés, laquele
sera ounie a l'aire du triangle.[6]

En la seconde maniere, se tu fais une quadrangle   15
duquel li 1 des costes soit de 6 piés, ounis a livel, et
li autres costes de 7 piés, ounis au coste du triangle,
keure li 1 sor l'autre, la somme fera l'aire d'un quadrangle

---

2 devise: devisee R
3 saches: saces R / ke: que R
4 cascuns: cascun R
7 ou: u R
9 che: ce R
10 petit: peti R
16 a: au R

---

[5]"Se tu veus . . . de 21 piés" (I, 2.1-8). Cf. ACC I,
2.1-13; the reference to Gerbert is omitted.

[6]"Che pues prover . . . l'aire du triangle (I, 2.9-14).
Cf. ACC I, 2.14-31; the Pratike is more concise and its proof
is less complete.

de 42 piés, laquele dobble a l'aire du triangle. Dont l'aire

du triangle sera de 21 piés.[7]   20

Et saches bien k'il sont 3 manieres des piés: piés

lons, piés quarrés, piés cras. Piés lons a 4 paumes de lonc

pour mesurer lignes, piés cras pour hautesches et les per-

fondesces, piés quarrés a mesurer les planeces.[8]

---

19 dobble: double R

20 de om.R

21 saches: saces R / k'il: qu'il R / des: de R

22 lons: lonc R

23 hautesches: hauteces R / perfondesces: perfondeces R

---

[7]"En la seconde maniere . . . de 21 piés" (I, 2.15-20).
Cf. ACC I, 2.32-49; again the Pratike's justification is in-
complete. Also the excursus on the Boethian solution in
ACC I, 2.50-61 is omitted from the translation.

[8]"Et saches . . . planeces" (I, 2.21-24). Cf. ACC I,
2.62-65. The author of the French version is confused about
the purpose of solid or cubic feet.

[I, 3] Se tu veus trover l'aire du triangle ki a 2 costes

ounis. Tant seulement keure la moitiés de la basse sor le

livel; la somme fera l'aire. Si com la basse est 14 piés,

<li liviaus 24 piés>, l'aires sera de <168> piés. Che pues

prover par cele devant en 2 manieres par les quadrangles.   5

   Troveras le livel en cel maniere: se tu mesures le

moitié de le basse par soi misme, et tu sostrais cele somme

du coste, <la rachine de> la remamance fera le livel. Si

comme se la basse est de 14 piés et cascuns costes de 25,

li liviax sera de 24 piés selonc ceste rieule.[9]   10

---

4 li liviaus 24 piés (ed.) om.RS / l'aires om.S /
   168 (ed.): 164 RS / che: ce R
6 cel: tel R / le (2): lel ?R
7 misme: meisme R / sostrais: sousterais R
8 la rachine de (ed.) om.RS / remanance: remaneniee R
9 basse: besse R
10 rieule: riule R

---

[9]Se tu veus . . . selonc ceste rieule" (I, 3.1-10). Cf.
ACC I, 3.1-14; note that the proofs are treated in summary.

[I, 4] Se tu veus savoir l'aire du triangle ki a tous les 3 costes divers. Keure li basse le moitié del livel; la somme fera l'aire. Si cum se li basse est de 25 piés, li liviax est de 12, l'aire sera de 150. Che pues prover par le seconde en une maniere par le grant quadrangle. 5

Troveras le livel en tel maniere: se tu mesures le menu coste et tu le multiplies par soi misme et tu devises cele somme par le basse, la denominations sera le menour partie de le basse. Si comme li menres costes est de 15 piés, la basse de 25, et graindres costes de 20, la 10 menre partie de le basse sera de 9. Laquele se tu multiplies par soi, et tu soustrais cele somme de le somme du menour coste multiplié par soi misme, la rachine de la remanance du nombre sera li liviax. Dont li liviax sera de 12 piés selonc ceste raison.[10] 15

---

1 ki: qui R
2 li: le R
3 cum: com R
4 150: 15 S / che: ce R
7 multiplies: multeplies R / misme: meisme R
8 sera: fera R / menour: menor R
9 comme: com R
11 post partie add. sera RS
11-12 multiplies: multeplies R
12 le: la R
13 menour: menor R / multiplié: multeplie R / misme: meisme R / rachine: racine R
14 remanance: ramanance R

---

[10]"Se tu veus . . . selonc ceste raison" (I, 4.1-15). Cf. ACC I, 4.1-20, except that the derivation of the "precisura" of ACC I, 4.9-16 is distorted beyond recognition in the Pratike.

[I, 5] Se tu veus savoir l'aire de l'ortogone, keure li

liviax sor le basse; li moitiés de la somme fera l'aire.

Che pues prover par le seconde par le quadrangle.

Tu troveras les costes par le livel par ceste rieule:

se tu multiplies le moitié del livel par soi et tu abas   5

1 de cele somme, s'averas la basse. Se tu i aioustes 1,

s'averas de 17. Che pues prover par le premiere.[11]

---

2 la: le R
3 che: ce R / par (2): part S
4 le om.S
5 multiplies: multeplies R
7 che: ce R

---

[11]"Se tu veus . . . par le premiere" (I, 5.1-7). Cf.
ACC I, 5.1-9.

[I, 6] Se tu veus trover le dyametre du cercle escrit en

l'ortogone. Se tu assambles le livel a se basse et sous-

trais le fause ligne de cele somme, la remanance fera

le dyametre du cercle. Si com se li liviax est de 6, la

basse de 15, la fause ligne de 17, li diametres sera   5

de 5. Che pues.prover par le premiere.[12]

---

5 diametres: dyametres <u>R</u>
6 che: ce <u>R</u>

---

[12]"Se tu veus . . . par le premiere" (I, 6.1-6). Cf.
<u>ACC</u> I, 1.1-6; the reference to Archytas is omitted.

[I, 7]   Se tu veus trover du triangle ambigoine l'aire,
keure li liviax sor le moitié de la basse; la somme fera
l'aire.   Che pues prover par le seconde et par le quadrangle.[13]

---

1 veus: vels R
2 liviax: livax R / basse: besse R
3 che: ce R

---

[13]"Se tu veus . . . par le quadrangle" (I, 7.1-4).   Cf.
ACC I, 7.1-3.

[I, 8] Se tu veus del triangle ortogone l'aire, keure le
moitié de le basse sor le livel; la somme fera l'aire.
Che pues prover par le seconde partie del triangle.

Or t'avons dit et essengnié les figures des triangles
par ordene.[14]   5

---

1 veus: vels R / le: li R
3 che: ce R
4 essengnié: ensegnié R

---

[14]"Se tu veus . . . par ordene" (I, 8.1-5). Cf. ACC I,
8.1-6; Euclid is not mentioned in the Pratike's version.

[I, 9]   Se tu veus savoir l'aire del quarré, keure le menres

sor le gregnour; la somme fera l'aire.   Che pues prover

par le seconde.[15]

---

1 veus: vels <u>R</u> / quarré: quarer <u>R</u> / le: li <u>R</u>
2 gregnour: greignor <u>R</u> / che: ce <u>R</u>

---

[15]"Se tu veus . . . par le seconde" (I, 9.1-3). Cf. <u>ACC</u>
I, 9.1-3.

[I, 11] Se tu veus trover l'aire del combe equilatere,

keure li liviaus sor le dyagonal; la somme fera l'aire.

Che pues prover par le seconde.

Le livel troveras en tel maniere: se tu multiplies

<li costes par soi et> le moitié de le dyagonal par soi   5

et tu soustrais l'une somme de l'autre, la rachine de

le remanance du nombre fera li livel. Se cum se cascuns

costes del combe de 10, la dyagonal de 12, li liviaus

de <8>, et l'aire sera de 96. Ce pues prover par le

premiere.[16]   10

---

1 veus: vels R
2 liviaus: liviax R
3 che: ce R
4 multiplies: multeplies R
5 li costes par soi (ed.): om.RS
6 rachine: racine R
7 li: le R / se (1): si R / cum: com R / se (2): om.R /
   cascuns: cascons R
8 liviaus: liviax R
9 8 (ed.): 6 RS

---

[16]"Se tu veus . . . par le premiere" (I, 11.1-10). Cf.
ACC I, 11.1-12. The Pratike omitted an important part of
the procedure; the side of the rhumbus squared is the product from which the square of half the diagonal is subtracted.

[I, 12] Se tu veus trover l'aire del combe non equilatere,
tu auras les costes contraires. Keure li 1 costes sor
l'autre; la somme fera l'aire. Si cum se li graindres
costes est de <8> et ses contraires de 6, et li costes de
4 et ses contraires de 2; se tu la ounies: li 1 sera de   5
3, li autres de 7, <l'aire> de 21. Che pues prover par le
seconde.[17]

---

1 veus: vels <u>R</u>
3 cum: com <u>R</u> / li: il <u>R</u>
4 8 (<u>ed.</u>): 6 <u>RS</u>
5 la: las <u>R</u>
6 l'aire (<u>ed.</u>): li autres <u>RS</u> / che: ce <u>R</u>

---

[17]"Se tu veus . . . par le seconde" (I, 12.1-7). Cf. <u>ACC</u>
I, 12.1-6. The <u>Pratike</u> does not mention the averaging of
the opposite sides as the beginning of the procedure. It
also omits the question-begging justification of <u>ACC</u> I,
12.6-9.

[I, 13] Se tu veus l'aire trover du trapezire ortogone,
se tu assambles le basse a le vertis, et la moitiés de
la somme keure sor le livel, la somme fera l'aire. Si
cum se li basse est de 45 piés, et la vertis de 15, et
li liviax est de 30, l'aire sera de <600>. Che pues   5
prover par la quadrangle.[18]

---

1 veus: vels R / l'aire trover tr.R / trapezire: ?S
   trabezire R / ortogone: orgone R
4 cum: com R
5 est om.R / 600 (ed.): 90 RS / che: ce R
6 la: le R

---

[18]"Se tu veus . . . par la quadrangle" (I, 13.1-6). Cf.
ACC I, 13.1-8.

[I, 14]   Se tu veus trover du quadrangle ortogoine la

dyagonal, multeplie le greingnour coste par soi <et le

menre coste par soi>, et se tu assambles ces ii sommes,

la racine du nombre fera le dyagonal.   Si comme se li

graindres costes est de <8>, li menres de 6, la dyagonal   5

sera de 10.   Che pues prover par le premiere.[19]

---

1 veus: vels R / post trover add. l'aire RS / du quadrangle
   om. S
2 greingnour: grignour R
2-3 et le menre coste par soi (ed.) om.RS
3 sommes: somme R
5 8 (ed.): 4 RS
6 che: ce R

---

[19]"Se tu veus . . . par le premiere" (I, 14.1-6).   Cf.
ACC I, 14.1-6.   Note that the numerical values follow those
of H, the revised version of ACC.   One step of the procedure
was left out of the French version.

[I, 15 (alt.)] Se tu veus trover l'aire du quintangle,

soit li pentagones devisés en <5> costes ounis, keure li

1 des costes sor le moitié de sen livel et multiplie

cele somme par 5; la somme fera l'aire. Che pues prover

par le seconde.[20]   5

---

1 veus: vels R
2 5 (ed.): 2 RS
3 le: la R / multiplie: monteplie R
4 che: ce R

---

[20]"Se tu veus . . . par le seconde" (I, 15.1-5). This
is not a translation of ACC I, 15. The method is correct
if "sen livel" refers to the altitude of any of the five
equal isosceles triangles making up the pentagon.

[I, 16 (alt.)] Se tu veus trover l'aire du sixteangle,

soit la sextagones, devisés en 6 costes ounis, keure li 1

des costes sor le moitie de sen livel et multiplie cele

somme par 6. La somme fera l'aire. Ce pues prover par

cele devant.[21]   5

---

1 veus: vels R

3 le om.S / multiplie: monteplie R

---

[21]"Se tu veus . . . par cele devant" (I, 16.1-5). This
is substituted for a translation of ACC I, 16. The procedure needs to be corrected or understood in a way analogous
to the preceding.

[I, 17 (alt.)]   Se tu veus trover l'aire du septangle,

soit li piagones, devisés en 7 costes ounis, keure li 1

costes sor le moitié de sen livel et multeplie cele

somme par 7; cele somme fera l'aire.   Che pues prover par

cele devant.   5

Or t'ai ensengié des toutes les figures angulers.[22]

Aprés che dirons des figures circulers.

---

1 veus: vels R / du: dou R
3 post sen add. li R / multeplie: monteplie R
4 cele: la R / fera: ferai R / che: ce R
6 ensengie: ensignie R / des: de R / toutes: totes R
7 che: ce R

---

[22]"Se tu veus . . . figures angulers" (I, 17.1-6). This
replacement for I, 17 will work if corrections similar to
those needed in I, 15 and I, 16 are applied.

[I, 21]   Se tu veus trover la circonference del compas,
multeplie le dyametre du compas par 3, et si aiouste la
septisme partie de soi, si auras la circonference.[23]

---

1 veus: vels R
2 par: pax R / si: se R / la: le R
3 si auras: saras R / circonference: circonferrence R

---

[23]"Se tu veus . . . la circonference" (I, 21.1-3). Cf.
ACC I, 21.1-4; the Pratike does not justify this by an appeal
to the rectification of a curve, as did ACC.

[I, 22] Se tu veus trover l'aire del compas, multeplie

la moitié <de la circonference par la moitié> de sen

dyametre; la somme fera l'aire. Si comme se li dyametres

est de 7, l'aire sera de <38> et demi. Ce pues prover

en autre maniere. Se tu multeplies le dyametre par soi   5

et tu multeplies cele somme par <11> et tu devises cele

somme par 14, la denominations fera l'aire de la circon-

ference. Ce pues prover par le seconde.[24]

---

1 veus: vels <u>R</u> / multeplie: monteplie <u>R</u>

2 de la circonference par la moitié (<u>ed.</u>) <u>om.RS</u>

3 comme: com <u>R</u>

4 l'aire: l'aires <u>R</u> / 38 (<u>ed.</u>): 36 <u>RS</u>

5 multeplies: monteplies <u>R</u>

6 multeplies: monteplies <u>R</u> / 11 (<u>ed.</u>): 13 <u>RS</u>

---

[24]"Se tu veus . . . par le seconde" (I, 22.1-8). Cf.
<u>ACC</u> I, 22.1-7. The lengthy proof of <u>ACC</u> is omitted here.

[I, 23] Se tu veus trover l'orneure du cercle, multeplie

le dyametre par soi et cele somme multeplie par 22 et cele

somme devise par 7; la denominations fera l'orneure du

cercle car elle est quadruple a l'aire du cercle. En

autre maniere le pues prover se tu multeplies la circon-   5

ference par sen dyametre. Si com li dyametres est de 7,

l'orneure sera de 154. Ce pues prover par raison de

nombre.[25]

---

1 veus: vels <u>R</u> / multeplie: monteplie <u>R</u>

2 multeplie: monteplie <u>R</u>

4 elle: ele <u>R</u>

5 multeplies: monteplies <u>R</u>

---

[25]"Se tu veus . . . par raison de nombre" (I, 23.1-8).
Cf. <u>ACC</u> I, 23.1-7.

[I, 24] Et par che pues tu trover ke 1 dyametres qui

doubble a 1 autre fait l'aire quadruble a l'aire de celui.

Che pues prover par raison de nombre.[26]

---

1 che: ce R / trover: prover R

2 doubble: double R

3 che: ce R

---

[26]"Et par che . . . par raison de nombre" (I, 24.1-3).
Cf. ACC I, 24.1-6.

[I, 26] Se tu veus trover le quarré du cercle, la rachine

de cele aire fera le coste du quarré ouni au cercle. Ce

pues prover en tel maniere: car cil costes multepliés par

soi fera l'aire du quarré ouni a l'aire du cercle. Si

comme se li dyametres est de 7, li costes del quarré   5

sera de 6 et une quinte, laquele se multeplie par soi se

fera l'aire de l'un et de l'autre. Et note ke se li nombres

de l'aire est seus et non entendables, tu prenderas la

rachine par minuces.[27]

---

1 veus: vels <u>R</u> / rachine: racine <u>R</u>

2 quarré: quaré <u>R</u>

3 multeplies: monteplies <u>R</u>

5 comme: com <u>R</u> / se <u>om.R</u>

6 multeplie: monteplie <u>R</u>

7 ke: que <u>R</u>

9 rachine: racine <u>R</u>

---

[27]"Se tu veus . . . par minuces" (I, 26.1-9). Cf. <u>ACC</u> I,
26.1-10. Note the interesting double translation of <u>surdus</u>
in the French as "seus et non entendables," which might be
translated as "surd and incomprehensible." The more accurate fractional value for the side of the square is not
given in the <u>Pratike</u>.

[I, 27] Se tu veus trover la circonference del quarré, se tu multeplies le coste du quarré par soi et tu multeplies cele somme par 14 et tu devises cele somme par 11, la rachine de la denomination fera la dyametre du cercle ouni au quarré. Si comme se li costes du quarré 5 est de 6 et une quinte, li dyametres du cercle sera de 7 piés. Ce pues prover par le premiere et par le seconde.[28]

---

2 multeplies: monteplies R
2-3 multeplies: monteplies R
4 rachine: racine R / la (2): le R
5 li om.R

---

[28]"Se tu veus . . . par le seconde" (I, 27.1-7). Cf. ACC I, 27.1-7. The enunciation of the Pratike's version is misleading; it wants the circling of the square not its circumference.

[I, 28]   Se tu veus trover le sorcrois du quarré au cercle
escrit dedans le quarré, se tu proves l'aire du cercle par
celes devant, et tu proves l'aire du quarré par sen coste
et tu soustrais l'une somme del l'autre, tu sarras le
sorcrois de l'une a l'autre. Si comme se li dyametres   5
du cercle est de 7, li sorcrois del quarré sera de 10 et
demi.[29]

---

1 veus: vels R
2 proves: prueves R
3 et om.R / proves: preves R
4 del: de R / sarras: saras R
5 li om.R

---

[29]"Se tu veus . . . sera de 10 et demi" (I, 28.1-7). Cf.
ACC I, 28.1-6.

[I, 29]   Se tu veus trover le sorcrois du cercle au
quarré escrit dedens de dele cercle; se tu ses l'aire
de le cercle et tu soustrais de cele somme par 4 fois
14, la remanance fera l'aire del quarré escrit dedens
le cercle. Si comme se li dyametres est de 14, l'aire   5
sera de <154>, de laquele se tu sostrais par 4 fois 14,
l'aire du quarré sera de 98. Che pues prover par le
seconde.[30]

---

1 veus: vels R
3 de le: du R
5 comme: com R / se om.R
6 154 (ed.): 145 RS / sostrais: soustrais R
7 che: ce R

---

[30]"Se tu veus . . . par le seconde" (I, 29.1-8). Cf.
ACC I, 29.1-7. The French version of this confused proposition has "4 fois 14" instead of the fraction 4/14. The
last few lines of ACC, which contain a converse operation,
are not translated.

[I, 30] Se tu veus trover le sorcrois du quarré ki contient
le cercle <du> petit quarré ki est contenus el cercle; se
tu ses l'aire de greignor quarré et l'aire du cercle et
l'aire du petit quarré, tu sarras le sorcrois du gregnour
quarré au menor, car l'aire del 1 est doble a l'autre.   5
Car l'aire du greignour est de 196, de quel li costes est
de 14. Et li dyametres del cercle est ausi de 14. Che
pues prover par celes devant.[31]

---

1 ki: qui R
2 du (ed.) om.RS
3 de: du R / greignor: gregnor R
4 sarras: saras R / gregnour: gregnor R
5 doble: double R
6 greignour: gregnor R
7 che: ce R

---

[31]"Se tu veus . . . par celes devant" (I, 30.1-8). Cf.
ACC I, 30.1-9.

[I, 31] Se tu veus trover la contenance d'une plaine de
200 piés de lonc et de <100> piés de léeche, ke 1 laterculers tient 5 piés de lonc et 4 piés de lei. Devise
de lonc le planece par 5, c'est en 40, et le lei par 4,
c'est en 25. Keure 25 sor 40; sauras le nombre de   5
laterculers. Che pues prover par nombre.[32]

---

1 veus: vels R / la: le R
2 100 (ed.): 200 RS / léeche: léece R
5 sauras: saveras R
6 che: ce R

---

[32]"Se tu veus . . . pues prover par nombre" (I, 31.1-6).
Cf. ACC I, 31.1-8.

[I, 32]   Se tu veus trover le nombre des arpens d'un grant
camp conneu.   Se li frons del camp est de 50 et li costes
contraires est de 60, li doi coste de lonc soient de 100
cascuns, assambles le coste de 50 a celi de 60 et le
devises par mi en 55, <keure 55> sor 100.   La somme fera   5
le nombre des perces, laquele tu devises par les perces
d'un arpent; la denominations fera le nombre des arpens.[33]

---

1 veus: vels <u>R</u>
5 keure 55 (<u>ed.</u>) <u>om.</u><u>RS</u>

---

[33]"Se tu veus . . . le nombre des arpens" (I, 32.1-7).
Cf. <u>ACC</u> I, 32.1-10. <u>ACC</u> has some specific metrological material, including the number of perches in an arpent and
several other equivalences, which is omitted here.

[I, 33] Se tu veus trover le nombre des arpens d'un camp
quadruple conneu. Soit li premiers costes de 30, et ses
contraires de 32, et li tiers de 33, li quars de 34.
Assambles les costes dou lonc, et si les devises par <mi>.
Assambles les costes dou lé, et si les devises par <mi>.   5
Keure li l sor l'autre; la somme fera le nombre des perces,
laquele tu devises par ces d'un arpent; la denominations
fera le nombre des arpens. Che pues prover par raison de
nombre.[34]

---

1 veus: vels <u>R</u>
4 dou: du <u>R</u> / mi (<u>ed.</u>): iiii <u>RS</u>
5 mi (<u>ed.</u>): iiii <u>RS</u>
8 che: ce <u>R</u>

---

[34]"Se tu veus . . . par raison de nombre" (I, 33.1-9).
Cf. <u>ACC</u> I, 33.1-7. The numerical values here are those of
MS <u>E</u> of <u>ACC</u>.

[I, 34]   Se tu veus trover d'une planete conneue le nombre
des maisons a faire. Soit li planece de 1000 piés de lonc
en 1 coste et en l'autre coste contraire de 1100, en cascun
coste de lé, 600. et cascune maison ait 40 perces de
lonc et 30 de lei. Assambles les costes dou lonc et si   5
les devises par mi. Si devise cele moitié par 40. Assamble
les costes du lei et si le devises par mi. Si devise cele
moitié par <30>. Keure l'une devisions sor l'autre; le
somme fera le nombre des maisons a faire. Che poes prover
par raison de nombre. Et par ceste raison pues tu prover 10
les contenances de toutes planeces en lonc et en lei.[35]

---

1 veus: vels <u>R</u> / conneue: conneus <u>R</u>
3 contraire: contraires <u>R</u>
5 dou: du <u>R</u>
7 le: les <u>R</u> / devise: devises <u>R</u>
8 30 (<u>ed.</u>): 40 <u>RS</u> / devisions: devisons <u>R</u> / le: la <u>R</u>
9 che: ce <u>R</u> / poes: pues <u>R</u>
11 toutes: totes <u>R</u>

---

[35]"Se tu veus . . . en lonc et en lei" (I, 34.1-11). Cf.
<u>ACC</u> I, 34.1-13.

[I, 35] Se tu veus trover d'une planece trianguler et
le nombre des laterculers. Tu troveras l'aire de le
planece trianguler par sen livel. Lakele se tu la devises
par l'aire des laterculers, la denominations fera [le
nombre] des laterculers. Si comme se cascuns costes   5
est de 70 piés, la somme des laterculers sera de 105.
Che pues prover par raison de nombre.[36]

---

1 veus: vels <u>R</u> / trianguler: triangle <u>R</u>
3 trianguler: triangler <u>R</u> / lakele: laquele <u>R</u>
4 l'aire: l'ai <u>R</u>
4-5 le nombre (<u>ed.</u>): l'aire <u>RS</u>
6 105: 145 <u>?R</u>
7 che: ce <u>R</u>

---

[36]"Se tu veus . . . par raison de nombre" (I, 35.1-7).
This is not a translation but a loose adaptation of <u>ACC</u> I,
35. Since the size of the small rectangles (<u>laterculers</u>)
is not given here I have no idea where the numbers come
from.

[I, 36] Se tu veus trover le nombre des maisons d'une
cité reonde, tu troveras l'aire de la cité par le moitié
de son dyametre mené en <la moitie de> se<n> circonference.
Dont se cascune [des maisons] est de 30 piés de lonc et
de 20 de lei, keure li 1 sor l'autre, la somme fera 600.   5
Se tu devises l'aire de le cité par ces sommes, la denominations fera le nombre des maisons.

Or t'avons enseingnié le mesure des planeces.[37]

---

1 veus: vels R
2 la: le R
3 la moitié de (ed.) om.RS / sen (ed.): se RS
4 se om.R
8 enseingnié: ensegnié R

---

[37]"Se tu veus . . . le mesure des planeces" (I, 36.1-8).
Cf. ACC I, 36.1-8.

[III, 1] Or te dirons des crasses mesures.

Se tu veus trover le mesure del combe quarré, keure

li costes del quarré sor le combe, la somme fera l'aire.

Derechief keure li costes sor l'aire, la somme fera la

combe du quarre. Si comme se li costes est de 7, li   5

conbles del quarré sera de 343. Che pues prover par

raison de nombre.[1]

---

2 veus: vels R

4 derechief: derekief R / li: le R / costes: coste R /

  la (2): le R

6 che: ce R

---

[1]"Or te dirons . . . par raison de nombre" (III, 1.1-7).
Cf. ACC III, 1.1-8. The alternate technique of cubing the
side and the justification of the procedure are omitted from
the Pratike. There may be some confusion about the meaning
of combe; combe or conbe are not misreadings of coube. The
word conbles or combles in line 6 makes the author's intention clear. Comble in modern French is a heaped measure;
conbles del quarré, literally then a heaped-up square, is
the translation for the Latin cubus. It seems that combles,
the nominative form, has combe for its oblique cases. Thus
combles quarrés, found in its oblique form combe quarré in
line 2 seems to translate the Latin corpus cubici.

[III, 2] Se tu veus trover le conbe d'un pilier reont,
tu troveras l'aire par le moitié de sen dyametre en sa
circonference. Keure l'aire sor le lonc; la somme fera
la conbe du pilier. Che pues prover ausi ke devant.[2]

---

1 veus: vels R / conbe: combe R
4 conbe: combe R / che: ce R / ke: que R

---

[2]"Se tu veus . . . ausi ke devant" (III, 2.1-4). Cf. ACC
III, 2.1-4. ACC's proof by drawing parallel lines is omitted
here.

[III, 3] Se tu veus trover la mesure del espere reonde,
tu troveras le dyametre du cercle par le premier livre,
delquel tu quiers la crasse mesure. Tu cuberas che
dyametre en son quarré. Et si note bien ke li combes du
quarré sorcroist le contenance del combe reont, lekel   5
reont il te convient soustraire. Se tu veus trover le
contenance del combe reont, che feras tu en tel maniere:
tu deviseras le combe del quarré par 21, et multeplieras
cele devision par 10; la somme fera le sorcrois del combe
quarré au combe reont. Derekief se tu multeplies cele   10
misme devision par <11>, la somme fera le contenance del
combe reont. Si comme se li dyametres du cercle est de
7 piés, se tu le combes, la somme fera 343. Se tu devises
cele somme par 21, la devision fera 16 piés et le tierce
part d'un piét. Se tu multeplies cele devision par 10,   15

---

1 veus: vels R
3 la: le R / cuberas: comberas R / che: ce R
4 ke: que R
5 combe: coube ?S / lekel: lequel R
6 veus: vels R
7 combe: coube ?S
8 combe: coube ?S / multeplieras: monteplieras R
9 combe: coube ?S
10 combe: coube ?S / multeplies: monteplies R
11 misme: meisme R / 11 (ed.): 2 RS
12 combe: coube ?S
15 multeplies: monteplies R

la somme fera 163 piés et le tierce part d'un pié. C'est

li sorcrois del combe quarré au combe reont. Derekief

se tu multeplies cele misme devision par <11>, la somme

fera <179> piés et 2 parties d'un pié. Che sera la

contenance del combe reont a lequele tu assambleras le   20

sorcrois; saveras le coumbe del quaré, 343. Ce pues

prover par celes devant.[3]

---

17 combe (1): coube ?S / combe (2): coube ?S
18 multeplies: monteplies R / misme: meisme R /
    11 (ed.): 2 RS
19 179 (ed.): 176 RS / pié: piét R / che: ce R
20 combe: coube ?S
21 coumbe: combe R / quaré: quarré R

---

[3]"Se tu veus . . . par celes devant" (III, 3.1-22). Cf.
ACC III, 3.1-19. The scribe of MS S seems not to have understood the word combles (combe in object case); he seems
to have written coube in most of its occurrences, although
he may have been writing conbe. For the reasons explained
in note 1, I have chosen to use combe, the reading of R.

[III, 7] Se tu veus trover le contenanche du mui. Se

tu mesures le dyametre du fons et la dyametre deseure

ki atouche le vin, et tu les devises par mi. Se tu

troeves l'aire du fons par le dyametre et tu ses de quans

piés l'aire soit. Se tu multeplies cele aire par le   5

haut dou mui, la somme te dira de quans piés li muis sera.

Se tu devises cele somme par le sestier, la devision fera

le nombre des sestiers. Che pues prover par raison de

nombre.[4]

---

1 veus: vels <u>R</u> / contenanche: contenance <u>R</u>

2 la: le <u>R</u>

3 ki: qui <u>R</u> / atouche: atouce <u>R</u>

4 troeves: trueves <u>R</u> / l'aire du fons: le fons de l'aire <u>R</u>

5 multeplies: monteplies <u>R</u>

6 dou: du <u>R</u>

7 devision: devisions <u>R</u>

8 che: ce <u>R</u>

---

[4]"Se tu veüs . . . par raison de nombre" (III, 7.1-8).
Cf. <u>ACC</u> III, 7.1-8. This text seems to include material
from some other proposition, perhaps <u>ACC</u> III, 8.

[III, 8] Se tu veus trover la contenanche du tonel. Se
tu ses l'aire du tonel par le dyametre du fons et tu
ses le lonc du tonel, se tu multeplies l'aire par le lonc,
la somme fera la contenance dou tonel. Se tu devises cele
somme par le contenance du mui, la devisions fera <le   5
nombre des muis>.[5]

---

1 veus: vels R / contenanche: contenance R / du: d'un R
   tonel: tonnel R
2 tonel: tonnel R
3 tonel: tonnel R / multeplies: monteplies R
4 la (2): le R / dou: du R / tonel: tonnel R
5-6 le nombre des muis (ed.): l'aire RS

---

[5]"Se tu veus . . . <le nombre des muis>" (III, 8.1-6).
Cf. ACC III, 8.1-7. The further conversion from modii to
sextarii and from sextarii to numerata is omitted from the
Pratike.

[III, 9] Saches ke 2 tonel d'un lonc se li dyametres del

1 est dobbles au dyametre del autre, la contenance del 1

sera quadruble a le contenance del autre. Che pues prover

par les dyametres et par les circonferences.[6]

---

1 saches: saces <u>R</u> / ke: que <u>R</u> / d'un: du <u>R</u>
2 dobbles: doubles <u>R</u>
3 quadruble: quadruple <u>R</u> / che: ces <u>R</u>

---

[6]"Saches ke 2 tonel . . . par les circonferences" (III,
9.1-4). Cf. <u>ACC</u> III, 9.1-9.

[III, 11]   Se li dyametres du tonel est dobles au dyametre del autre, liquels soit quadruples au lonc del autre, les contenances de 2 seront ounies.   Che pues prover par celes devant et par raison de nombre.[7]

---

1 tonel: tonnel R / dobles: doubles R
2 liquels: lequels R
3 de: des R / ounies: onnies ?R / che: ce R

---

[7]"Se li dyametres . . . par raison de nombre" (III, 11.1-4).   Cf. ACC III, 11.1-4.   The fuller explanation of ACC III, 11.5-8 is omitted here.

[III, 12] Se tu veus trover l'aire d'un vaissel reont ou

quarré. Se tu ses l'aire, multiplie l'aire par le haut;

la somme la rechevance, si comme nous avons prové.[8]

---

1 veus: vels R / ou: u R

2 multiplie: multeplie R

3 rechevance: recevance R

---

[8]"Se tu veus . . . nous avons prové" (III, 12.1-3). Cf.
ACC III, 12.1-4.

[III, 14] Saches ke de 2 huces quarrées se li costes del
une est dobbles au coste del autre en lonc et en lei, la
contenance del une fera 8 tans ke l'autre.   Ce pues prover
par raison de nombre.[9]

---

1 saches: saces R / huces: hauces R
2 dobbles: doubles R
3 ke: que R

---

[9]"Saches ke de 2 huces . . . par raison de nombre" (III,
14.1-4). Cf. ACC III, 14.1-2; the rest of ACC III, 14 is
left out here.

[III, 13] Se tu veus trover le contenance d'une huce.

Se tu multiplies l'aire par le haut et tu devises la

somme par le contenance de la huce. Che pues prover par

nombre.[10]

---

1 veus: vels R

2 multiplies: multeplies R

3 che: ce R

---

[10]"Se tu veus . . . prover par nombre" (III, 13.1-4).
Cf. ACC III, 13.1-5. In both MSS R and S, this proposition
follows III, 14. The author scrambled the sense of the original Latin in abridging it.

[III, 15] Se tu vels faire une huce onnie a 1 tonel. Se
tu ses l'aire du tonel par le dyametre et tu quiers la
rachine de cele aire de laquele tu faiches une aire quarré
onnie a l'aire del fons, et tu fais le lonc de le huce onni
au lonc du tonel. Si ke les aires soient onnies et   5
keurent sor 1 lonc, les contenances seront onnies. Che
pues tu prover par nombre.[11]

---

1 onnie: ounie <u>R</u> / tonel: tonnel <u>R</u>
2 tonel: tonnel <u>R</u>
3 rachine: racine <u>R</u> / cele: cel <u>R</u> / faiches: faices <u>R</u> /
   quarré: quarrée <u>R</u>
4 onnie: ounie <u>R</u> / fais: faces <u>R</u>
5 tonel: tonnel <u>R</u> / ke: que <u>R</u>
6 che: ce <u>R</u>
7 tu <u>om.R</u>

---

[11]"Se tu vels . . . prover par nombre" (III, 15.1-7).
Cf. <u>ACC</u> III, 15.1-9. The reference to the quadrature of the
circle is omitted from the <u>Pratike</u>.

[III, 16] Se tu vels trover 1 tonel ouni a 1 huce. Se tu
ses l'aire de le huce et tu quiers le dyametre du tonel,
duquel tu faiches une aire ounie a l'aire del huce, et tu
fais le lei dou tonel ouni au lonc de le huce ke les aires
soient ounies et keurent sor 1 lonc, les contenances   5
seront onnies.[12]

---

1 tonel: tonnel R
2 tonel: tonnel R
3 duquel: douquel R / faiches: faces R / une: 1 R /
   del: de le R
4 dou: du R / tonel: tonnel R / ke: que R
6 onnies: ounies R

---

[12]"Se tu vels . . . seront onnies" (III, 16.1-6). Cf.
ACC III, 16.1-8; the proof of ACC is changed a bit here.

[III, 17]   Se tu vels trover les contenances d'un puis.

Se tu mesures l'aire d'un puis et tu le multeplies par sen

haut et tu devises cele somme par le contenances des

vaisseaus, teus com tu vauras, la divisions fera le nombre

des vaisseaus.[13]   5

---

2 tu (1) om.R
3 haut: halt R / le: les R
4 vaisseaus: vaissiaus R / divisions: devisions R
5 vaisseaus: vaissiaus R

---

[13]"Se tu vels . . . le nombre des vaisseaus" (III, 17.1-5).
Cf. ACC III, 17.1-4.

[V, 1] Se tu mesures 1 quarré de 4 costes, cascune coste
de 14 piés, multeplie l'un des costes par soi, c'est a
dire par 14 fois 14. La somme fera 196.

Se tu fais 1 compas dedens cest quarré si grant com
tu porras, il tenra en l'aire du compas <154> piés. Don-   5
ques li sorcrois del quarré au reont monte <42> piés.

Donques encor quier 1 quarré dedens cel compas si
grant com tu pues: cil quarrés tenra [?] enseure <98> piés,
car il vaut le moitié du premier. Donques li sorcrois del
compas au petit quarré monte 56 piés.   10

Aprés si tu trueves dedens toutes ces mesures, c'est
a dire dedens le premier quarré: 1 triangle equilatere,
cascuns costes sera de 14 piés, et l'aire du triangle sera
de 160 piés et 8. Toutes ces mesures puent estre dedens
le premier quarré. Et saches ke cascuns des 4 angles du   15
premier quarré monte en soi <11> piés et cascuns des 4
angles du compas par defors le petit quarré 14 piés. Or
as tu <9> parties de mesure dont cascune est nombrée et

---

1 cascune: cascun R
2 multeplie: montcplie R
4 com: ke R
5 154 (ed.): 156 RS
6 42 (ed.): 49 RS
8 tenra ?S / enseure: deseure R / 98 (ed.): 96 RS
9 vaut: valt R
10 compas: compos R
11 si tu: se R / toutes: totes R / post ces add. figures S
14 toutes: totes R
15 saches: saces R / ke: que R
16 11: 2 RS
18 9 (ed.): ci R 12 S / parties: paties S / nombrée: nombré R

V, 1. If you [want to] measure a square of four sides, each
one of 14 feet, multiply one of the sides by itself, that is,
14 times 14. The product will be 196.

If you make a circle as large as you can inside this
square, the area of the circle will be 154 [square] feet.
Thus, the excess of the square over the circle is 42 [square]
feet.

Then, to find a square as big as you can inside this circle: the square will be 98 [square] feet, for it is half the
first square. Thus the excess of the circle over the small
square is 56 [square] feet.[1]

Then, if you [want to] find the measure of all these figures [inscribed] within the first square: [for] an equilateral triangle, each side will be of 14 feet, and the area
of the triangle will be 168 [square] feet. All these measures can be within the first square.[2] And know that each
of the four corners of the first square is 11 [square] feet
and each of the four corners [i.e., segments] of the circle
outside of the small square is 14 [square] feet. Then you
have these parts of measure, each one of which is numbered

---

[1]"Se tu mesures . . . 56 piés" (V, 1.1-10). Proposition
V, 1 occurs in the manuscripts after the French version of
I, 30. This part treats material closely related to that of
ACC I, 28-I, 30, which deal with inscribed and circumscribed
squares and circles. The rest of the proposition extends
its concerns to other figures. See fig. 84.

[2]"Aprés si tu . . . dedens le premier quarré" (V, 1.11-15).
This is, in effect, the calculation of the area of an equilateral triangle with a given side; here it happens to equal
that of the given square. See fig. 85. The calculation uses
the approximation 7:6 for the side to altitude ratio in the
equilateral triangle. The ratio 7:6 is very near the correct
$\sqrt{3}$:2. The same value is used in ACC I, 2.8-9.

mesurée par soi et toutes sont contenues dedens le premiere.

Se tu fais le triangle dedens le quarré si a cascuns   20
costes 14 piés, ausi com i'ai dit devant, et li liviax en
a 12, la somme est de 84 piés et les clés en coste cascune
<42> piés et la lisiere deseure doit avoir 2 piés de lé et
14 de lonc, che sont 28. En ces 4 figures est contenus
tous li quarrés.   25

Se tu fais 1 compas dedens le triangles si grant
com tu pues, il tenra en soi 50 piés et le quart d'un pié
et les 7esme du quart d'un pié. Et les 3 cornes dou triangle par defors le compas; cascune tient en soi <11> piés
et les <4> d'un pié, li 21esme d'un quarte mains.   30

Derechief se tu fais un triangle dedens ce compas,
il terra en soi <21> piés. Et es 3 angles du compas par

---

22 84: 88 <u>R</u>
23 42 (<u>ed.</u>): 4 <u>RS</u>
24 14: 114 <u>R</u> / che: ce <u>R</u> / <u>post</u> contenus <u>add.</u> li qua <u>?R</u>
28 dou: du <u>R</u>
29 11 (<u>ed.</u>): 2 <u>RS</u>
30 4 (<u>ed.</u>): 3 <u>RS</u> / quarte: quarre <u>R</u>
31 derechief: derekief <u>R</u> / un: 1 <u>R</u>
32 terra: tenra <u>R</u> / 21 (<u>ed.</u>): 31 <u>RS</u>

and measured by itself, and all are contained in the first.[3]

If you make the triangle inside the square 14 feet on a side, as I said before, and its altitude 12, the area is 84 feet, and the keys on each side, 42, and the border above should be 2 feet in width and 14 in length, which makes 28 [square feet].   In these 4 figures the whole square is contained.[4]

If you make a circle as big as you can inside the triangle, it will have an area of 50 and a quarter and a seventh part of a quarter of a [square] foot. And each of the three corners of the triangle outside of the circle have an area of 11 and a fourth [square] feet, less a twenty-first part of a fourth [of a square foot].[5]

Then if you make a triangle inside this circle, it will have an area of 21 [square] feet, and for the three corners

[3]"Et saches . . . dedens le premiere" (V, 1.15-19). The differences between the circle and its circumscribed and inscribed squares was found in lines 5-6 and 9-10 above. When those excesses are each divided by four, 10 1/2, rounded to 11, and 14 result. The area of the eight "corners" plus the smaller square make up the area of the larger square, as in fig. 84.

[4]"Se tu fais le triangle . . . tous li quarrés" (V, 1.20-25). Here the author considers the equilateral triangle inscribed in the square, as in lines 11-15 above. The various figures into which the square is divided may be seen in fig. 85.

[5]"Se tu fais 1 compas . . . d'un quarte mains" (V, 1.26-30). The computation of the area of a circle inscribed in an equilateral triangle of side 14 is correct if certain assumptions are made. The radius of the circle is taken as 4, where it should be $\frac{7\sqrt{3}}{3}$. This approximation is analogous to that in note 2 above. The standard medieval approximation of π as 22/7 is also used here to compute the area of the circle. The difference between the area of the circle, 50 2/7, and the area of the triangle, 84, is divided by 3 to produce the area of the corners, 11 5/21 each. Note that the author gives his results in unit fractions, that is, in fractions having one as a numerator.

dehors le petit triangle, cascuns tient en soi 9 piés et

les <3> pars d'un pié et le 21esme del quart d'un pié.

Donques se tu gardes 1 des angles dou compas et le   35

triangle ensamble, ches 2 coses font une figure ki a anom [?]

cone; se tient en soi 30 piés et 3 pars d'un pié et le

21esme du quart d'un piét.

---

34 3 (ed.): 2 RS
35 dou: du R
36 ches: ces R / ki: qui R

[i.e., segments] of the circle outside of the little triangle, each one has 9 and three-fourths and a twenty-first
part of the fourth of a [square] foot. Thus if you keep
one of the corners [i.e., segments] and the triangle together, these two things make up a figure that is called [?]
a cone; it has an area of 30 and three-fourths and a twenty-
first part of a fourth of a [square] foot.[6]

[Fig. 84]   [Fig. 85]

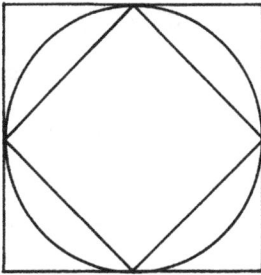   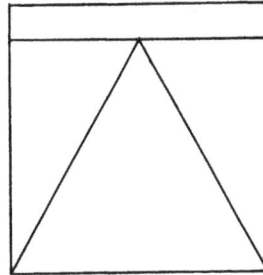

[Fig. 86]

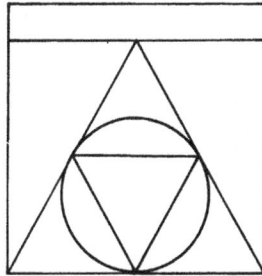

---

[6]"Derechief . . . du quart d'un piét" (V, 1.31-38). The
height of an equilateral triangle inscribed in a circle of
radius 4 will be 6 (see Fig. 86). The author assumes its
side will be 7, making the same presumption about the side-
altitude ratio as above. (See notes 2 and 5.) Again he computes one-third of the difference between the two figures to
find the area of each of the three corner segments. By adding one of those segments to the inscribed triangle, he gets
the area of the "ice-cream cone" shape.

[V, 2] Li toniaus de droite muison doit estre tels: de 2
piés et demi doit estre li fons, et le bouge doit avoir
le quart d'un piét et si doit avoir de lonc 4 piés et le
12es d'un piét. Et li piés est de 12 pauls, et ensi aura
li toniaus de gros 31 paus et demi et li lons 49 paus   5
dedens.

   Se tu veus savoir par art combine cis toniax tient,
tout avant multeplie le fons par sen dyametre, et donc
remaine [?] le quarré el reont par les rieules de geometrie.
Si averas le contenance dou fons. Et cele somme multe-   10
plie par 49, et ensi auras le contenance de tout le tonel
per paus. Et ces paus remanras en piés. Et saches ke

---

1 toniaus: tonnials R / tels: cels R
2 le bouge: li bouce R
4 pauls: paus R / aura: avera R
5 toniaus: tonnials R
7 veus: vels R / toniax: tonniax R
8 tout: tot R / multeplie: monteplie R / sen: sens R /
   ante dyametre add. de R
9 remaine: remainne R / el: en R / rieules: riules R /
   geometrie: gyometrie R
10 dou: du R
10-11 multeplie: muleplie S
11 auras: averas R / tout: tot R
12 remanras: ramenras R / saches: saces R / ke: que R

V, 2. A barrel of correct measure should be as follows:
the bottom should be two and a half feet [across], the
mouth should be a quarter of a foot [across], and it
should be 4 and a twelfth feet long. And a foot is of 12
inches, and thus the [average] width of the barrel will be
31½ inches and its internal [?] length 49 inches.[7]

If you want to know "by art" how much this barrel holds,
first of all multiply the [average] base by its diameter
and then the square is brought to roundness by the rules
of geometry. You will then have the area of the base. Multiply the area of the base by 49 and then you will have the
volume of the whole barrel in [cubic] inches.[8] And you will
convert these inches into feet. Know that each [cubic] foot

---

[7]"Li toniaus . . . 49 paus dedens" (V, 2.1-6). This material and everything up through V, 8 follows the French version of ACC III, 11, which deals with the volume of barrels.
This section is metrological, as are parts of V, 7 and V, 8
below. The author gives us the dimensions of a barrel, which
seems to have the shape of the frustum of a cone. The foot
here contains 12 inches. The author has averaged the top and
bottom diameters to find the width of the barrel.

[8]"Se tu veus savoir . . . le tonel per paus" (V, 2.7-12).
The author tells us to square the average base, yielding a
square measurement, which we are then to "round," presumably
by taking 11/14 of the square. Next we are to multiply that
value by the length of the barrel. The procedure amounts to
using the formula $V = \frac{11}{14}(\frac{d+D}{2})^2 h$, where d and D are the diameters of the two bases and h is the height of the barrel.
If all of the measurements are in inches, the result will
be in cubic inches.

cascuns piés tient en soi 1700 paus et 28, et li <140> paus

et <4> font le quarré, et ensi tient li piés 12 quarrés.

Et li toniax tient 22 et li viii isme d'un piét.   15

---

13 140 (ed.): 180 RS
14 et (1) om.R / 4 (ed.): 8 RS / quarré: quarrés R
14-15 et ensi . . . quarrés. Et om.R
15 toniax: tonnials R / viii isme: viiiesme R

contains 1728 [cubic] inches, and 144 [square] inches make

a square foot. And thus the [cubic] foot contains 12

"squares." And the barrel holds 22 and an eighth [cubic]

feet.[9]

---

[9]"Et ces paus . . . li viii isme d'un piét" (V, 2.12-15)
The author is not clear about the distinction between linear,
square, and cubic feet. In order for his comment about a
cubic foot equaling 12 "squares" to make sense, the "square"
must be taken to mean 144 cubic inches; this value will be
needed again below. The value of 22 1/8 cubic feet is very
close to what would be computed using the procedure in note
8; the resulting volume would be 30201 5/8 in.[3] or very
nearly 22 1/8 ft.[3]

[V, 3]   Li piés reons doit estre de <12> pax de dyametres.
Quant il est multepliés par soi et ramenés en reont si
contient l'aire.

   Si com l'aire du fons [est] 114 paus, et cele somme
multeplié par 12 si fait <1368>. Cest somme devisé par   5
le quarré, c'est a dire par 144 ke la quarré tient; si
troveras on pié reont 9 quarrés et demie. Et de teus piés
tient le toniax 28. Or as les rieules de droite muison.
Et sachies ke cis seus tonniaus soit 1 pauc plus lons,
si seus paus le croisteroit 5 quarrés et les 3 pars de   10
demie quarré; et s'il estoit 1 pauc plus gros, cis paus
le croisteroit 17 quarrés.

---

1 12 (<u>ed.</u>): 22 <u>RS</u> / pax: paus <u>R</u>
4 paus: pals <u>R</u>
5 1368 (<u>ed.</u>): 1376 <u>RS</u>
6 ke: que <u>R</u>
7 on: 1 <u>R</u> / pié: piét <u>R</u> / teus: tels <u>R</u>
8 le: li <u>R</u> / toniax: tonniax <u>R</u> / rieules: riules <u>R</u>
9 sachies: sacies <u>R</u> / ke: que <u>R</u> / tonniaus: tonniax <u>R</u> /
   lons: lonc <u>R</u>
11 paus: pals <u>R</u>

V, 3. A round foot should be of 12 inches in diameter.
When [12] is multiplied by itself and brought back to
roundness, the area is contained in it.[10]

So that if the area of the base is 114 inches and that
sum multiplied by 12, 1368 is produced. Divide that sum
by the "squares" the [cubic] foot contains, that is by 144;
you will find that one [solid] round foot contains 9½ squares.
And the barrel contains 28 of such feet. Now you have the
rules of correct measure.[11]

And know that if this same barrel was one inch longer,
this single inch would enlarge it 5 and 3/8 "squares";
and if it was an inch bigger across, this inch would enlarge it 17 "squares."[12]

---

[10]"Li piés reons . . . contient l'aire" (V, 3.1-3). The
author introduces the very odd measure of the round foot; it
is the area of a circle inscribed in a one-foot square. If
the usual 11:14 ratio of circle to circumscribing square is
used, the round foot would contain 113 1/7 square inches.
"Brought back to roundness" must mean multiply by 11 and
divide by 14.

[11]"Si com l'aire du fons . . . droite muison" (V, 3.4-8).
Multiplying the area of a round base by a height would yield
the volume of the cylinder based on that area. The numerical
cal values suggest that the author is giving us the dimensions of the "cylindrical foot" based on his round foot.
This solid round foot would contain 1368 in.^{3} or 9½ "squares"
or nearly 11/14 ft.^{3} The barrel containing 28 solid round
or "cylindrical" feet is very close to the same size as that
containing 22 1/8 ft.^{3} described in V, 2.14.

[12]"Et sachies . . . 17 quarrés" (V, 3.9-12). The author's
values for the amount of change in the barrel's volume are
very nearly right for the barrel and method in the preceding
proposition. The author's method for calculating volume is to
presume the barrel is a cylinder having as its diameter the
mean diameter and its length the length of the barrel. Increasing the length by one inch would enlarge the barrel by
779 5/8 in.^{3} or by 5 53/128 or 5.414 "squares" of 144 in.^{3}
One more inch in diameter would add 2464 in.^{3} or 17 1/9
"squares."

[V, 4] Or se tu veus multeplier le dyametre par soi misme, cele somme tu deviseras par 14 et tantes fois com tu troveras 14, el quarré avera ii el reont. Et saches ke ce sorcroist 780, le quart d'un pauc l'uitisme mains, ansi com par l'autre rieule devant dite. Et cele somme 5 dois tu multeplier par le lonc du tonel, c'est a dire par 49 paus, si troveras en le somme de tout le tonel <38202> pauchiers.

---

1 veus: vels R / multeplier: monteplier R / misme: meisme R
3 saches: saces R / ke: qui ?R
4 ce: ces R
5 rieule: riule R
6 tonel: tonnel R
7 paus: pauc R / tout: tot R / tonel: tonnel R /
   38202 (ed.): 34202 RS

V, 4. Now if you want to multiply the diameter by itself
and divide that sum by 14, then each time you find 14, the
square will have 2 [times] the circle.[13] And know that the
excess is 780, less a quarter and an eighth, inches [i.e.,
779 and 5/8 inches], just as by the aforesaid rule. And
you should multiply that sum by the length of the barrel,
that is, by 49 inches; you will find in the sum of the
whole barrel 38202 inches.[14]

---

[13]"Or se tu veus . . . el reont" (V, 4.1-3). It is not
clear what the author intends here. Perhaps he means something that amounts to saying that areas of circles are to
one another as the squares on their diameters.

[14]"Et saches ke ce sorcroist . . . pauchiers" (V, 4.3-8).
As described in note 12, enlarging the barrel--the cylinder
really--by one inch in length would increase the volume by
779 5/8 in.^{3} For a cylinder, the volume is proportional to
the length; therefore, the volume of the whole cylinder is
38201 5/8 in.^{3}

[V, 5] Et se tu veus en autre maniere, multeplier le

dyametre par soi misme et cele somme multeplie par le lonc,

si as le contenance del vassel quarré. Se tu veus ramener

en reont multeplie toute cele somme par 11, et la somme

ki en istra devisée par 14. Cele devisions te monstera   5

quant paus cis toniaus tient.

---

1 veus: vels <u>R</u>
2 misme: meismes <u>R</u>
3 vassel: vaissel <u>R</u> / veus: vels <u>R</u>
4 multeplie: multeplier <u>R</u> / toute: tote <u>R</u>
5 ki: qui <u>R</u> / istra: istera <u>R</u> / devisée: devise <u>R</u> /
    monstera: mousterra <u>R</u>
6 quant: quans <u>R</u> / toniaus: tonniax <u>R</u>

V, 5. And if you want [it] in another way, multiply the
diameter by itself and multiply the product by the length,
then you will have the contents of the square vessel.[15]
If you want to convert it into round, multiply the whole
product by 11 and divide the resulting product by 14. This
division will show you how many inches the barrel holds.[16]

---

[15]"Et se tu veus . . . vassel quarré" (V, 5.1-3). The
author gives a method for finding the volume of a square
column: $V = d^2h$, where d is the side of the square base
and h the height of the column.

[16]"Se tu vels ramener . . . toniaus tient" (V, 5.3-6).
A round column or cylinder will have a volume $\pi/4$ times that
of a square column with the side of its base equal to the
diameter of the round base.   The common medieval value of
11/14 is used here for $\pi/4$.

[V, 6]   Et ki auroit 1 tonel dont li fons fust des 3 piés

et demi et li lons de 6 piés, li toniaus terra 7 m^o et <5>

sestiers et 1 quarte.   Et ki auroit 1 autre tonel dont li

fons eust 7 piés et li lons 12, si terroit 58 muis, une

quarte mains.   Car c'est droite rieule en gyometrie ke   5

se li 1 vassels se double en lonc et en lei, li secons

tient les 8 des premiers.

---

1 ki: qui R / auroit: averoit R / tonel: tonnel R / des: de R
2 toniaus: tonniax R / terra: tenra R / 5 (ed.): 6 RS
3 ki: qui R / tonel: tonnel R
4 terroit: tenroit R
5 ke: que R
6 vassels: vaissaus R

V, 6. And if you have a barrel, the base of which is 3 and a half feet and the length 6 feet, it will hold 7 muids, 5¼ setiers. And if you have another barrel, the base of which is 7 feet and the length 12, it would hold 57 3/4 muids.[17] For this is a correct rule in geometry: if one vessel is double another in length and width, the second holds 8 times the first.[18]

---

[17]"Et ki auroit . . . 58 muis, une quarte mains" (V, 6.1-5). The author seems to be dealing with a cylindrical barrel in this and the next proposition. The material in these propositions yields some valuable insights into the metrology of the author's time and region. Since one barrel, containing 57 3/4 muids, is 8 times as large as one holding 7 muids plus 5¼ setiers, then one muid must equal 24 setiers. The emendation in V, 6.2 from 6 to 5 is possible because of the further information in V, 7.5 that each "piés quarré" or cubic foot contains 3 setiers.

[18]"Car c'est droite . . . 8 des premiers" (V, 6.5-7). The author's rule is correct for columnar vessels; the statement is an extension of the concerns of III, 9 - III, 12.

[V, 7] Se tu veus savoir de 1 tonel combien il tient,
multiplie par soi le dyametre--s'il est de 2 piés et demi,
si aras 6 piés et le quart d'un pié; aprés multiplie cele
somme par le lonc--s'il est de 4 piés, che sont 25 pié
quarré. Et saces ke cascuns piés quarré tient 3 sestiers.   5
Aprés multiplie ces 25 piés par 3; si aras 75 sestiers.
Et sacun sestier tient 4 quartes. Donc multiplieras ces
sestiers par 4; ce sont 300. Aprés le remenras en reont
en tel maniere: multiplie cele somme par 11--s'aras 3300--
<et devise 3300 par 14--s'aras 235 quartes et 3 pars, le   10
vii esme partie d'un quart mains>.

---

1 veus: vels R̲ / tonel: tonnel R̲
2 multiplie: multeplie R̲
3 multiplie: multeplie R̲
4 che: ce R̲
6 multiplie: multeplie R̲
7 sacun: cascuns R̲ / sestier: sestiers R̲ / multiplieras:
    multeplieras R̲
8 post 4 add. fois R̲
9 multiplie: multeplie R̲
10-11 et devise . . . mains (ed.): dont cascuns (cascons R̲)
    fait l'onsisme (onsime R̲) partie d'une quarte mains R̲S̲

V, 7.   If you want to know how much a barrel holds, multiply

its diameter by itself--if it is 2½ feet, you will have 6¼

[square] feet; then multiply that sum by the length--if it

is 4 feet, there will be 25 [cubic] feet.[19]   And know that

each [cubic] foot contains 3 setiers.   Then multiply these

25 feet by 3, you will have 75 setiers.   And each setier

contains 4 quarts.   Then multiply those setiers by 4, there

will be 300.[20]   Then convert to round as follows: multiply

that sum by 11, you will have 3300; and divide 3300 by 14.

You will have 235 3/4 - 1/28 quarts [i.e., 235 5/7].[21]

---

[19]"Se tu veus . . . 25 pié quarré" (V, 7.1-5).   Once again
the author is applying the procedure of V, 5.1-3, squaring
the diameter and multiplying the product by the height to
find the volume of the square column circumscribing the cylindrical barrel.   Again, as in note 17, "pié quarré" must be
understood as cubic feet.

[20]"Et saces . . . ce sont 300" (V, 7.5-8).   The author
yields important metrological information here; a "sestier"
is one-third of a cubic foot, and a "quarte" is one-fourth
of a "sestier."

[21]"Aprés le remenras . . . <d'un quart mains>" (V, 7.8-
11).   Converting the volume of the square column to that of
a cylinder requires a technique like that in note 16.   I have
freely changed the reading from that of the manuscripts (see
critical apparatus) to produce the required numerical values.
The corruption of the text may have stemmed from the misreading of vii esme as unsisme.

[V, 8] Se tu veus savoir d'un tonel de 66 sestiers, c'est

a dire de 3 muids, une quarte mains, par pauchiers, li fons

a 30 pauchiers et 3 pauchiers de bouge, ce sont 31 et demi.

Dont multeplieras 31 et demi par soi misme, si aras 992

pauchiers et le quart d'un pauchier. Aprés multiplieras   5

cele somme par le lonc du tonel, ki est de 49 paus, che sont

<48620¼> pauchiers. Aprés multiplieras toute cele somme

encore par 11 et la somme ki en venra deviseras par 14;

si averas <38202> pauchier. Et cele somme deviseras par

les quartes, c'est a dire par <144>; si troveras <265>   10

quartes, ce sont 66 sestiers et <une> quartes.

Et se tu veus cel tonel engrossier 1 pauc, tu

troveras le sorcrois 171 quartes.

---

1 veus: vels R / tonel: tonnel R
4 misme: meisme R
5 multiplieras: multeplieras R
6 tonel: tonnel R / ki: qui R / che: ce R
7 48620¼ (ed.): 48602 RS / multiplieras: multepliera R /
    toute: tote R
8 ki: qui R
9 38202 (ed.): 38200 RS
10 144 (ed.): 36 RS / troveras: trovera R / 265 (ed.):
    1061 RS
11 une (ed.): 5 RS / quartes: quarres R
12 veus: vels R / tonel: tonnel R / engrossier: ergrossier R

V, 8. If you want to know [the dimensions] of a barrel of
66 setiers, or 2 3/4 muids, in inches,[22] [they are as follows:] the base is 30 inches [in diameter], the mouth is
3 inches [in diameter], [averaged] they are 31½ inches.
Then multiply 31½ by itself and you will have 992¼ inches.
Then multiply that sum by the length of the barrel, which
is 49 inches, there will be 48620¼ inches. Then multiply
that whole sum by 11 and divide the result by 14; you will
have 38202 [cubic] inches.[23] Divide that sum by the [cubic
inches in one] quart, that is by 144, you will find 265
quarts, which are 66 setiers and one quart.[24]

And if you want to enlarge this barrel by 1 inch, you
will find the increase to be 171 quarts.[25]

---

[22]"Se tu veus savoir . . . par pauchiers" (V, 8.1-2).
Further evidence of the author's metrology is the information that 2 3/4 muids equals 66 setiers. From this equivalence we have one muid equal to 24 setiers, just as in
note 17.

[23]"li fons a 30 pauchiers . . . si averas <38202> pauchier" (V, 8.2-9). This is a numerical example of the technique described in V, 5. The numbers here have been modified
slightly for consistency.

[24]"Et cele somme . . . 66 sestiers et <une> quartes" (V,
8.9-11). Here the numbers had to be emended significantly
because according to V, 7.5-8 (see note 20), there are 12
quarts in a cubic foot; thus 144 cubic inches make a quart.

[25]"Et se tu veus . . . 171 quartes" (V, 8.12-13). I cannot determine what way the author would have us enlarge the
barrel to increase its contents 171 quarts.

[V, 9] En tans piés comme li pauchiers a de lonc vaut 1 pauchiers 1 pié, et en tant piés comme li piés a pauchiers en l'aire de ses quarrie vaut 1 pauchiers tant piés comme li piés a pauchiers de lonc.

---

1 comme: que R
3 vaut: valt R / comme: com R

V, 9.   In as many feet as there are inches in length, one
inch is worth one foot; and in as many feet as a foot has
inches in the area of its square, one inch is worth as
many feet as the foot has inches in length.[26]

---

[26]"En tans piés . . . pauchiers de lonc" (V, 9.1-4).   This
and the next two propositions follow the French version of
III, 17.   I cannot understand this proposition.   Perhaps the
author is saying something like "when you add an inch to the
length of a rectangle, you add as many square inches as there
are inches in its width," or $(h+x)w = hw + wx$ where x is the
number of inches added to the length h, and w is the width.

[V, 10] Derechief en tantes perces comme la perce a piés
de lonc, vaut 1 piés 1 perce. Et en tantes perces comme
la perce a piés en l'aire de quarrie, vaut li piés tantes
perces comme la perce a piés de lonc. Ausi ravient des
perces a arpens. Se tu as mesuré une droite partie et tu   5
vels assambler demi pié, saches ke chis demis piés vaut
tant piés comme la quarrie avoit devant lonc, et plus le
quart, demi pié mains. Et se li quadrangles est lons et
estrois par le moitié dou lonc et dont le fera ausi.

---

1 derechief: derekief R / comme: com R
3 de: des R / vaut: valt R / tantes: tante R
4 comme: com R
6 saches: saces R / ke: que R / chis: cis R
9 dou: du R

V, 10. Then in as many rods as the rod has feet in length, one foot is worth one rod. And in as many rods as the rod has feet in the area of its square, the foot is worth as many rods as the rod has feet in length. Similarly rods are reduced to arpents. If you have measured one whole [?] part and you want to add a half foot, know that these half-feet are worth as many feet as the square had before in length plus a fourth, less a half foot. And if the rectangle is long and narrow, [add as many square feet as] half the length, and that will do it too.[27]

---

[27]"Derechief en tantes perces . . . le fera ausi" (V, 10.1-9). The author seems to be extending the technique of the previous proposition.

[V, 11] Se tu vels en droite quarrie croistre, alonge ta
lisiere le quart et le siste, et c'est li lons de le ligne
dyagonal. Si com on le puet plus prendre pres, c'est a
dire ke la desraine valt les 2 pars et plus en 72 un quart.

---

4 ke: que R / desraine: daesraine R

V, 11. If you want to enlarge in a correct square, lengthen
your side a fourth and a sixth; then that is the length of
the diagonal line. One can take it closer as well; that is,
the next [approximation] is worth the two fractions plus
1/288 [extra] [i.e., subtract 1/288 from their sum].[28]

---

[28]"Se tu vels . . . en 72 un quart" (V, 11.1-4). This is
a technique for doubling a square. The method is to lengthen
the side to make it equal to the diagonal of the original
square. That diagonal will be the side of a square having
twice the area of the first one. The author's first approximation for the diagonal d amounts to $d/a = 17/12$ where a is
the side of the first square. This value yields a new square
with area 2 and 1/144 times the first. The closer approximation is $d/a = [(17/12) - (1/288)]$, producing an area 2 minus
5/1728 times the first.

[VI, 1]   Se tu as le fort monoie, multeplie par cange tant
saus com tu vels cangier, et cele devise par 12. Et se
tu vels le foible monoie tant saus com tu as, multeplie
par 12, et cele somme devise par le cange. Che sont o.
che ke tu fais devant par 12, fai par 24, et des pugoces   5
par 48, et cel cange a cele misme mesure.

---

1 monoie: monnoie R
3 monoie: monnoie R / com: comme R
4 che: ce R
5 che: ce R / pugoces: pugoises R
6 misme: meisme R

VI, 1. If you have strong money, multiply by the exchange
as many <u>sous</u> as you want to change, and divide that by 12.
And if you want to change weak money, multiply the <u>sous</u> you
have by 12, and divide that sum by the exchange.[1] [If the
rate is in] <u>oboles</u>, do by 24 what you did before by 12; for
<u>pugoises</u>, by 48; and that exchange has the same value.[2]

---

[1]"Se tu as . . . par le cange" (VI, 1.1-4). In order for
this procedure to make sense, the exchange rate must be given
in <u>deniers</u> per <u>sou</u>, with the <u>deniers</u> always in the weaker
money. Thus, the exchange rate must normally have been the
number of <u>deniers</u> of the weaker money needed to buy one <u>sou</u>
of the stronger money. The factor of 12 used in the operations converts, in effect, the <u>deniers</u> of the exchange rate
to <u>sous</u>. The conversion works, of course, because there are
12 <u>deniers</u> in one <u>sou</u>.

[2]"Che sont . . . mesure" (VI, 1.4-6). An <u>obole</u> is one-
half a <u>denier</u>; a <u>pugoise</u>, also called <u>poitevine</u> or <u>pite</u>, is
one-quarter of a <u>denier</u>. In these instances the conversion
is based on the equivalences of 24 <u>oboles</u> per <u>sou</u> and 48
<u>pugoises</u> per <u>sou</u>.

[VI, 2] Des 6 freires ki avoient 7560 sen: ot li i le
quart, che fu 1890; li autres le quint, che fu 1512; li
autres le siste, che fu <1260>; li autres le septisme,
che fu 1080; li autres le witisme, che fu <945>; li autres
le viiii isme, che fu 840, et si remest 33.   5

---

1 des: es R / freires: freres R / 7560: 7566 R / ot: ont R
2 che (1): ce R / che (2): ce R
3 che: ce R / 1260 (ed.): 1056 RS
4 che (1): ce R / witisme: viii esme R / che (2): ce R /
    fu om.S 945 (ed.): 985 RS
5 viiii isme: viiii esme R / che: ce R / 840: 880 R

VI, 2.   Six brothers had 7560 <u>sous</u>.   The first took a fourth,
which was 1890; the next took a fifth, which was 1512; the
next took a sixth, which was <1260>; the next, a seventh,
which was 1080; the next an eighth, which was <945>; the
next a ninth, which was 840; and 33 remained.[3]

---

[3]"Des 6 freires . . . remest 33" (VI, 2.1-5).   This is
perhaps an exercise in fractional parts.   It can be written
as $\frac{1}{4}n+\frac{1}{5}n+\frac{1}{6}n+\frac{1}{7}n+\frac{1}{8}n+\frac{1}{9}n+33=n$, where n = 7560.

[VI, 3] Li xiii en monte li xii ismes et descent li xiii
ismes. A ii s. la livre monte li x ismes et descent li
onsismes. A xiii ein et o. monte li viii ismes et descent
li viiii ismes. A xiiii ein monte li vi ismes et descent
li vii ismes. A iiii s. la libre monte li v ismes et   5
descent li sisimes. A xv ein monte li quars et descent li
v ismes. A xv ein et o., ii d. plus la libre, monte li
quars et li xx ismes. A xvi ein, iiii plus la libre, monte
li quars et li x ismes. A xvi ein monte li iii et descent
li quars. A xvi ein et o. monte li quars et li viii   10
ismes. A xvii en, iiii mains la libre, monte li tiers et
li quinsismes et descent li quars et li <xxviii> ismes.
A xvii em et o. monte li tiers et li viii ismes. A xviii
em monte la moitiés et descent li tiers. A xix, pugoise
mains, monte li moitiés et li xvi ismes. A xviiii ein   15

1-2 xiii ismes: xiii isme R
3 onsismes: xi ismes R / ein: em R
4 viiii: ix R / ein: em R
6 sisimes: vi ismes R / ein: em R
7 ein: em R
8 ein: em R / la: li R
9 ein: em R
10 ein: em R
11 en: em R
12 quinsismes: xv ismes R / xxviii (ed.): xviii RS
13 et (1) om.R
14 em: en R / xix: xxx R
15 xviiii: xviii R / ein: em R

VI, 3.   [Part of text presented as a table]

value	"monte"	"descent"[4]
13	1/12	1/13
2s./lb.	1/10	1/11
13½	1/8	1/9
14	1/6	1/7
4s./lb.	1/5	1/6
15	1/4	1/5
15½ + 2d./lb.	1/4 + 1/20	
16 + 4d./lb.	1/4 + 1/10	
16	1/3	1/4
16½	1/4 + 1/8	
17 - 4d./lb.	1/3 + 1/15	1/4 + 1/28
17½	1/3 + 1/8	
18	1/2	1/3
18 3/4	1/2 + 1/16	

---

[4]"Li xiii . . . et li vi ismes" (VI, 3.1-24). Although
it seems a great mystery, and may remain so after my explanation, this section seems to relate the cost of a commodity,
probably money, to a standard cost of 12 deniers for that commodity. If this material is related to the exchange rates in
VI, 1, then the commodity must be one sou of a stronger money.
Thus, taking the first example, given that it costs 13
deniers for a sou of a stronger money, the "monte" figure
means that it will cost 1/12 more to buy the sou than it
would a sou of the weaker money. The "descent" figure means
that one can buy 1/13 less of the stronger money with weaker
money of the same face value. The only cases where the reference is not to the cost of a sou are the examples of 2 sous
and 4 sous per livre, where the reference is clearly to a
livre. Thus "2 s. la livre" means a pound of the stronger
money costs 2 sous more of the weaker money, i.e. 22 sous.
In other words it will cost 1/10 more to buy stronger money
of the same face value, or one could get 1/11 less of the
stronger money. With only a few minor changes, the values
given in VI, 3 are correct, as may be seen in the table which
I have presented instead of a translation. The "monte" and
"descent" may be computed or verified using the following

monte li moitiés et li xii ismes. A xix em, iiii plus le
libre, monte li moitiés et li x ismes et descent li quars
et li viii ismes. A xix em et o. monte li moitiés et li
viii isme. A xx em monte les ii pars et descent li
<tiers> et li xv ismes. A xx em et o., ii mains la libre, 20
monte li moitiés et li v ismes. A xxi em montent les iii
pars. A xxii montent les iii pars et li xii ismes. A
xxii et o. monte les iii pars et li viii ismes. A xxiii
em monte les iii pars et li vi ismes.

. . . quant li xii valent xxiiii, c'est a dire i por ii.   25

   Autant valent xx libres xxvi ein com xl libres xiii ein;
autant valent xvi libres xxvii em com xxxii libres xiii
ein et o.; autant valent xviii libres xxviii ein com xxxvi
libres xiiii ein; autant valent xiiii libres xxx ein com
xxviii libres xv ein; autant valent x libres xxx et   30

---

16 xii ismes: xii isme R
19 viii isme: viii ismes R / em: en R
20 tiers (ed.): quars RS
21 moitiés om.S
22 A xxii montent les iii pars om.S
24 em: en R
26 ein (1): em R / com: monte R / xiii: iii S /
   ein (2): em R
27 em com xxxii libres xiii om.S
28 ein (1): em R / ein (2): em R
29 ein (1): ain R / autant valent tr.S / ein (2): em R
30 ein: em R

19	1/2 + 1/12	
19 + 4d./lb.	1/2 + 1/10	1/4+1/8
19½	1/2 + 1/8	
20	2/3	1/3+1/15
20½ - 2d./lb.	1/2 + 1/5	
21	3/4	
22	3/4 + 1/12	
22½	3/4 + 1/8	
23	3/4 + 1/6	

. . . when 12 are worth 24, that is one for two.[5]

Twenty lbs. at 26 of them are worth as much as 40 lbs. at 13 of them; 16 lbs. at 27 of them are worth as much as 32 lbs. at 13½ of them; 18 lbs. at 28 of them are worth as much as 36 lbs. at 14 of them; 14 lbs. at 30 of them are worth as much as 28 lbs. at 15 of them; 10 lbs. at 32 are worth

---

formulas: "monte" = k-1, and "descent" = $1-\frac{1}{k}$, where k depends on the form in which the values are given. If the value v is understood as in deniers per sou then k = $\frac{v}{12}$. If another adjustment a, in deniers per livre, is made then k = $\frac{v}{12} \pm \frac{a}{240}$. in the case of exchange values s given in extra sous to be paid per livre then k = $\frac{s+20}{20}$. This same kind of material is found below in VI, 3.37-43.

[5]". . . quant li xii . . . i por ii" (VI, 3.25). This line makes no sense at all in its context. Perhaps material like that in the next paragraph, but with the ratio of one to two, preceded this line but was lost in copying.

doubliaus com xx libres xvi ein; autant valent viii libres

xxxiii ein com xvi libres xvi ein et o.; autant valent <vii>

libres xxxvi ain comme xiiii libres xviii ain; autant valent

c s. xl ain comme x libres xx ain; autant valent xx libres

xl et doubliaus comme xl libres xxi ain, quant li xii   35

valent xlviii, c'est a dire i por iiii.

A xxvi en descent li moitiés et li xxvi ismes. A xxvii

en descent li moitiés et li xviii ismes. A xxviii en

descent li moitiés et li <xiiii> ismes. A xxx en descent

li moitiés et li x ismes. A xxx et doubliaus descent li 40

moitiés et li viii ismes. A xxxvi en descent les ii pars.

A xl descent li moitiés et li v ismes. A xlviii en descent

les iii pars.

Autant valent xvii libres l [i.e. 50] com l [i.e. 50]

libres xvii ain.   45

---

31 xvi: xv <u>S</u> / ein: em <u>R</u>
32 ein (1): em <u>R</u> / ein (2): em <u>R</u> / vii (<u>ed.</u>): viii <u>RS</u>
33 comme: come <u>R</u>
34 ain (1): em <u>R</u> / comme: com <u>R</u>
35 et <u>om.S</u> / comme: com <u>R</u>
37 en: ain <u>R</u>
37-38 xxvi ismes. A xxvii en descent li moitiés et li <u>om.R</u>
39 xiiii (<u>ed.</u>): xii <u>R</u> xiii <u>S</u>
40 doubliaus: doubliax <u>R</u>
41-42 les ii pars. A xl descent <u>om.R</u>

as much as 20 <u>lbs.</u> at 16 of them, 8 <u>lbs.</u> at 33 of them are worth as much as 16 <u>lbs.</u> at 16½ of them. 7 <u>lbs.</u> at 36 of them are worth as much as 14 <u>lbs.</u> at 18 of them; 100 <u>sous</u> at 40 of them are worth as much as 10 <u>lbs.</u> at 20 of them; and 20 <u>lbs.</u> at 42 are worth as much as 40 at 21 of them, when 12 is worth 48, that is one for four.[6]

value	"descent"[7]
26	1/2 + 1/26
27	1/2 + 1/18
28	1/2 + 1/14
30	1/2 + 1/10
32	1/2 + 1/8
36	2/3
40	1/2 + 1/5
48	3/4

Seventeen <u>lbs.</u> at 50 are worth as much as 50 <u>lbs.</u> at 17 of them.[8]

---

[6]"Autant valent . . . i por iiii" (VI, 3.26-36). This section seems to be little more than a table of reciprocal values to show that if something quadruples in price, one can get only half as much for twice as much money. If this material has some relation to exchange rates I have not determined what it is.

[7]"A xxvi . . . les iii pars" (VI, 3.37-43). This is a continuation of the material found in VI, 3.1-24 above. The explanation given in note 4 applies equally here. This section, however, gives only the "descent" values omitting those for "monte."

[8]"Autant valent . . . xvii ain" (VI, 3.44-45). This appears to be part of another reciprocal values table like that in VI, 3.26-36.

[VI, 4] Quant li mars d'or sera vendus xx libres et ie
vaurai savoir ke vaura li d. pesans de <xxiiii> d. en l'once,
ie prendrai xx d. et xx p., che sont xxv d. ke li d.
pesans vaura de xxiiii en l'once.

Quant li mars d'or sera vendus xvi libres et ie   5
vaurai savoir ke vaura li esterlins pesans de xx en l'once,
je prendrai xvi d. et xvi o., c'est a dire xxiiii d. ke
li esterlins pesans vaura. Et a cascune libre ke li
mars d'or vaura, je prendrai iii o. por le valor del
esterlin pesant de xx en l'once.   10

Quant li mars d'or sera vendus xii libres et ie
vaurai savoir ke vaura li esterlins pesans de xl en l'once,
je prendrai xii o. et xii poitevines, c'est a dire ix d.
ke li esterlins pesans vaura de xl en l'once.

---

2 vaurai: vaura R / ke: que R / vaura om.S / xxiiii (ed.):
  xiiii RS
3 che: ce R / ke: que R
6 vaurai: vaura R / ke: que R / esterlins: esterlin R
7 je: ie R / ke: que R
8 a om.R / ke: que R
12 vaurai: valra R / ke: que R / vaura: valra R
14 ke: que R

VI, 4. When a _marc_ of gold is sold for 20 _livres_ and I want to know the value of a _denier_ weighing 24 _deniers_ to the _once_, I will take 20 _d_. and 20 _p_., which are 25 _d_., which the _denier_ weight will be worth at 24 _d_. to the _once_.[9]

When a _marc_ of gold is sold for 16 _livres_ and I want to know the value of an _esterlin_ weighing 20 to the _once_, I will take 16 _d_. and 16 _o_., that is 24 _d_., which the _esterlin_ weight will be worth. And [in general] for each _livre_ that the _marc_ of gold is worth, I will take 3 _o_. [i.e. 1½ _d_.] for the value of the _esterlin_ weighing 20 to the _once_.[10]

When a _marc_ of gold is sold for 12 _livres_ and I want to know the value of an _esterlin_ weighing 40 to the _once_, I will take 12 _o_. and 12 _p_., that is 9 _d_., which the _esterlin_ at 40 to the _once_ will be worth.[11]

---

[9]"Quant li mars . . . de xxiiii en l'once" (VI, 4.1-4). This and the following are simple problems in conversion of units. Here the procedure amounts to multiplying the value n of _livres_ per _marc_ by 1¼ d. to find the value of a _denier_ weighing 1/24 of an _once_. The only "hidden" part in the conversion is the fact that one _marc_ equals 8 oz. The multiple 1¼ may be derived in this way for a _denier_ weighing 24 per _once_. Let _D_ be that _denier_, then

$$\frac{1 \text{ oz.}}{24 \text{ D}} \times \frac{1 \text{ marc}}{8 \text{ oz.}} \times \frac{n \text{ lb.}}{\text{marc}} \times \frac{240 \text{ d.}}{\text{lb.}} = \frac{240 \cdot n \text{ d.}}{24 \cdot 8 \text{ D}} = \frac{1\frac{1}{4} \cdot n \text{ d.}}{D} .$$

The example in the text has the value of the _marc_ as 20 lbs., i.e. n is 20. The resulting value of d./D is 20x1¼ or 25 d./D.

[10]"Quant li mars . . . xx en l'once" (VI, 4.5-10). This is another conversion of units problem. This time the cost n of a _marc_ in lbs. is to be multiplied by 1½ d. to find the value of an _esterlin_ E weighing 1/20 of an _once_. The conversion factor of 1½ d. is derived as follows:

$$\frac{1 \text{ oz.}}{20 \text{ E}} \times \frac{1 \text{ marc}}{8 \text{ oz.}} \times \frac{n \text{ lb.}}{\text{marc}} \times \frac{240 \text{ d.}}{\text{lb.}} = \frac{240 \cdot n \text{ d.}}{20 \cdot 8 \text{ E}} = \frac{1\frac{1}{2} \cdot n \text{ d.}}{E}$$

The example in the text has n as 16, and thus the value for d./E is 1½·16 or 24.

[11]"Quant li mars . . . xl en l'once" (VI, 4.11-14). This conversion exercise has us multiply n by 3/4 d. to get the

Quant li once d'or sera vendue xl s. et ie vaurai   15

savoir ke vaura li esterlins pesans du marc de flandres

de xvi en l'once, je prenderai xl o. et xl poit., c'est a

dire xxx d. ke li esterlins pesans vaura. Et tout cist

conte sont contable par ceste misme raison.

Quant li mars d'argent sera vendus xlviii s. et   20

ie vaurai savoir ke li mars de billons a v d. vaura, je

prendrai par v fois xlviii, c'est a dire xx s., ke

li mars de billons a v d. vaura. Et billons a iiii d.,

xlviii d., c'est a dire iiii s. [por le d. de billons].

Et billons a une o., xlviii o., c'est a dire ii s. Et   25

billons a une poit., xlviii poit., c'est a dire xii d.

Et billons a 1 grain, xlviii grains ki font ii d., a le

---

15 vaurai: valrai R
17 poit.: p. R
18 ke: que R / pesans: pensans R / tout om.R / cist: cis R
19 conte: contes R / misme: meisme R
21 vaurai: vaura R
22 prendrai: prenderai R / ke: que R
23 vaura: valra R
26 poit. (1): poi. R / poit. (2): poi. R / a dire om.S
27 ki: qui R

When an <u>once</u> of gold is sold at 40 <u>s</u>. and I want to know
the value of an <u>esterlin</u> weighing 16 to the <u>once</u> of the <u>marc</u>
of Flanders, I will take 40 <u>o</u>. and 40 <u>p</u>., that is, 30 <u>d</u>.
which the <u>esterlin</u> weight is worth. And all these problems
may be computed in the same way.[12]

When a <u>marc</u> of silver is sold at 48 <u>s</u>. and I want to know
the value of a <u>marc</u> of bullion at 5 <u>d</u>., I will take 5 [<u>d</u>.]
times 48, that is 20 <u>s</u>. which the <u>marc</u> of bullion at 5 <u>d</u>. is
worth. And for bullion at 4 <u>d</u>. I take 48 <u>d</u>., that is 4 <u>s</u>.
for each <u>denier</u> of bullion. And for bullion at one <u>o</u>. I
take 48 <u>o</u>., that is 2 <u>s</u>. And for bullion at 1 <u>poitevine</u> I
take 48 <u>poitevines</u>, that is 12 <u>d</u>. And for bullion at 1
<u>grain</u> I take 48 <u>grains</u> which makes 2 <u>d</u>., with a <u>marc</u> of

value of an <u>esterlin</u> E weighing 1/40 of an <u>once</u>:

$$\frac{1 \text{ oz.}}{40 \text{ E}} \times \frac{1 \text{ marc}}{8 \text{ oz.}} \times \frac{n \text{ lb.}}{\text{marc}} \times \frac{240 \text{ d.}}{1\text{b.}} = \frac{240 \cdot n \text{ d.}}{40 \cdot 8 \text{ E}} = \frac{(3/4) \cdot n \text{ d.}}{E} \ .$$

Here 3/4 of 12 produces the text's 9 d./E.

[12]"Quant li once . . . misme raison" (VI, 4.15-19). Aside
from the fact that here we have the price of gold in <u>sous</u> per
<u>once</u> instead of in <u>lbs.</u> per <u>marc</u>, this problem is like the
ones above. The conversion factor, 3/4 d. arises from:

$$\frac{1 \text{ oz.}}{16 \text{ E}} \times \frac{k \text{ s.}}{\text{oz.}} \times \frac{12 \text{ d.}}{\text{s.}} = \frac{12 \cdot k \text{ d.}}{16 \text{ E}} = \frac{(3/4) \cdot k \text{ d.}}{E} \ , \text{ where } \underline{k} \text{ is the}$$

price of gold in <u>sous</u> per <u>once</u>. In the example <u>k</u> is 40,
which when multiplied by 3/4 d. yields 30 d.

raison del marc d'argent de xlviii s.

Quant li mars de fin argent sera vendus xxxiiii s.

et ie vaurai savoir ke li mars de billons a iii d. vaura, 30

je prendrai par iii fois xxxiiii d., c'est a dire viii s.

et demi ke li mars de billons a iii d. vaura; et billons

a 1 d. xxxiiii d.; et billons a 1 grain xxxiiii grains dont

li <xxiiii> grains poisent 1 d. de xxiiii d. en l'once.

Et tout cist conte sont contable par cest misme raison.   35

---

28 marc: mars R
30 vaurai: vaura R
31 prendrai: prenderai R
32 ke: que R
34 xxiiii (ed.): xxxiiii RS
35 tout: tot R / contable: contables R / misme: meisme R

silver worth 48 s.[13]

When a <u>marc</u> of fine silver is sold for 34 s. and I want
to know the value of a <u>marc</u> of bullion at 3 d., I will take
3 times 34 d., that is 8½ s. which the <u>marc</u> of bullion at
3 d. will be worth. And bullion at 1 d. will be worth 34
d. and bullion at 1 <u>grain</u> will be worth 34 <u>grains</u>, such that
24 <u>grains</u> weigh 1 d. at 24 d. in the <u>once</u>. And all these
problems may be computed in the same way.[14]

---

[13]"Quant li mars . . . de xlviii s." (VI, 4.20-28). Here
again we have a conversion, but this time the purpose is to
know the value of <u>marcs</u> of silver bullion minted at different
purities, given in <u>deniers</u> or parts of <u>deniers</u> per <u>sou</u>. If
the mintage rate in <u>deniers</u> per <u>sou</u> is m and the cost of a
marc of pure silver in <u>sous</u> is q, then the value of the <u>marc</u>
of bullion is $\frac{m \text{ d.}}{\text{s.}} \times \frac{q \text{ s.}}{\text{marc}}$ or $\frac{m \cdot q \text{ d.}}{\text{marc}}$. This value is easily
converted to <u>sous</u> by dividing by 12. In the example given
m is 5 and q is 48; 5 d. times 48 is 240 d. or 20 s. The
rest of this paragraph deals with mintage rates in fractions
of <u>deniers</u> per <u>sou</u>. The values computed are all for silver
at 48 s. per <u>marc</u>.

[14]"Quant li mars . . . misme raison" (VI, 4.29-35). This
is simply another example of the procedure in the previous
paragraph. Here m is 3 and q is 34; the value of the <u>marc</u>
of bullion is 102 d. or 8½ s.

[VI, 5]   Tous alouemens de monoie va par xvi s. le marc.
Et se i'ai une monoie a v d. et ie voil savoir combien
il a de fin argent el marc, je conterai par xvi fois v,
c'est a dire vi s. et viii d., ki font iii onces et le
tierc d'une once; et billons a iii d., par xvi fois   5
iii d., c'est a dire iiii s. ki font 1 firton; et billons
a 1 d. par xvi fois 1 d., c'est a dire xvi d. ki font <les
ii pars d'une> once; et billons a 1 maille par xvi fois
1 o., c'est a dire viii d. ki font le tierc d'une once;
et billons a 1 poit. par xvi fois 1 poit., c'est a dire   10
iiii d. ki font le tierc demie once; et billons a 1 grain
par xvi fois 1 grain, c'est a dire xvi grains dont li
xxiiii grain poisent 1 d. de xxiiii en l'once. Et tout
cist conte sont contable par ceste misme raison.

---

1 alouemens: aloemens R / xvi om.R
3 conterai: contera R
4 onces: onnces R
5 d'une: de 1 R
6 ki: qui R / firton: ferton R
7 fois om.R / ki: qui R
7-8 les ii pars d'une (ed.): demi RS
8 maille: o. R
9 tierc: tiers R
10 poit. (1): poi. R / poit. (2): poi. R
12 xvi: cxvi(?) R
13 poisent: font R / tout: tot R
14 misme: meisme R

VI, 5. All alloys of money are reckoned by 16 s. to the
marc. And if I have money at 5 d. and I want to know how
much fine silver there is in a marc of it, I compute 16
times 5, that is 6 s. and 8 d., which makes 3 and a third
onces. And for bullion at 3 d. I take 16 times 3 d., that
is 4 s., which makes a firton; and for bullion at 1 d. I
take 16 times 1 d., that is 16 d. which makes a half-once.
And for bullion at 1 maille I take 16 times 1 o., that is
8 d., which makes a third of an once. And for bullion at
1 poitevine I take 16 times 1 poitevine, that is 4 d.,
which makes a sixth of an once. And for bullion at 1 grain
I take 16 times 1 grain, that is 16 grains such that 24
grains weigh 1 d. at 24 to the once. And all these problems
are computable in the same way.[15]

---

[15]"Tous alouemens . . . misme raison" (VI, 5.1-14). This
section is a conversion from mintage rates given in deniers
or their fractions per sou to the value of the silver in the
alloy and then to the weight of silver in onces per marc.
Taking m as the mintage rate again, the value of silver in
a marc of 16 sous is $\frac{16 \text{ s.}}{\text{marc}} \times \frac{m \text{ d.}}{\text{s.}} = \frac{16 \cdot m \text{ d.}}{\text{marc}}$. Since there
are 24 d. per once this value need only be divided by 24 to
get the number of onces of silver per marc. Evidently a
firton is a weight of 2 onces.

[VI, 6]   Conte a xii d. est appellée fine; draant de Damas
sont a xii d.   Venicien et grousenois et roial d'Acre et
roial de Marseilles a xii d., poi. mains, c'est a dire ke
li mars tient iiii d. ki sont le tierc de demie once.
Esterlin et hollandois sont a xi d. et o., c'est a dire   5
ke li mars tient viii d. ki font le tierc d'une once.
Artisien noef et messaig sont a xi d. et poi., c'est a
dire ke li mars tient demie once.   Trossien et liegois
sont a xi d.   Brouselois sont a x d. et o., c'est a dire
<une once> en 1 marc de fin argent.   Frizacois sont a   10
x d.   Ramnés borgois sont a ix d.   Bialvoisien et toulousain de Toulouse et molain de Gascoigne sont a viii d.
Blanc de Valencienes sont a vii d. et o.   Mansois sont
a vii d., poi. mains.   Cambrisien sont a vi d. et poi.
Fors de Laon et fors de Suse sont a vi d., c'est a dire   15
le moitié de keuvre et le moitié d'argent.   Masconois
et clingnisien et tunitsois et scalonge d'Ardane et
moritien et li d. de Losane sont a v d. et o.   Vienois

---

1 conte: toute (?) S / appellée: apelée R
2 d. om.S
3 Marseilles: Marseille R / ke: que R
5 xi d. et o.: xi o. S
6 ki: qui R / d'une: d'un R
10 une once (ed.): vii onces RS
11 ramnés: ravivés (?) R
13 Valencienes: Valenchienes R / vii d. et o.: viis d. S
15 Laon: Loon R
18 v: vi R / et om.S

VI, 6. Account money at 12 <u>d</u>. is called fine.

The "draant" of Damascus is at 12 <u>d</u>.

The money of Venice, Grousenois [?], the "royal" of Acre
and the "royal" of Marseilles are at 11 3/4 <u>d</u>., that is, the
<u>marc</u> contains 4 <u>d</u>. [of alloy], which is a sixth of an <u>once</u>.

Sterling and the money of Holland are at 11½ <u>d</u>., that is,
the <u>marc</u> contains 8 <u>d</u>. [of alloy], which makes a third of
an <u>once</u>.

The new money of Arras and the money of Metz are at 11¼
<u>d</u>., that is, the <u>marc</u> contains [12 <u>d</u>. or] a half-<u>once</u> [of
alloy].

The monies of Troyes [?] and Liège are at 11 <u>d</u>.

The money of Brussels is at 10½ <u>d</u>., that is, there is
one <u>once</u> [of alloy] in one <u>marc</u> of fine silver.

Frizacois [?] is at 10 <u>d</u>.

The reduced [?] money of Bourges is at 9 <u>d</u>.

The money of Beauvais, the "toulousain" of Toulouse and
the "molain" of Gascony are at 8 <u>d</u>.

The "blanc" of Valenciennes is at 7½ <u>d</u>.

The money of le Mans is at 6 3/4 <u>d</u>.

The money of Cambrai is at 6¼ <u>d</u>.

The strong monies of Laon and Suse [?] are at 6 <u>d</u>., that
is, half copper and half silver.

The monies of Macon, Cluny [?], Tunis [?], the "scalonge"
of Ardennes, Moritien [?], and the <u>denier</u> of Lausanne are at
5½ <u>d</u>.

et valencenois et susain petit et emperial et princois

d'Antioches et marbotin de Saint Quentin sont a v d. poi.   20

Viés et les vieuses monoies paresies sont a v d., poi.

mains. Raencien et paresis noef sont a xiiii d. et iii

esses, c'est a dire ke en xii d. pesant a iiii d. et o.

et iiii grains. Certain et bloisois et bordelois et

sanserois et clermontois et doisien sont a iiii d., demie   25

poit. mains. Nantois et vendomois et parcerain et dunoi-

sien et barbarin de Saint Tirre sont a iiii d., poi. mains.

Fors de Soignie et aucerrois et fors de Provins et li

monoie ki court ensamble sont a xi d. et iii esses, c'est

a dire en xii d. pesant a iii d. et o. et iiii grains.   30

Genevois et li viés noiret de Soisons et burgalois et

fors de Navers et l'angoine et estievenent de Besencon

sont a iii d. et o. Lonisien sont a iii d. Diounois

sont a viii d. et iii esses, c'est a dire k'en xii d.

pesans a ii d. et o. et iiii grains. Caorsin et   35

---

20 Antioches: Antioce R / v: vi R

21 vieuses: vielles R / paresies: paresis R

23 ke: que R / et om.S

24 certain: chartain R

25 clermontois: clementois R

26 poit.: poi. R

27 Tirre: Tire R

29 ki: qui R / court: cort R / ensamble: emsamble R /
    esses: esse R

31 li om.R / noiret: noirois R

33 et om.S / diounois: dijounois R / pesans: pesant R

35 et om.S

The monies of Vienne, Valence, the small money of Suse, the "imperial," the "princois" of Antioch, and the "marbotin" of St. Quentin are at 5¼ d.

The "viés" and the old money of Paris are at 4 3/4 d.

The money of Reims and the new money of Paris are at 14 d. in 3, that is, there are 4 2/3 d. [of silver] in 12 d. weight.

The monies of Chartres, Blois, Bordeaux, Sancerre, Clermont, and Douay [?] are at 3 7/8 d.

The monies of Nantes, Vendôme, le Perche, Châteaudun, and the "barbarin" of Saint Tirre [?] are at 3 3/4 d.

The strong money of Soignies, the money of Auxerre, the strong money of Provins and the money that runs with them are at 11 d. in 3, that is, there are 3 2/3 d. [of silver] in 12 d. weight.

The money of Geneva, the old "noiret" of Soissons, Burgalois [?], the strong money of Nevers, the money of Anjou [?], and the "estievenent" of Besançon are at 3½ d.

The money of Laon [?] is at 3 d.

The money of Dijon is at 8 d. in 3, that is, there are 2 2/3 d. [of silver] in 12 d. weight.

roedenois et sevois et vienois sont a ii d. et o. Parmesain

et venisien petit sont a ii d. Nacois et barselevois [?]

sont a iii o. Papiois de Tollete sont a v poi. Quinunois

d'Auscorgne et besant de Babilonie et venicien de Venice

sont a xi d. et o. de fin or. Li droite o. de Muce [?]   40

est a xi d. Cerin et marin et moilekin et li maille

ke li rois marrois fist faire sont a x d. et o. Eufosin

de Tollete sont a x d. Leonnois et portigalois sont a

---

36 et o̅m.S
38 Tollete: Tolete R
39 Babilonie: Babiloine R
40 et o̅m.S / de fin or. Li droite o. o̅m.S
41 moilekin: molekin R / maille: o. R
42 ke: que R / marrois: marros R / et o̅m.S
43 Tollete: Tolete R

The monies of Cahors, Rodez, Savoy [?], and Vienne are at 2½ d.

The money of Parma and the small money of Venice are at 2 d.

Nacois [?] and the money of Barcelona are at 1½ d.

The "papiois" of Toledo is at 1¼ d.[16]

The "quinunois" of Auscorgne [?], the "besant" of Babilonia [? Cairo], the "venicien" of Venice are at 11½ d. of fine gold.

The correct "obole" of Muce [?] is at 11 d.

The "cerin," marin, moilekin, and the "maille" that the moorish king [? the King Marrois] had made are at 10½ d.

The "eufosin" [? alphonsine] of Toledo is at 10 d.

The monies of Leon [?] and Portugal are at 9 d.

---

[16]"Conte a xii d. . . . a v poi." (VI, 6.1-38). This section gives the mintage rate or title of silver coinages of a variety of authorities. The conversion procedures of the previous section are applied to determine the value and weight of silver per marc given the mintage rate in deniers and their fractions per sou. Some of the values are given in grains; a grain is 1/24 of a denier. As far as possible I have identified the localities of the various coinages. I have not compared this text with other materials to check the accuracy of the values given. The only way the values for the monies of Reims, Paris (new) (VI, 6.22-24), Soignies, Auxerre, Provins, etc. (VI, 6.28-30), and Dijon (VI, 6.33-35) can make sense is if "et iii esses" is construed as "in three (sous)"; this reading is possible if the scribal ampersand, normally read et, is taken as é, one old-French form of the preposition "en" and if a "troisesse" is a threesome or three sous.

ix d. Li perpe de Constantinoble ke li emperreres Manasces

fist faire est a ix d. Li perpe d'Aquilee est a viii d.   45

et poi. Boison d'Aicre sont a viii d. et poi. et lor

parfais a vi d. Terin de Poelge sont a vii d. Scenas-

pois sont a vi d. Terin de Cecille sont a iiii d. de fin

or et a iiii d. d'argent et a iiii d. de keuvre.

---

44 Constantinoble: Constentinoble R / ke: que R /
   emperreres: empereres R
46 d'Aicre: d'Achre R
48 Cecille: Sesile R

The "perpe" of Constantinople that the emperor Manasces
[?] had made are at 9 d̲.

The "perpe" of Aquilea is at 8½ d̲.

The "boisan" [? "besant"] of Acre is at 8½ d̲. and sometimes at vi d̲.

The "terin" of Poelge [?] is at 7 d̲.

Scenaspois ₍?] is at 6 d̲.

The "terin" of Sicily is at 4 d̲. of fine gold, 4 d̲. of
silver and 4 d̲. of copper.[17]

---

[17]"Quinunois d'Auscorgne . . . de keuvre" (VI, 6.38-49).
This section gives mintage rates for gold monies. They also
seem to be given in deniers and their parts per sou. There
was no Roman emperor in the East named Manasces (VI, 6.44-45).

[VI, 7]   Or commencerons de l'art d'arismetike. Si commence
en tel forme car tout li nombre de 1 dusqua x sont appellé
d'Egipte por chu ke li 1 se gete sor l'autre. Car par 1
fois 1 fait 1, par ii fois ii font iiii, par iii fois iii
font ix, par iiii fois iiii font xvi, par v fois v font   5
xxv, par vi fois vi font xxxvi, par vii fois vii font
xlviiii, par viii fois viii font lxiiii, par ix fois
ix font iiii^{xx} et 1, par x fois x font c. Cist petit nombre
sont profitable a multeplier le disisme desus, par xi fois
xi font vi^{xx} et 1, par xii fois xii font vii^{xx} et iiii, par   10
xiii fois xiii font <viii^{xx} et ix, par xiiii fois xiiii font>
ix^{xx} et xvi, par xv fois xv font xi^{xx} et v, par xvi fois
xvi font xii vins et xvi, par xvii fois xvii font xiiii
vins et ix, par xviii fois xviii font xvi vins et iiii, par
xviiii fois xix font <xviii^{xx}> et 1.   15

---

1 commencerons: commenceront *R* / arismetike: arismetique *R*
2 tout: tot *R* / de: des *R* / dusqua: iusques a *R* /
   appellé: apelé *R*
3 chu: ce *R* / gete: giete *R*
4 ii om.*R*
7 xlviiii: xlix *R*
9 disisme: disime *R*
11 viii^{xx} et ix, par xiiii fois xiiii font (ed.) om.*RS*
12 ix^{xx}: ix vins *R* / xi^{xx}: xi vins *R*
13 xii vins: xii^{xx} *R*
13-14 xiiii vins: xiiii^{xx} *R*
14 xviii fois xviii font xvi vints et iiii, par iter.*R*
15 xviiii: xix *R* / xviii^{xx} (ed.): xix^{xx} *RS*

VI, 7. Now we begin the art of arithmetic. It begins in this way because all the numbers from 1 to 10 are called "of Egypt" because one is cast ("se gete") upon the other,

for
  $1 \times 1 = 1$
  $2 \times 2 = 4$
  $4 \times 4 = 16$
  $5 \times 5 = 25$
  $6 \times 6 = 36$
  $7 \times 7 = 49$
  $8 \times 8 = 64$
  $9 \times 9 = 81$
  $10 \times 10 = 100.$

These little numbers are convenient for multiplying higher

than 10, for
  $11 \times 11 = 121$
  $12 \times 12 = 144$
  $13 \times 13 = 169$
  $14 \times 14 = 196$
  $15 \times 15 = 225$
  $16 \times 16 = 256$
  $17 \times 17 = 289$
  $18 \times 18 = 324$
  $19 \times 19 = 361.$ [18]

---

[18]"Or commencerons . . . <xviii> et 1" (VI, 7.1-15). This section claims to be an introduction to arithmetic; however, it merely gives a list of numbers and their squares. "Egyptian numbers" is a curious name for the Hindu-Arabic numerals; the author's explanation for the name is more curious yet. Although not uncommon, indicating multiples of twenty by a superscript xx over the number deserves notice.

Et se ie voil savoir ke font par xx fois xx, je garderai
en xx quantes x ismes il i a, c'est <ii> fois x et conterai
par ii fois ii^c font cccc ke par xx fois xx font, car cascune disaine doublée en soi valt c. Par xxx fois xxx
font ix^c, et par xl fois xl font <xvi^c, et par l fois l   20
font xxv^c, et par lx fois lx font> xxxvi^c, par lxx fois lxx
font xlix^c, par iiii^{xx} fois iiii vins font lxiiii^c, par iiii
vins et x fois iiii^{xx} et x font viii^m et c, par c fois c
font x^m.

Et se ie voil savoir ke font par c et x fois cent   25
et x, c'est a dire xi disaines, et conterai par xi fois
xi^c font xii mile et i^c, par vi vins fois <vi vins> font
xiiii mile et iiii^c, par vi vins et x fois vi vins et x
font xvi mile et viiii^c, par <vii> vins fois vii vins
font xix^m et vi^c, par vii^{xx} et x fois vii^{xx} et x font xxii mile   30
et v^c, par viii vins fois viii vins font xxv mile et vi^c,
par viii vins et x fois viii^{xx} et x font xxviii mile et ix^c,

16 ke: que R / je: ie R
17 ii (ed.): xi RS / conterai: contera R
18 cccc: iiii^c R / ke: que R
19 doublée: doublé R
20-21 xvi^c, et par l fois l font xxv^c, et par lx fois lx
    font (ed.) om.RS
22 iiii vins: iiii^{xx} R
22-23 iiii vins: iiii^{xx} R
25 voil: vel R / cent om.S
27 vi vins (ed.) om.RS
28 vi vins (2): vi^{xx} R
29 viiii^c: ix^c R / vii (ed.): xvii^m R xvii S
30 x (2): dis R
31 viii vins (1): viii^{xx} R / viii vins (2): viii^{xx} R
32 viii vins: viii^{xx} R / et x (1) om.R

And if I want to know how much 20 x 20 makes, I will
keep the number of tens there are in 20, that is 2 times
10, and I will compute 2 times 200, which makes 400, which
is 20 x 20, for each ten multiplied by itself is worth 100.

$$30 \times 30 = 900$$
$$40 \times 40 = 1600$$
$$50 \times 50 = 2500$$
$$60 \times 60 = 3600$$
$$70 \times 70 = 4900$$
$$80 \times 80 = 6400$$
$$90 \times 90 = 8100$$
$$100 \times 100 = 10,000.$$

And if I want to know what 110 times 110 makes, that
is, 11 tens, I will compute 11 times 1100, which makes
12,100,

$$120 \times 120 = 14,400$$
$$130 \times 130 = 16,900$$
$$140 \times 140 = 19,600$$
$$150 \times 150 = 22,500$$
$$160 \times 160 = 25,600$$
$$170 \times 170 = 28,900$$
$$180 \times 180 = 32,400$$
$$190 \times 190 = 36,100$$
$$200 \times 200 = 40,000 \quad [19]$$

---

[19]"Et se ie voil . . . font xl mile" (VI, 7.16-35). Here
we have a continuation of the list of numbers and their
squares but with an explanation of the procedure for multiplying products of ten. The author suggests considering 20
as 2 times 10; then 20 x 20 = 20 x 10 x 2. Also 110 is to
be considered 11 times 10; the same procedure applies. A
superscript c is used to indicate multiples of 100 and a
superscript m for thousands.

par ix vint fois ix vins font xxxii mile et iiii^{c},

par ix vins et x fois ix vins et x font xxxvi mile et i^{c},

par ii^{c} fois ii^{c} font xl mile.   35

Et se ie voil savoir ke font par xxv fois xxv, je conterai par v fois v font xxv et garderai quantes disaines a en xx, c'est a dire ii fois x, et conterai par ii fois v font x, et par v fois ii font x, c'est a dire xx di-saines ki font ii^{c}; iiii^{c} et ii^{c} et xxv font vi^{c} et xxv ke   40 par xxv fois xxv font.

Par xxx fois xl font par iii fois iiii^{c}: xii^{c}. Et se ie voil savoir ke font par lxx fois vii^{c}, je garderai quantes fois x a en lxx, c'est a dire vii fois x, et conterai par vii fois vii mile font, car toutes disaines ietées   45 par deseure cent valent mil. Et tot cist conte, sont contable par ceste misme raison.

---

33 ix vint: ix^{xx} R / ix vins: ix^{xx} R
34 ix vins (1): ix^{xx} R / x (1): dis R / ix vins (2): ix^{xx} R
35 fois ii^{c} om.S
36 ke: que R
37 conterai: contera R / disaines: x aines R
38 conterai: contera R
40 ki: qui R
43 voil: veul R / post fois add. lxx S / je: ie R
44 conterai: contera R
46 cent: c R
47 misme: meismes R

And if I want to know what 25 x 25 makes, I will compute
5 x 5, which makes 25, and I will keep as many tens as
there are in 20, that is, 2 times ten, and I will compute
2 x 5, which makes 10, and 5 x 2 which makes 10, that is,
20 tens which makes 200. 400 and 200 and 25 makes 625,
which is 25 x 25.[20]

30 x 40 is done as 3 x 400, which is 1200. And if I
want to know what 70 x 700 is, I will keep the number of
tens there are in 70, that is, 7 times ten, and I will
compute 7 x 7000 [which makes 49,000], because every ten
cast on 100 is worth 1000. And all these computations
may be done in the same way.[21]

---

[20]"Et se ie voil . . . xxv fois xxv font" (VI, 7.36-41).
The procedure for squaring a number that is not an even
multiple of 10 is described in this section. The description is incomplete, leaving out the derivation of some of
the subproducts. What the author gives is part of the pro-
cess of squaring a binomial, viz. $25 \times 25 = (20+5)(20+5) = (2 \cdot 10+5)(2 \cdot 10+5) = (2 \cdot 5+5 \cdot 2) \cdot 10 + 20 \cdot 20+5 \cdot 5 = 200 + 400 + 25 = 625$.

[21]"Par xxx . . . misme raison" (VI, 7.42-47). This continuation of the rules for finding products of multiples of
10 gives the procedure for multiples of 100.

[VI, 8] Les ix figures d'augorismes se multeplient en tel forme: car la premiere, quant cifre le maine, valt par x fois soi misme, se font x; li seconde par x fois x, font c; la tierce par x fois c, fait mil; la quarte par x fois mil, fait $\overset{m}{x}$; li v isme par x fois $<x$, fait$> \overset{m}{c}$; la   5 vi isme par x fois $\overset{m}{c}$, font mil mile; la vii isme par x fois mil mile, font $\overset{m}{x}$ mile; la viii isme par x fois x millier mile, font c mil mile; li ix isme par x fois $\overset{m}{c}$ mile, font mile mile mile.

---

1 les: des R / d'augorismes: d'augorisme R
3 misme: meisme R
4 c: x R / fait: font R
5 $\overset{m}{x}$ fait (ed.) om.RS / la: li R
7 $\overset{m}{x}$: x mil R
8 millier mile: mil mile R / c mil: $\overset{m}{c}$ R / par x fois: fois par R
9 post font add. mile RS

VI, 8.   The 9 figures of algorism multiply in that way because the first figure when zero leads it (i.e. is to its
right) is worth 10 times itself making 10; the second by
10 times 10 making 100; the third by 10 times 100 making
1000; the fourth by 10 times 1000 making 10,000; the fifth
by 10 times 10,000 making 100,000; the sixth by 10 times
100,000 making 1,000,000; the seventh by 10 times 1,000,000
making 10,000,000; the eighth by 10 times 10,000,000 making
100,000,000; the ninth by 10 times 100,000,000 making
1,000,000,000.[22]

---

[22]"Les ix figures . . . mile mile mile" (VI, 8.1-9). The
author seems to be using the word "figure" to mean both
figure and position. Thus he talks about nine figures but
gives examples of positional notation only for powers of
ten. This material seems to fit with the explanation of
multiplying products of ten in VI, 7.

[VI, 9]   Quant la libre de poivre sera vendue xx d., et
je vaurai savoir ke vauront les ccc libres, je conterai
ke la libre et li quarterons vauront xxv d.; les ccc
libres valront xxv libres, car autant d. cum li livre
et li quarterons valra, tante libres valront les iii^c   5
libres.

    Quant c libres d'avoir de pois seront vendues xx s.,
la libre valra ii d. et ii v ismes de d. Quant cc libres
seront vendues xx s., la libre vaura x s. et le v isme
de i d. Quant ccc libres de gimgembre seront vendues   10
xxv libres et ie vaurai savoir ke valra li libre, ie
prendrai de xxv libres xxv d., des queus xxv ie osterai
le v isme, c'est a dire v d., si demoront xx d. ke la
libre de gimgembre valra. Et tout cist conte sont contable par ceste misme raison.   15

---

1 et om.R
2 vaurai: valrai R / ke: que R / vauront: valront R /
   ccc: iiii^c R
3 ke: que R / vauront: valront R / ccc: iii^c R
4 cum: com R / livre: libre R / post livre add. vaura R
5 valra om.R
9 vaura: valra R / x s. om.S
10 ccc: iii^c R / gimgembre: gyngembre R / post vendues add.
   xxxii S
11 vaurai: valra R / ke: que R
12 prendrai: prenderai R / de: des R / des queus:
   desquels R / ie osterai: aioustera R
13 demoront: demonront R
14 gimgembre: gymgembre R / valra: vaura R / tout: tot R
15 misme: meisme R

VI, 9. When pepper is sold for 20 d. a pound and I want to know the value of 300 pounds, I compute that a pound and a quarter will be worth 25 d. then the 300 pounds (weight) will be worth 25 livres, because the 300 pounds (weight) will be worth in livres just as much as the pound and a quarter (weight) will be worth in deniers.[23]

When 100 pounds avoirdupois are sold for 20 s., the pound will be worth 2 and 2/5 d. When 200 pounds are sold for 20 s., the pound will be worth 1 and 1/5 d.[24] When 300 pounds of ginger are sold for 25 livres and I want to know the value of a pound of ginger, I will take for the 25 livres 25 d., from which I will take away a fifth, that is 5 d.; 20 d., which the pound of ginger is worth, will remain. And all these problems may be computed in the same way.[25]

---

[23]"Quant la libre . . . iii^{c} libres" (VI, 9.1-6). The justification for the author's procedure comes from the relation between the denier and the livre, as follows:

$$\frac{20 \text{ d.}}{\text{lb.(wt.)}} = \frac{25 \text{ d.}}{1\frac{1}{4} \text{ lb.(wt.)}} \times \frac{1 \text{ lb.}}{240 \text{ d.}} = \frac{25 \text{ lb.}}{300 \text{ lb.(wt.)}}$$

[24]"Quant c libres . . . v isme de i d." (VI, 9.7-10). Converting sous to deniers establishes these proportions:

$$\frac{20 \text{ s.}}{100 \text{ lb.(wt.)}} = \frac{2 \text{ } 2/5 \text{ d.}}{1 \text{ lb.(wt.)}} \text{ and } \frac{20 \text{ s.}}{200 \text{ lb.(wt.)}} = \frac{1 \text{ } 1/5 \text{ d.}}{1 \text{ lb.(wt.)}} \text{ .}$$

[25]"Quant ccc libres . . . misme raison" (VI, 9.10-15). The factor of 4/5 involved in this calculation arises as follows:

$$\frac{25 \text{ lb.}}{300 \text{ lb.(wt.)}} \times \frac{240 \text{ d.}}{\text{lb.}} = \frac{4}{5} \times \frac{25 \text{ d.}}{\text{lb.(wt.)}} = \frac{20 \text{ d.}}{\text{lb.(wt.)}} \text{ .}$$

Quant li esterlins sera vendus iiii d., poi. mains,
et ie valrai savoir ke valt li mars de xiii s. iiii d.,
je conterai par iiii fois iiii d., poi. mains, c'est a
dire, xv d. Et quant iiii d. valront xv d., iiii libres,
ki font vi mars, valent xv libres. De queus xv libres   20
ie prenderai les vi isme, c'est a dire, l [i.e. 50] s. ke
li mars de xiii s. et iiii d. vaura.

Quant li esterlinc sera vendus iii d. et o., <demie>
poi. mains, et je valrai savoir ke valra li mars de xiii
s. et iiii d., je conterai par viii fois iii d. et o.,   25
demie poi. mains, c'est a dire xxvii d. Et quant viii d.
valent xxvii d., viii esterlins qui font xii mars ki valent
xxvii libres, des quels xxvii libres je prenderai le xii
isme, c'est a dire xlv s. ke li mars valra.

Quant li esterlins sera vendus iiii d. et o. et ie   30
vaurai savoir ke li mars de xiii s. et iiii d. vaura, je

---

17 valt: valra R / post xiii s. add. et R
18 je: ie R
20 ki: qui R / queus: ques R
21 prenderai: prendrai R / les: le R / ke: que R
22 vaura: valra R
23 esterlinc: esterlins R / et om.S
24 je: ie R / post ke add. li R
25 iii d. et o.: iiis d. S
27 esterlins: esterlin R / ki om.R
30 esterlins: esterlin R / iiii d. et o.: iiiis d. S
31 ke: que R / vaura om.S

When the esterlin is sold for 3 and 3/4 underline{d}. and I want to
know the value of a mark of 13 underline{s}. and 4 underline{d}., I compute 4
times 3 and 3/4 underline{d}., that is 15 underline{d}., and when 4 esterlins are
worth 15 underline{d}., 4 pounds [sterling], which make 6 marks, are
worth 15 livres.   Then I take one-sixth of those 15 livres,
that is, 50 underline{s}., which the mark of 13 underline{s}. and 4 underline{d}. will be .
worth.[26]

When the esterlin is sold for 3 and 3/8 underline{d}. and I want to
know the value of a mark of 13 underline{s}. and 4 underline{d}., I compute 8
times 3 and 3/8 underline{d}., that is, 27 underline{d}.   Then when 8 underline{d}. (sterling)
are worth 27 underline{d}., 8 [pounds] sterling, which make 12 marks
(sterling), are worth 27 livres.   I then take one-twelfth
of those 27 livres, that is 45 underline{s}., which the mark (sterling)
will be worth.[27]

When the esterlin is sold for 4½ underline{d}. and I want to know
the value of a mark of 13 underline{s}. and 4 underline{d}., I compute 2 times

---

[26]"Quant li esterlins . . . et iiii d. vaura" (VI, 9.16-
22).   The author's procedure is apparently as follows:

$$\frac{3\ 3/4\ \text{d.}}{\text{E}} = \frac{15\ \text{d.}}{4\ \text{E}} = \frac{15\ \text{lb.}}{4\ \text{lb.(st.)}} = \frac{15\ \text{lb.}}{6\ \text{mark(st.)}} = \frac{50\ \text{s.}}{\text{mark(st.)}}.$$

[27]"Quant li esterlinc . . . li mars valra" (VI, 9.23-29).
This calculation is very similar to the preceding one:

$$\frac{3\ 3/8\ \text{d.}}{\text{E}} = \frac{27\ \text{d.}}{8\ \text{E}} = \frac{27\ \text{lb.}}{8\ \text{lb.(st.)}} = \frac{27\ \text{lb.}}{12\ \text{mark(st.)}} = \frac{45\ \text{s.}}{\text{mark(st.)}}.$$

conterai par ii fois iiii d. et o., font ix d. Et

quant ii d. valent <ix> d., xl s. ki valent <iii> mars

valent ix libres. Je prenderai le tierc de ix libres,

c'est a dire lx s. ke li mars vaura.   35

    Quant li esterlins sera vendus iiii d. et ie vaurai

savoir ke li mars de xiii s. et iiii d. vaura, je pren-

drai por iiii d. iiii libres; des queus iiii libres ie

osterai <le tierc>, c'est a dire <si demoront> liii s.

et iiii d. ke li mars vaura.   40

    Quant li esterlins sera vendus iii d. et o. et

ie valrai savoir ke li mars de x s. valra, ie conterai

par x fois iii d. et o., c'est a dire xxxv d. Et quant

x d. valent xxxv, x s., ki font i marc, valent xxxv s.

car cascun d. ke li x valent, ie prendrai tant s. por le 45

---

32 iiii d. et o.: iiiis d. S
33 ix (ed.): viii RS / ki: qui R / iii (ed.): xii RS
35 ke: que R
36 esterlins: esterlin R
37 ke: que R
38 des queus: de ques R
38-39 ie osterai: iousterai R
39 le tierc (ed.): les ii pars RS
40 ke: que R / vaura: valra R
41 esterlins: esterlin R
42 valrai: valra R / ie (2): je R
43 iii d. et o.: iiis d. S
43-44 et quant x d. valent xxxv om. S
44 ki: qui R
45 ie: je R

$4\frac{1}{2}$ d. which makes 9 d. And when 2 d. (sterling) are worth
9 d., 40 s. (sterling), which are worth 3 marks, are worth
9 livres. I then take a third of 9 livres, that is 60 s.,
which the mark will be worth.[28]

When the esterlin is sold for 4 d. and I want to know
the value of a mark of 13 s. and 4 d., I take for 4 d., 4
livres; from those 4 livres, I take away one-third, that
is 53 s. and 4 d., which the mark will be worth.[29]

When an esterlin is sold for $3\frac{1}{2}$ d. and I want to know
the value of a mark of 10 s., I compute 10 times $3\frac{1}{2}$ d.,
that is 35 d. And when 10 d. (sterling) are worth 35 d.,
10 s. (sterling), which make a mark, are worth 35 s., because for each denier that the 10 d. (sterling) are worth,
I take that many sous for the value of the mark, and for

---

[28]"Quant li esterlins . . . li mars vaura" (VI, 9.30-35).
Here the author's conversion amounts to:

$$\frac{4\frac{1}{2} \text{ d.}}{\text{E}} = \frac{9 \text{ d.}}{2 \text{ E}} = \frac{9 \text{ lb.}}{2 \text{ lb.(st.)}} = \frac{9 \text{ lb.}}{40 \text{ s.(st.)}} = \frac{9 \text{ lb.}}{3 \text{ mark(st.)}} = \frac{60 \text{ s.}}{\text{mark(st.)}}.$$

[29]"Quant li esterlins . . . li mars vaura" (VI, 9.36-40).
The method given here is equivalent to:

$$\frac{4 \text{ d.}}{\text{E}} = \frac{4 \text{ lb.}}{\text{lb.(st.)}} = \frac{2/3 \ 4 \text{ lb.}}{\text{mark(st.)}} = \frac{2}{3} \times \frac{80 \text{ s.}}{\text{mark(st.)}} = \frac{53 \text{ s. 4 d.}}{\text{mark(st.)}}.$$

valeur du marc, et por le maille vi d., et por le tierc
d'un d. iiii d., et por le poi. iii d.

Quant li d. de xvi s. le mar sera vendus ii d. et
o., et je valrai savoir ke valra li mars; ie conterai
par xvi fois ii d. et o., c'est a dire xl d.; et xvi s.   50
ki font i mars valent xl s.

Quant i esterlins et li <ix> ismes d'un esterlin
sera vendus iiii d., et ie valrai savoir ke li mars de xiii
s. et iiii d. vaura, je conterai par ix fois iiii d., font
xxxvi d. et conterai ke ix d. et ix ix ismes de d. font   55
x d. Et quant x d. valent xxxvi d., x libres ki valent
xv mars valent xxxvi libres. De queus xxxvi libres ie
prendrai le xv isme, c'est a dire xlviii s. ke li mars
valra. Et quant li mars valt xlviii s. et ie vaurai savoir

---

46 maille: male R
48-49 ii d. et o.: iis d. S
49 je: ie R / ke: que R / ie: je R
50 et (1) om.S / et (2) om.R
51 ki: qui R / i: le R
52 et om.R / ix (ed.): viii RS / ismes: isme R
54 vaura: valra R / d. (2) om.R / font: ce sont R
55 ix (2) om.R / ismes: isme R / de: d'un R
57 de queus: desquels R / ie: je R
58 prendrai: prenderai R

each maille [=½ d.] , I take 6 d., and for each third of a

denier, 4 d., and for each poitevine, 3 d.[30]

When the denier of a mark of 16 s. is sold for 2½ d.,

and I want to know the value of a mark, I compute 16 times

2½ d., that is 40 d., and the 16 s. that make a mark are

worth 40 s.[31]

When 1 and 1/9 esterlin are sold for 4 d., and I want to

know the value of a mark of 13 s. and 4 d., I compute 9

times 4 d., which is 36 d., and I figure that 9 and 9 ninths

of a denier makes 10 d. And when 10 d. (sterling) are worth

36 d., 10 livres (sterling), which are worth 15 marks (ster-

ling), are worth 36 livres. I then take one-fifteenth of

those 36 livres, that is 48 s., that the mark (sterling) will

be worth. And when a mark is worth 48 s. and I want to know

---

[30]"Quant li esterlins . . . por le poi. iii d." (VI,
9.41-47). Again we have a simple conversion based on the
standard relations between deniers, sous and livres:

$$\frac{3\frac{1}{2} \text{ d.}}{E} = \frac{35 \text{ d.}}{10 \text{ E}} = \frac{35 \text{ s.}}{10 \text{ s.(st.)}};$$ the general procedure is based

on these proportions: $\frac{\text{d.}}{10 \text{ E}} = \frac{\text{s.}}{10 \text{ s.(st.)}} = \frac{12 \text{ d.}}{10 \text{ s.(st.)}}.$

Taking one-half, one-third, and one-fourth of each side of
the proportion yields the values of the text.

[31]"Quant li d. . . . valent xl s." (VI, 9.48-51). If we
let D be the denier of a mark of 16 sous and let S be that
sou, then the relation in the text is simply:

$$\frac{2\frac{1}{2} \text{ d.}}{D} = \frac{40 \text{ d.}}{16 \text{ D}} = \frac{40 \text{ s.}}{16 \text{ S}}.$$

ke valront <v> mars, je conterai par v fois xlviii s.,   60

font xii libres.

    Quant v marc valent xii libres, $\overset{c}{v}$ mars valent

xi$\overset{c}{i}$ libres car por cascune libre ke li v marc valront,

li $\overset{c}{v}$ mars valront tante c libres. Et por xiii s. iiii d.,

les ii pars de c libres; et por x s., le moitié de c   65

libres; et por vi s. et viii d., le tierc de c libres;

et por v s., le quart de c libres; et por iiii s., le

v isme de c libres. Et tout cist conte sont contable

par ceste misme raison.

---

60 valront: vauront <u>R</u> / v (<u>ed.</u>): $\overset{c}{v}$ <u>RS</u>

62 marc: mars <u>R</u>

63 marc: mars <u>R</u>

64 valront: vauront <u>R</u> / c <u>om.R</u> / <u>post</u> xiii s. <u>add.</u> et <u>R</u> /
    d. <u>om.R</u>

65 le moitié <u>iter.S</u>

68 v isme: vi isme <u>R</u> / tout: tot <u>R</u>

69 misme: meisme <u>R</u>

the value of 5 marks, I compute 5 times 48 s., which makes 12 livres.[32]

When 5 marks are worth 12 livres, 500 marks are worth 1200 livres, because for each livre that 5 marks is worth, 500 marks will be worth that many 100 livres. And for a mark of 13 s. and 4 d., I take two-thirds of 100 livres; and for a mark of 10 s., I take half of 100 livres; and for a mark of 6 s. and 8 d., a third of 100 livres; and for a mark of 5 s., a quarter of 100 livres; and for a mark of 4 s., a fifth of 100 livres. And all these calculations may be done in the same way.[33]

---

[32]"Quant i esterlins . . . font xii libres" (VI, 9.52-61). The author's procedure involves these conversions:

$$\frac{4 \text{ d.}}{1\,1/9 \text{ E}} = \frac{36 \text{ d.}}{10 \text{ E}} = \frac{36 \text{ lb.}}{10 \text{ lb.(st.)}} = \frac{36 \text{ lb.}}{15 \text{ mark(st.)}} = \frac{48 \text{ s.}}{\text{mark(st.)}}.$$

[33]"Quant v marc . . . misme raison" (VI, 9.62-69). The conversions for the different kinds of marks are based on the equivalence 20 s. = 1 lb. Then 13 s. 4 d. = 2/3 lb., 10 s. = ½ lb., 6 s. 8 d. = 1/3 lb., 5 s. = ¼ lb., and 4 s. = 1/5 lb.

Quant li mars de fin or sera vendus xxx libres   70

et ie vaurai savoir le xii isme de xxx libres, c'est a

dire l [i.e. 50] s. Et conterai par x fois l [i.e. 50]

s. font <xxv> libres ke ors a <x> d. valra; et ors

a vii d. par vii fois l [i.e. 50] font xvii libres et

x s.; et ors a i d., l [i.e. 50] s.; et ors a une maille, 75

xxv s.; et ors a une poi., xii s. vi d.; et ors a i

grain xxv d.

---

71 vaurai: valrai R / xii isme: xii S

72 conterai: contera R

73 xxv (ed.): xv RS / ke: que R / post ke add. li R /
   x (ed.): ii RS

75 une maille: i o. R

76 post xii s. add. et R / d. om.R

When a mark of fine gold is sold for 30 livres and I
want to know the twelfth of 30 livres, that is, 50 s.,
I will compute 10 times 50 s. which makes 25 livres, that
[a mark of] gold [minted] at 10 d. will be worth. And for
gold at 7 d., I take 7 times 50 which makes 17 livres and
10 s. And for gold at one denier I take 50 s.; and for gold
at one maille, 25 s.; and for gold at one poitevine, 12 s.
and 6 d.; and for gold at one grain, 25 d.[34]

---

[34]"Quant li mars . . . a i grain xxv d." (VI, 9.70-77).
Taking one-twelfth of the value of a mark of fine gold facilitates the computation of the value of a mark of gold
alloy or gold coins, where the purity is measured in deniers
and their fractions per sou. The examples are computed by
multiplication using the standard relations between lb., s.,
d. and the parts of a denier, viz. 1 denier = 2 mailles (or
oboles) = 4 poitevines = 24 grains.

[VI, 10] Se ie voil savoir le droite rieule a savoir le
somme et a trover les missions des viles, et ie voil avoir
de le libre, d., ie prendrai le xii isme; et de le libre,
xvi d., le xv isme; et de le libre, le xv d., le xvi isme;
et de le libre, xii d., le xx isme; de la libre, x d.,   5
le xxiiii isme; et de libre, viii d., le xxx isme; de la
libre, vii et o., le <xxxii> isme; de la libre, vi d.,
le xl isme; de la libre, v d., le xlviii isme; de la
libre iiii d., le <lx> isme; de le libre, iiii d., poi.
mains, le lxiiii isme; de le libre, iii d., le iiii$\overset{xx}{}$;   10
de le libre, iis d. [i.e. 2 d. et o.], le iiii$\overset{xx}{}$ et xvi isme;
de la libre, ii d., le vi$\overset{xx}{}$ isme; de la libre, iii o., le
viii$\overset{xx}{}$ isme.

---

1 voil: vel R / rieule: riule R
2 viles: villes R / voil: voel R
3 prendrai: prenderai R / et om.R / post libre (2) add. le R
4 xvi d. le xv isme; et de le libre le om.R /
   xvi isme: xv isme S
5 le (1): la R
6 et de libre om.R
7 vii et o.: viis d. S / xxxii (ed.): xii RS / la: le R
8 la (2): le R
9 lx (ed.): xl RS
11 de le libre, iis d., le iiii$\overset{xx}{}$ om.R
12 iii o.: iiis d. S

VI, 10. If I want to have the correct rule for knowing the
sum and for finding the coinages of cities, and I want to
know deniers per livre, I will take a twelfth; and if I want
to know 16 d. per livre, I will take a fifteenth; and for
15 d. per livre, a sixteenth; and for 12 d. per livre, a
twentieth; for 10 d. per livre, a twenty-fourth, for 8 d.
per livre, a thirtieth; for 7½ d. per livre, a thirty-second;
for 6 d. per livre, a fortieth; for 5 d. per livre, a forty-
eighth; for 4 d. per livre, a sixtieth; for 3 and 3/4 d. per
livre, a sixty-fourth; for 3 d. per livre, an eightieth; for
2½ d. per livre, a ninety-sixth; for 2 d. per livre, a one-
hundred-twentieth; for 1½ d. per livre, a one-hundred-
sixtieth.[35]

---

[35]"Se ie voil . . . le viii isme" (VI, 10.1-13). This
section is in effect a table relating fractional parts of
a livre to their expressions in deniers per livre. The only
exception is the first value which should probably be amend-
ed to deniers per sou. I do not understand how the author
means to use these values to find sums and coinages.

[VI, 11]   Li contemens d'alouement de monoie va par xvi s.

le marc.   Et si se conte en tel forme ke se i'ai une

monoie a ii d. et poi. et une autre a iiii d., et ie

le voil i a faire a iii d., si garderai quantes poi.

a de <ii> d. et poi. iusqua iii d., c'est a dire iii   5

pugoises, et de iii d. iusca iiii d. a iiii poi.   Si

prendrai por iii poi. ki sont en bas iii mars d'escange

ki est a iiii d., et por iiii d. ki sont en haut prendrai

iiii mars d'escange ki est a ii d. et poi.   Car li nombres

ki est d'aloiement de monoie va de haut en bas et de bas   10

en haut.   Et se ie voil ke cis aloiemens de monoie soit

bien fais, je conterai quant mars de xvi s. a ii d. et

poi., a xxxvi d., et a iiii mars par ceste misme raison

a xii s.; <et quant mars de xvi s. a iiii d., a lxiiii d.,

et a iii mars a xvi s. xvi s.> et xii s. font xxviii s.   15

de fin argent ke en iiii mars et en iii mars meslé en-

---

1 d'alouement: aloiement R / monoie: moinoie R

5 ii (ed.): iii RS / iusqua: iusca R

7 prendrai: prenderai R / ki: qui R

8 ki (2): qui R / sont: est S / prendrai: prenderai R

9 d'escange: des cange S / post poi. add. ki est en bas RS

10 ki: qui R / d'aloiement: aloiement R

11 ke: que R

12 quant: quans R / ii: deus R

13 misme: meisme R

VI, 11. The alloy content of money is reckoned by 16 $\underline{s}$. per mark. It is computed as follows: if I have one coinage at 2¼ $\underline{d}$. and another at 4 $\underline{d}$., and I want to make coins at 3 $\underline{d}$., I will keep as many $\underline{pites}$ [or $\underline{pougoises}$] as there are from 2¼ $\underline{d}$. up to 3 $\underline{d}$., that is, 3 $\underline{pites}$, and from 3 $\underline{d}$. up to 4 $\underline{d}$. there are 4 $\underline{pites}$. I will then take 3 marks of the exchange money at 4 $\underline{d}$. for the 3 $\underline{pites}$ below [i.e., between 2¼ $\underline{d}$. and 3 $\underline{d}$.] and 4 marks of exchange money at 2¼ $\underline{d}$. for the 4 $\underline{pites}$ above [i.e., between 3 $\underline{d}$. and 4 $\underline{d}$.]. This is done because numbers for alloying money go from high to low and from low to high.[36] And if I want these alloys to be made well, I will compute how many marks of 16 $\underline{s}$. there are for the 2¼ $\underline{d}$.; there are 36 $\underline{d}$. [of fine silver in 16 $\underline{s}$. at 2¼ $\underline{d}$.], and in 4 marks there are 12 $\underline{s}$. [of fine silver] by the same reason; <and for the marks at 4 $\underline{d}$. there are 64 $\underline{d}$. [of fine silver in 16 $\underline{s}$. at 4 $\underline{d}$.], and in 3 marks there are 16 $\underline{s}$. 16 $\underline{s}$.> and 12 $\underline{s}$. together make 28 $\underline{s}$. of fine silver, which there is in 4 marks and 3 marks mixed together. And

---

[36]"Li contemens . . . de bas en haut" (VI, 11.1-11). The formula implicit in this discussion of alloying is: $m_1 p_1 + m_2 p_2 = (m_1 + m_2) \cdot p_3$, where $\underline{p}$ is the mintage rate such that $p_1 > p_3 > p_2$ and m_1 and m_2 are the quantities of the coinages to be alloyed. In this example, $p_1 = 4$ $\underline{d}$., $p_2 = 2$¼ $\underline{d}$., and $p_3 = 3$ $\underline{d}$., per $\underline{sou}$. The author tells us that to find the quantities, we first find $e = p_1 - p_3$ and $f = p_3 - p_2$; here $\underline{e}$ equals 4 $\underline{pites}$ and $\underline{d}$ equals 3 $\underline{pites}$. Then $m_1 \cdot e = m_2 \cdot f$; that is, the differences in value between the old monies and the one to be minted are inversely proportional to their quantities.

samble. Et au marc a iii d., iiii s.; par vii fois
iiii s. font xxviii s.

Se i'ai d'one monoie a iii d. et ie'l voil faire
a iiii d., je conterai combien a de iii dusca iiii, c'est   20
a dire i d., et de iiii dusca xii, ki est appellés fins
argens, si a viii d. Si prendrai de i d., i marc d'escange
a xii d., et por viii d., viii marc de celi a iii d.,
c'est a dire ix mars, mellé ensamble, sont a iiii d.

---

17 ante Et add. a RS / d. om.R / post d. add. et R
19 ie'l: ie le R
20 dusca: iusca R
21 dusca: iusca R / ki: qui R / appellés: apelés R
22 prendrai: prenderai R
24 mellé: meslé R

in the mark at 3 $\underline{d}$. there are 4 $\underline{s}$. of silver; 7 times 4 $\underline{s}$. makes 28 $\underline{s}$.[37]

If I have some money at 3 $\underline{d}$. and I want to make it into money at 4 $\underline{d}$., I will count how many there are from 3 to 4, that is, 1 $\underline{d}$., and from 4 to 12, which is called fine silver, there are 8 $\underline{d}$. If I take one mark of the exchange money at 12 $\underline{d}$. for the 1 $\underline{d}$., and 8 marks of that at 3 $\underline{d}$. for the 8 $\underline{d}$.; that is 9 marks, mixed together, and they are at 4 $\underline{d}$.[38]

---

[37]"Et se ie voil . . . font xxviii s." (VI, 11.11-18). The author verifies his calculation of the quantities by calculating $m_1 p_1$, $m_2 p_2$ and $(m_1 + m_2) \cdot p_3$ and checking that the sum of the first two equals the third, according to the formula in note 36. The calculation uses the conversion factor of 16 $\underline{s}$. per mark.

[38]"Se i'ai d'one . . . sont a iiii d." (VI, 11.19-24). Using the formulas of note 36, p_1 = 12 $\underline{d}$. (fine silver), p_2 = 3 $\underline{d}$. (the money in hand), and p_3 = 4 $\underline{d}$. (to be minted). Then e = 12 d. - 4 d. = 8 d. and f = 4 d. - 3 d. = 1 d. Thus one mark of fine silver and 8 marks of silver at 3 d. will produce 9 marks at 4 $\underline{d}$.

Abbott, T. K. 1900. Catalogue of the Manuscripts of Trinity College, Dublin . . . (Dublin).

Abu Bekr. See Busard, 1968.

Adelard of Bath. See Willner, 1903.

Alonso Alonso, M., ed. 1954. Dominicus Gundissalinus. De scientiis: Compilación a base principalmente de la Kitab Ihsā al-ᶜulūm de al-Fārābī (Madrid).

Alessio, Franco. 1965. "La filosofia e le artes mechanicae nel secolo XII." Studi medievali, 6: pp. 71-161.

Anaritius. See Curtze, 1899a.

Arrighi, Gino, ed. 1966. Pisano (Fibonacci). La pratica di geometria, volgarizzata da Cristofano di Gherardo di Dino cittadino pisano (Pisa).

Augustine. See Migne, 1844-1864, v. 34.

Bacon, Roger. See Bridges, 1897-1900; Burke, 1928; and Steele, 1940.

Baldelli, Ignazio. 1965. "Di un volgarizzamento pisano della 'Practica geometrie'." Rivista di cultura classica e medioevale, 7: pp. 74-92.

Bandinius, A. M. 1775. Catalogus codicum latinorum Bibliothecae Mediceae Laurentianae, v. 2 (Florence).

Baron, Roger. 1955a. "Hugues de Saint-Victor, auteur d'une 'Practica geometriae'." Mediaeval Studies, 17: pp. 107-116.

------ 1955b. "Sur l'introduction en Occident des termes 'geometria theorica et practica'." Revue d'histoire des sciences, 8: pp. 298-302.

------, ed. 1956. Hugh of St. Victor. De ponderibus et mensuris. In: "Hugues de Saint-Victor lexicographe-- Trois textes inédits." Cultura neolatina, 16: pp. 132-137.

------ 1957. "Note sur les variations au XIIe siècle de la triade géométrique: altimetria, planimetria, cosmimetria." Isis, 48: pp. 30-32.

------ 1962. "Note sur la succession et la date des écrits de Hugues de Saint-Victor." Revue d'histoire ecclésiastique, 57: pp. 88-118.

------, ed. 1966. Hugh of Saint-Victor. Opera propaedeutica (Notre Dame, Ind.).

------ 1969. "L'Insertion des arts dans la philosophie chez Hugues de Saint-Victor." Actes du Quatrième Congrès international de philosophie médiévale: Arts libéraux et philosophie au Moyen Age (Montreal and Paris), pp. 551-555.

al-Battani. See Nallino, 1903-1907.

Baur, L., ed. 1903. Dominicus Gundissalinus. De divisione philosophiae. Beiträge zur Geschichte der Philosophie des Mittelalters, 4, 2-3.

Beaujouan, Guy. 1954. "L'enseignement de l'arithmétique élémentaire à l'Université de Paris aux XIIIe et XIVe siècles." Homenaje a Millás-Vallicrosa (Barcelona).

------ 1975. "Réflexions sur les rapports entre théorie et pratique au moyen âge." Murdoch and Sylla, eds. 1975. The Cultural Context of Medieval Learning. Dordrecht and Boston, pp. 437-484.

Beresford, M. 1967. New Towns of the Middle Ages (London).

al-Biruni, Abu'l-Rayhan Muhammad ibn Ahmad. 1934. The Book of Instruction in the Elements of the Art of Astrology. tr. R. Ramsay Wright (London).

Blume, F., K. Lachmann, and A. Rudorff, eds. (1848-1850). Die Schriften der römischen Feldmesser. v. 1: Texte und Zeichnungen, v. 2: Erläuterungen zu den Schriften der römischen Feldmesser (Berlin). Both vols. repr. Rome (?), n.d.

Boethius. See Friedlein, 1867.

Boinet, A. 1921. Les Manuscrits à peintures de la Bibliothèque Sainte-Geneviève de Paris (Paris).

Boncompagni, Baldassarre. 1851. "Della vita e delle opere di Gherardo Cremonese, traduttore del secolo duodecimo." Atti dell'Accademia Pontifica de' Nuovi Lincei, 4: pp. 387-493.

------ 1852. "Della vita e delle opere di Leonardo Pisano, matematico del secolo decimo terzo." Atti dell'Accademia Pontifica de' Nuovi Lincei, 5: pp. 1-91, 208-246.

------ 1854. *Intorno ad alcune opere di Leonardo Pisano, matematico del secolo decimoterzo* (Rome).

------, ed. 1857. *Trattati d'aritmetica* (Rome).

------, ed. 1857-1862. *Scritti di Leonardo Pisano.* v. 1: Liber abbaci. v. 2: *Practica geometriae ed opusculi* (Rome).

Bowie, Theodore, ed. 1959. *The Sketchbook of Villard de Honnecourt* (Bloomington and London).

Branner, Robert. 1957. "Three Problems from the Villard de Honnecourt Manuscript." *Art Bulletin*, 39: pp. 61-66.

Bridges, J. H., ed. 1897-1900. *The "Opus majus" of Roger Bacon* (3 vols., Oxford and London).

Britt, Nan Long, ed. 1972. *A Critical Edition of Tractatus Quadrantis* (Emory University, Ph.D. diss.).

Bubnov, Nicolaus, ed. 1899. *Gerberti postea Silvestri II papae Opera mathematica (972-1003) . . .* (Berlin; repr. Hildesheim, 1963).

Burke, R. B., tr. 1928. Roger Bacon. *Opus majus* (Philadelphia).

Busard, Hubert L. L., ed. 1965. "The Practica Geometriae of Dominicus de Clavasio." *Archive for History of Exact Sciences*, 2: pp. 520-575.

------, ed. 1968. "L'Algèbre au Moyen Age: le 'Liber mensurationum' d'Abū Bekr." *Journal des Savants*, 1968: pp. 65-124.

Buttimer, Charles Henry, ed. 1939. Hugh of Saint-Victor. *Didascalicon de studio legendi, a critical text* (Washington, D. C.).

Cantor, Moritz. 1875. *Die römischen Agrimensores und ihre Stellung in der Geschichte der Feldmesskunst* (Leipzig).

------ 1894-1900. *Vorlesungen über Geschichte der Mathematik.* v. 1: Von den ältesten Zeiten bis zum Jahre 1200 n. Chr. v. 2: Von 1200-1668 (2v., 2nd ed. Leipzig).

Catalogue des manuscrits français [de la Bibliothèque nationale]. 1868. v. 1 (Paris).

Catalogus codicum manuscriptorum Bibliothecae Regiae. 1744. v. 4 (Paris).

Chenu, M.-D. 1940. "Arts 'mécaniques' et oeuvres serviles."
Revue des sciences philosophiques et théologiques, 29:.
pp. 313-315.

Clagett, Marshall. 1953. "The Medieval Latin Translations
from the Arabic of the Elements of Euclid, with Special
Emphasis on the Versions of Adelard of Bath." Isis, 44:
pp. 16-42.

------ 1954. "King Alfred and the Elements of Euclid."
Isis, 45: pp. 269-277.

------ 1964. Archimedes in the Middle Ages. v. 1: The
Arabo-Latin Tradition (Madison, Wis.).

Curtze, Maximilian. 1896. "Über die im Mittelalter zur
Feldmessung benutzten Instrumente." Bibliotheca Mathematica, 10: pp. 65-72.

------, ed. 1897. Petri Philomenus de Dacia in Algorismum
vulgarem Johannis de Sacrobosco commentarius (Copenhagen).

------ 1898. "Über eine Algorismus schrift des XII. Jahr-
hunderts," Abhandlungen zur Geschichte der Mathematik, 8:
pp. 2-27.

------, ed. 1899a. Anaritius. In decem libros priores
Elementorum Euclidis commentarii. In: Supplementum to
Euclid, Opera omnia (Leipzig).

------, ed. 1899b. "Der Tractatus Quadrantis des Robertus
Anglicus in Deutscher Übersetzung aus dem Jahre 1477."
Abhandlungen zur Geschichte der Mathematik, 9: pp. 41-63.

------ 1900a. "Aus den 'Canones sive regule super tabulas
Toletanas' des Al-Zarkālī." Bibliotheca Mathematica,
3. F., 1: pp. 337-347.

------ 1900b. "Urkunden zur Geschichte der Trigonometrie
in christlichen Mittelalter." Bibliotheca mathematica,
3. F., 1: pp. 321-416.

------, ed. 1902a. "Der Liber Embadorum des Abraham bar
Chiyya Savasorda in der Übersetzung des Plato von Tivoli."
Abhandlungen zur Geschichte der mathematische Wissen-
schaften, 12: pp. 1-191.

------, ed. 1902b. "Die Practica Geometriae' des Leonardo
Mainardi aus Cremona." Abhandlungen zur Geschichte der
mathematische Wissenschaften, 13: pp. 336-434.

Daly, John F. 1964. "Mathematics in the Codices Ottoboni-
ani Latini." Manuscripta, 8: pp. 12-14.

De Smet, Antoine. 1949. "De l'utilité de recueillir les mentions d'arpenteurs cités dans les documents d'archives du Moyen Age." Annales du Congrès archéologique et historique de Tournai, 3: pp. 782-795.

------ 1966. "Landmeterstraditie en oude Kaarten van Vlaanderen." Verslagen en mededelingen van De Leiegouw, 8: pp. 209-218.

Dilke, O. A. W. 1971. The Roman Land Surveyors: an Introduction to the Agrimensores (Newton Abbot).

Dominicus de Clavasio. See Busard, 1965.

Drecker, E., ed. 1931. "Hermannus Contractus über das Astrolab." Isis, 16: pp. 200-219.

Du Colombier, Pierre. 1973. Les Chantiers des cathédrales: Ouvriers, architectes, sculpteurs (2nd ed., Paris).

Eneström, G., ed. 1913. "Der 'Algorismus de integris' des Meisters Gernardus." Bibliotheca mathematica, 3.F., 13: pp. 289-332.

------, ed. 1914a. "Der 'Algorismus de minutiis' des Meisters Gernardus." Bibliotheca mathematica, 3.F., 14: pp. 99-149.

------, ed. 1914b. "Das Bruchrechnen des Jordanus Nemorarius." Bibliotheca mathematica, 3.F., 14: pp. 41-54.

Euclid. 1926. The Thirteen Books of Euclid's Elements, tr. Sir Thomas L. Heath (3 v. 2nd ed. revised with additions. Cambridge; repr. New York, 1956).

Eynde, Damien van den. 1960. Essai sur la succession et la date des écrits de Hugues de Saint-Victor (Rome).

Fanfani, A. 1951. "La Préparation intellectuelle a l'activité économique en Italie du XIVe au XVIe siècle." Le Moyen Age, 57: pp. 327-346.

al-Farabi. See González Palencia, 1953.

Fisher, N. W. and Sabetai Unguru. 1971. "Experimental Science and Mathematics in Roger Bacon's Thought." Traditio, 27: pp. 353-378.

Folkerts, Menso, ed. 1970. "Boethius" Geometrie II: ein mathematisches Lehrbuch des Mittelalters (Wiesbaden).

Fournial, Etienne. 1970. Histoire monétaire de l'Occident médiéval (Paris).

Frankl, Paul. 1960. The Gothic: Literary Sources and In-
    terpretations through Eight Centuries (Princeton).

Friedlein, G., ed. 1867. Anicii Manlii Torquati Severini
    Boetii De institutione arithmetica, libre duo; De insti-
    tutione musica, libri quinque . . . (Leipzig).

Gagné, Jean. 1969. "Du Quadrivium aux Scientiae mediae."
    Actes du Quatrième Congrès international de philosophie
    médiévale: Arts libéraux et philosophie au Moyen Age
    (Montreal and Paris), pp. 975-986.

Gandz, S., ed. 1932. "The Geometry of Muhammad ibn Musa
    al-Khowarizmi." Quellen und Studien zur Geschichte der
    Mathematik, Astronomie und Physik, Abt. A, 2: pp. 61-89.

Gerbert. See Bubnov, 1899.

Gernardus. See Eneström, 1913 and 1914a.

Goldstein, Bernard R. 1967. Ibn al-Muthanna's Commentary
    on the Astronomical Tables of al-Khwarizmi (New Haven
    and London).

González Palencia, Angel, ed. and tr. 1953. al-Farabi.
    Catálogo de las ciencias. (2nd ed., Madrid).

Gossen, Charles Théodore. 1970. Grammaire de l'ancien
    picard (Paris).

Guenther, Siegmund. 1887. Geschichte des mathematischen
    Unterrichts im deutschen Mittelalter bis zum Jahre 1525,
    Monumenta germaniae paedagogica, 3 (Berlin).

Guilhiermoz, P. 1913. "De l'équivalence des anciennes
    mesures à propos d'une publication récente." Biblio-
    thèque de l'Ecole des Chartes, 74: pp. 267-328.

Gundissalinus, Dominicus. See Alonso Alonso, 1954 and
    Baur, 1903.

Gunther, R. T. 1929. Chaucer and Messahala on the Astro-
    labe, v. 5 of his Early Science in Oxford (Oxford).

Hahnloser, Hans R., ed. 1935. Villard de Honnecourt,
    Kritische Gesamtausgabe des Bauhüttenbuches ms. fr.
    19093 der Pariser Nationalbibliothek (Vienna).

Halliwell (-Phillips), James Orchard. 1841. Rara mathema-
    tica; or a Collection of Treatises on the Mathematics and
    Subjects Connected with Them, from Ancient Inedited Manu-
    scripts (2nd ed., London).

Haring, Nicholas M. 1964. "Thierry of Chartres and Dominicus Gundissalinus." Mediaeval Studies, 26: pp. 271-286.

Harper, Richard I. 1971. "Prophatius Judaeus and the Medieval Astronomical Tables." Isis, 62: pp. 61-68.

Hartner, W. 1960. "Astūrlab." Encyclopaedia of Islam (New ed., Leiden and London), v. 1: pp. 722-728.

Haskins, C. H. 1924. Studies in the History of Mediaeval Science (Cambridge, Mass.).

Henry, Charles, ed. 1880. "Prologus N. Ocreati in Helceph ad Adelardum Batensem magistrum suum, etc." Abhandlungen zur Geschichte der Mathematik, 3: pp. 129-139.

------, ed. 1882. "Les deux plus anciens traités français d'algorisme et de géométrie." Bulletino di bibliografia e di storia delle scienze matematiche e fisiche, 15: pp. 49-70.

Hermannus Contractus. See Drecker, 1931.

Hugh of St. Victor. See Baron, 1956 and 1966; Buttimer, 1939; Migne, 1844-1864, v. 176; and Taylor, 1961.

Hunt, R. W. 1948. "The Introductions to the Artes in the Twelfth Century." Studia mediaevalia in honorem admodum reverendi patris Raymundi Josephi Martin (Bruges).

Hyde, J. K. 1973. Society and Politics in Medieval Italy: The Evolution of Civil Life, 1000-1350 (New York).

Ibn al-Muthanna, Ahmad. See Goldstein, 1967, and Millás Vendrell, 1963.

Ibn 'Ezra, Abraham. See Millás Vallicrosa, 1940 and 1947.

James, Montague Rhodes. 1903. The Ancient Libraries of Canterbury and Dover (Cambridge).

John of Seville. See Boncompagni, 1857.

Jordanus Nemorarius. See Eneström, 1914b.

Judy, Albert G., ed. 1976. Robert Kilwardby, O. P. De ortu scientiarum. Auctores Britannici Medii Aevi, 4 (London and Toronto).

Juschkewitsch, A. P. 1964. Geschichte der Mathematik im Mittelalter (Leipzig).

Kennedy, E. S. 1956. "A Survey of Islamic Astronomical Tables." Trans. Amer. Philos. Soc. 46, 2: pp. 123-177.

al-Khwarizmi, Muhammed ibn Musa. See Gandz, 1932; Neugebauer, 1962; Suter, 1914; and Vogel, 1963.

Kibre, Pearl. 1969. "The Quadrivium in the Thirteenth Century Universities (with Special Reference to Paris)." Actes du Quatrième Congrès international de philosophie médiévale: Arts libéraux et philosophie au Moyen Age (Montreal and Paris), pp. 173-191.

Kilwardby, Robert. See Judy, 1976.

Klinkenberg, H. M. 1959. "Der Verfall des Quadriviums in fruhen Mittelalter." Joseph Koch, ed., Artes liberales (Leiden and Cologne).

Koch, Joseph, ed. 1959. Artes liberales, von der antiken Bilding zur Wissenschaft des Mittelalters (Leiden and Cologne).

Kohler, Charles. 1896. Catalogue des manuscrits de la Bibliothèque Sainte-Geneviève, 2 (Paris).

Lachmann, Karl, ed. 1848. Die Schriften der römischen Feldmesser, v. 1: Texte unde Zeichnungen (Berlin; repr. Rome, n.d.).

Leff, Gordon. 1968. Paris and Oxford Universities in the Thirteenth and Fourteenth Centuries (New York, London and Sydney).

Leonardo Pisano (Fibonacci). See Arrighi, 1966 and Boncampagni, 1857-1862.

Levey, Martin. 1970. "Abraham bar Hiyya ha-Nasi." Dictionary of Scientific Biography, ed. Charles Coulston Gillispie (14v., New York, 1970-1976), v. 1: pp. 22-23.

Little, A. G., ed. 1914. Roger Bacon Essays . . . (Oxford).

Macrobius. 1952. Commentary on the Dream of Scipio. tr. William Harris Stahl (New York).

Madan, F., H. H. E. Craster and N. Denholm-Young. 1937. A Summary Catalogue of Western Manuscripts in the Bodleian Library at Oxford, v. 2 (Oxford).

Mariétan, Joseph. 1901. Le problème de la classification des sciences d'Aristote à saint Thomas (Paris).

Marx, J. 1905. Verzeichnis der Handschriften-Sammlung des Hospitals zu Cues bei Bernkastel am Mosel (Trier).

Mendthal, H., ed. 1886. Geometria Culmensis: Ein agronom-
　　ischer Tractat aus der Zeit des Hochmeisters Conrad von
　　Jungingen (1393-1407) (Leipzig).

Michel, Henri. 1947. Traite de l'astrolabe (Paris).

Migne, Jacques Paul, ed. 1844-1864. Patrologiae cursus
　　completus, Series latina (221 v., Paris).

Millás Vallicrosa, José María. 1931. Assaig d'història de
　　les idees físiques i matemàtiques a la Catalunya medi-
　　eval (Barcelona).

------, ed. 1940. "Un neuvo tratado de astrolabio de R.
　　Abraham ibn'Ezra." Al-Andalus, 5: pp. 1-29.

------ 1942. Las traducciones orientales en los manu-
　　scriptos de la Biblioteca Catedral de Toledo (Madrid).

------ 1943-1950. Estudios sobre Azarquiel (Madrid).

------ 1944. "Un tratado de almanaque probablemente de R.
　　Abraham Ibn'Ezra." M. F. Ashley Montagu, ed., Studies
　　and Essays in the History of Science and Learning Of-
　　fered in Homage to George Sarton on the Occasion of the
　　Sixtieth Birthday (New York), pp. 421-432.

------, ed. 1947. Abraham Ibn'Ezra. El libro de los Fun-
　　damentos de las Tablas astronómicas (Madrid and Barce-
　　lona).

------ 1949. Estudios sobre historia de la ciencia es-
　　pañola (Barcelona).

------ 1960. Neuvos estudios sobre historia de la ciencia
　　española (Barcelona).

Millás Vendrell, Eduardo, ed. 1963. El comentario de Ibn
　　al-Mutanna a las Tablas Astronómicas de al-Jwarizmi
　　(Madrid and Barcelona).

Molinier, A. 1890. Catalogue des manuscrits de la Biblio-
　　thèque Mazarine, v. 3 (Paris).

Mortet, Victor. 1904. "Note historique sur l'emploi de
　　procédés matériels et d'instruments usités dans la géo-
　　métrie pratique au moyen âge (Xe-XIIIe siècles)." Con-
　　grès international de Philosophie, 2 (Geneva).

------, ed. 1909. "Le plus ancien traité français d'al-
　　gorisme." Bibliotheca mathematica, 3.F., 9: pp. 55-64.

------ 1910. "Le Mesure de la figure humaine et le canon des proportions d'après les dessins de Villard de Honnecourt, d'Albert Durer et de Léonard de Vinci." Mélanges offerts à M. Emile Chatelain (Paris), pp. 367-382.

Mortet, Victor and Paul Deschamps, eds. 1911-1929. Recueil de textes relatifs à l'histoire de l'architecture et à la condition des architectes en France au Moyen-âge. (2 v., Paris).

Murdoch, John E. 1968. "The Medieval Euclid: Salient Aspects of the translations of the Elements by Adelard of Bath and Campanus of Novara." Colloques, XIIe Congrès international d'Histoire des Sciences (Paris), pp. 67-94.

Murdoch, John E. and Edith Dudley Sylla, eds. 1975. The Cultural Context of Medieval Learning: Proceedings of the First International Colloquium on Philosophy, Science and Theology in the Middle Ages--September 1973. (Dordrecht, Holland and Boston).

Nallino, C. A., ed. 1903-1907. Al-Battani sive Albatenii Opus Astronomicum. Pubblicazioni del Reale Osservatorio di Brera in Milano, 40 (2 pts., Milan).

Nares, R. et al. 1808. A Catalogue of the Harleian Manuscripts in the British Museum, v. 1 (London).

Neugebauer, O., ed. and tr. 1962. The Astronomical Tables of al-Khwarizmi. Historisk-filosofiske Skrifter udgivet af Det Kongelige Danske Videnskabernes Selskab 4, 2: pp. 1-247.

Ocreatus, N. See Henry, 1880.

Panofsky, Erwin. 1946. Abbot Suger on the Abbey Church of St.-Denis and its Art Treasures (Princeton).

Paris, Paulin. 1841. Les Manuscrits français de la Bibliothèque du Roi, v. 4 (Paris).

Pedersen, Olaf. 1959. "Du quadrivium à la physique." Joseph Koch, ed., Artes liberales (Leiden and Cologne).

Peter Philomenus of Dacia. See Curtze, 1897.

Pingree, David. 1976. "The Indian and Pseudo-Indian Passages in Greek and Latin Astronomical and Astrological Texts," Viator, 7: pp. 141-195.

Pirenne, Henri. 1929. "L'Instruction des marchands au Moyen Age." Annales d'histoire économique et sociale, 1: pp. 13-28.

Podlaha, A. 1922. Soupis Rukopisů knihovny Metropolitní kapitoly Pražske, v. 2 (Prague).

Poullé, Emmanuel. 1954. "L'Astrolabe médiéval d'après les manuscrits de la Bibliothèque nationale." Bibliothèque de l'Ecole des Chartes, 112: pp. 81-103.

------ 1972. "Les Instruments astronomiques de l'Occident latin aux XIe et XIIe siècles." Cahiers de civilisation médiévale, 15: pp. 27-40.

Prior, W. H. 1924. "Notes on the Weights and Measures of Medieval England." Archivum latinitatis Medii Aevi, 1: pp. 77-97, 141-170.

Ptolemy, Claudius. 1952. Almagest, tr. R. Catesby Taliaferro. Great Books of the Western World, v. 16 (Chicago).

Richard of St. Victor. See Migne, 1844-1864, v. 196.

Robertus Anglicus. See Britt, 1972; Curtze, 1899b; and Tannery, 1897.

Savasorda, Abraham. See Curtze, 1902a.

Schum, W. 1887. Beschreibendes Verzeichnis der Amplonianischen Handschriften-Sammlung zu Erfurt (Berlin).

Sharp, D. E. 1934. "The 'De ortu scientiarum' of Robert Kilwardby." New Scholasticism (Washington) 8: pp. 1-30.

Shelby, Lon R. 1961. "Medieval Masons' Tools: The Level and Plumb Rule." Technology and Culture, 2: pp. 127-130.

------ 1965. "Medieval Masons' Tools. II: Compass and Square." Technology and Culture, 6: pp. 236-248.

------ 1969. "Setting Out the Keystones of Pointed Arches: A Note on Medieval 'Baugeometrie'." Technology and Culture, 10: pp. 537-548.

------ 1970. "The Education of Medieval English Master Masons." Mediaeval Studies, 32: pp. 1-26.

------ 1971. "Mediaeval Masons' Templates." Journal of the Society of Architectural Historians, 30: pp. 140-154.

------ 1972. "The Geometrical Knowledge of Mediaeval Master Masons." Speculum, 47: pp. 395-421.

Simson, Otto von. 1962. The Gothic Cathedral (New York).

Smalley, Beryl. 1964. The Study of the Bible in the Middle Ages. (2nd ed., Notre Dame, Ind.).

Spufford, P. 1963. "Coinage and Currency." Appendix in M. M. Postan, E. E. Rich, and Edward Miller, eds. The Cambridge Economic History of Europe, v. 3 (Economic Organization and Policies in the Middle Ages): pp. 576-602 (Cambridge).

Steele, Robert, ed. 1940. Bacon, Roger. Communia mathematica. Opera hactenus inedita Rogeri Baconis, 16 (Oxford).

Sudhoff, Karl. 1914. "Die kurze 'Vita' und das Verzeichnis der Arbeiten Gerhards von Cremona: von seinem Schülern und Studiengenossen kurz nach dem Tode des Meisters (1187) zu Toledo verabfasst." Archiv für Geschichte der Medizin, 8: pp. 73-82.

Suter, H., ed. 1914. Die astronomischen Tafeln des Muhammed ibn Musa al-Khwarizmi in der Bearbeitung des Maslama ibn Ahmed al-Madjriti und der latein. Uebersetzung des Athelhard von Bath. Det Kongelige Danske Videnskabernes Selskabs Skrifter, 7. Raekke, Historisk og Filosofisk Afdeling, 3, 1: pp. 1-255.

Tabulae codicum manuscriptorum praeter graecos et orientales in Bibliotheca Palatina Vindobonensi asservatorum. 1870, 1873. v. 4, v. 6 (Vienna).

Tannery, Paul, ed. 1897. "Tractatus quadrantis magistri Roberti Anglici." Notices et extraits des manuscrits de la Bibliothèque nationale, 35: pp. 561-640. (repr. in Paul Tannery, Mémoires scientifiques, 5: pp. 118-197).

Taylor, E. G. R. 1947. "The Surveyor." Economic History Review, 17: pp. 121-133.

Taylor, Jerome, ed. and tr. 1961. Hugh of Saint-Victor. Didascalicon: a medieval guide to the arts (New York).

Thorndike, Lynn. 1943 "Robertus Anglicus." Isis, 34: pp. 467-469.

------ 1947. "Who Wrote Quadrans vetus?" Isis, 37: pp. 150-153.

------ 1957a. "Cosmimetria or Steriometria?" Isis, 48: p. 458.

------ 1957b. "Notes on Some Astronomical, Astrological, and Mathematical Manuscripts of the Bibliothèque nationale, Paris." Journal of the Warburg and Courtauld Institutes, 20: pp. 112-172.

Thorndike, Lynn, and Pearl Kibre. 1963. A Catalogue of In-
    cipits of Mediaeval Scientific Writings in Latin (rev.
    and augmented ed., Cambridge, Mass.).

Thulin, C. 1911. Zur Überleiferungsgeschichte des Corpus
    Agrimensorum: Exzerptenhandschriften und Kompendien
    (Göteborg).

Toomer, G. J. 1968. "A Survey of the Toledan Tables."
    Osiris, 15: pp. 5-174.

Trabut-Cussac, Jean-Paul. 1961. "Date, fondation et iden-
    tification de la bastide de Baa." Revue historique de
    Bordeaux, n.s., 10: pp. 133-144.

Tropfke, Johannes. 1924-1940. Geschichte der Elementar-
    mathematik in systematischer Darstellung (v. 1: Rechnen.
    3d ed., Berlin and Leipzig, 1930. v. 4: Ebene Geom-
    etrie, ed. Kurt Vogel. 3d ed., Berlin, 1940. v. 7:
    Sterometrie, Verzeichnisse. 2d ed., Berlin and Leipzig,
    1924).

Ullman, B. L. 1964. "Geometry in the Mediaeval Quadri-
    vium." Studi di bibliografia e di storia in onore di
    Tammaro de Marinis (4 v., Verona) 4: pp. 263-285.

Victor, Stephen. 1970. "Johannes de Muris' Autograph of
    the De arte mensurandi." Isis, 61: pp. 389-395.

Villard de Honnecourt. See Bowie, 1959; Hahnloser, 1935;
    and Willis, 1859.

Vogel, Kurt, ed. 1963. Muhammed ibn Musa al Khwarizmi.
    Algorismus; das früheste Lehrbuch zum Rechnen mit indi-
    schen Ziffern (Aalen).

------. 1971. "Fibonacci." Dictionary of Scientific Bio-
    graphy, ed. Charles Coulston Gillispie (14 v., New York,
    1970-1976), v. 4: pp. 604-613.

Weisheipl, James A. 1965. "Classification of the Sciences
    in Medieval Thought." Mediaeval Studies, 27: pp. 54-90.

Weissenborn, Hermann. 1888. Gerbert: Beiträge zur Kenntnis
    der Mathematik des Mittelalters (Berlin).

Willis, Robert, ed. 1859. Facsimile of the Sketch-Book of
    Wilars de Honecourt, an Architect of the Thirteenth Cen-
    tury (London).

Willner, Hans, ed. 1903. Adelard of Bath. De eodem et
    diverso. Beiträge zur Geschichte der Philosophie des
    Mittelalters, 4, pt. 1: pp. 1-112.

INDEX OF LATIN TECHNICAL TERMS

This index includes most of the technical terms of <u>ACC</u>. The common terms for arithmetical operations and results, like addere, dividere, ducere, invenire, iungere, multiplicare, productum, subtrahere, sumere and summa, are not normally included. I have not indexed forms of numbers—cardinal, ordinal or fractional. Most of the other technical and unusual words of the text and variants are found here. Adjectives are listed by their masculine forms. Verb forms are listed under their infinitives, except for participles used principally as adjectives, which are listed as such.

Since the General Index lists terms discussed in the introduction and running commentary, they are not included here. A page number followed by "n" is a reference to a word appearing in the variants but not in the main text of that page. Words occuring in both the text and variants of a page are listed once, without an "n." Since the Latin text appears only on even-numbered pages, an inclusive notation means the word occurs on all the even-numbered pages included. Thus, the entry 282-286, 290n would mean that the word in question appears in the text (and perhaps in the variants) of pages 282, 284 and 286, and in the variants on page 290.

accipere, 204
activus, 196n
acumen, 246
adequatio, 276n, 278n
agregare, 134
aliquotus, 384, 386, 390, 392, 424
allidada, 114, 116, 294, 298, 310, 316
almagestans, 336n, 340n
Almagesti, 220, 242, 248, 254, 280, 282, 324, 334, 336, 342, 344
almucantarath, 238
altifa, 220
altimetra, 298
altimetria, 110, 218, 220n, 324

altitudo, 220-254, 258, 264-274, 292, 294, 298-304, 308, 310, 314-324, 348, 360-374
altrinsecus, 186n
altus, 316n, 318
altus locus, 276, 282-286, 290
ambligonus, 142
angularis, 170
angulus, 110, 118, 124, 150, 152n, 210n, 212n, 318, 326, 330, 380
anno domini, 282
anno verbi incarnati, 282n
annus, 282-286
anteriorare, 418
aprilis, 230

aqua, 300
aquarius, 256, 260
arare, 242
archa, 362-368
Archimenides, 176, 178
Archites, 140
arcus, 152n, 158n, 186n, 220, 226,
    242
area, 126-134, 138, 142-154, 162,
    166-170, 178-184, 190-202, 216n,
    218, 314, 326, 328, 348, 352-
    370, 376-380
argumentare, 450
aries, 230, 250, 256-260, 286, 288,
    292
arismetica, 108, 130
arpennius or arpemnius, 206n-210n
arpennus, 206n, 208n, 210. See
    also arpentus
arpentus, 206, 208
articus, 240
arundo, 304, 312
ascendens, 244, 246, 268
ascensus, 250-260
aspicere, 298
astrolabium, 114, 116, 120, 188,
    222, 224, 246, 264, 270-274,
    292, 294, 310
astrologus, 226, 402
astronomia, 110, 112, 324, 394,
    398n
astronomicus, 112, 226n, 382, 392,
    398, 448
auferre, 136
aux, 278, 284, 290
azimuth, 238

basis, 114, 116, 126, 132-144, 150,
    154, 224, 302
Boetius, 128, 130, 216n

cacumen, 298, 302
calculatio, 108
campus, 206, 210
cancer, 256, 262
canna, 116, 118, 300-304, 312
capacitas, 112, 204, 206n, 314,
    324, 348, 352-366, 370-374
capricornus, 256, 260
capud. See caput
caput, 114, 116, 126, 290, 318
caput drachonis, 290
cathetus, 114-118, 122-126, 132-
    144, 150, 154, 170n, 216n, 222-
    226, 294, 298, 302, 316, 320,
    322

cauda, 290
cema, 158n
cenith, 240
centrum, 178, 186, 188, 226, 276,
    278, 322
certificatio, 454
cifra, 442, 444, 460, 462
circinus, 186, 318, 322
circularis, 170, 322, 360n, 362
circulatio, 194, 310, 314, 368
circulus, 140, 172, 176-184, 186n,
    190, 194-198, 202, 226, 228, 232,
    238, 250, 258, 270, 278, 318,
    322, 330, 342, 348, 366, 460,
    462, 466, 468
circulus directus, 258
circulus rectus, 260
circumferentia, 152n, 176, 178,
    186n, 188, 218, 226, 318, 342
citius, 292
civitas, 218
clima, 230, 232, 238, 258, 260, 292
color, 322, 324
columpna, 328, 342
columpnaris, 346, 348, 360
communis, 128, 328n, 348n
comparare, 312
comparatio, 304
compositio, 430
comproportionalis, 396n
computare, 130, 274, 286, 288
computatio, 286
consonantia, 108, 110
constituere, 118
continentia, 352
continuus, 292
corausta, 140
corda, 158n, 220, 242, 244, 248
corpus, 110, 112, 294, 324, 326,
    346, 370
cosimetria, 380n
crassimetra, 364
crassimetria, 326n, 380, 430
crassitudo, 112, 130, 324-330, 334,
    370-376
crassus, 130, 314, 326
crescere, 278
cubicare, 326, 330, 418
cubicatus, 334
cubicatio, 422, 430
cubicus, 130, 326, 414n, 418, 422-
    426, 438
cubus, 326, 330, 332, 344, 412,
    414, 418, 428, 438
cubum, 338, 340

currere (super), 124, 132, 138, 142, 146, 152
curta piramis, 348n
curvare, 172
curvus, 172, 176

decagonus, 170
declinare, 212n, 226
declinatio, 266
declinus, 258
decrescere, 278
decuplare, 456n, 460n, 462n
deduplare, 444, 456, 460, 462, 466, 468
deduplatio, 458
demere, 138, 150, 244, 284
demonstrare, 196
demonstratio, 158n, 216n
denominare, 390, 392, 452
denominatio, 136, 176, 182, 194, 206, 210–220, 238, 242, 248–254, 258, 330, 332, 348, 352, 398, 424
depressio, 292
descendens, 244
describere, 318
dexter, 292
diagonalis, 150, 156
diagonaliter, 128
diameter. See dyameter
dies, 242, 244, 266, 272, 282, 286, 288, 292
differentia, 192, 442, 446, 456, 462, 464, 468, 470
difficilis, 374
difficultas, 392
digitus, 114, 116, 220–224, 248–254, 258, 260n, 298, 310, 316, 320–324, 348, 384, 414–422
directio, 272
directus, 258, 272
distantia, 234, 240, 276, 278, 284, 322
distare, 240
diversitas, 276, 372, 374
dividendus, 404
dividens, 404
divisio, 222, 392
diurnus, 242
doctrina, 348
dolius, 348n, 352–358, 364–368
domus capienda, 212
domus, 212, 218

ducere, 126, 132, 134, 138, 142, 144, 150, 154, 158, 186, 304, 318, 390
ductus, 126, 136, 178, 218
duplare, 442, 444, 456, 466n
duplatio, 458n
duplicare, 460n
dyagonalis. See diagonalis
dyameter, 140, 172, 176, 182, 184, 190, 194–198, 202, 210n, 310, 312, 314n, 322n, 330, 334–344, 348, 352–358, 364, 368, 372, 374, 430

ecentricitas, 270, 278
ecentricus, 276
eclipsis, 110
econtrarius, 286
embaldale, 122, 128
embaldum, 134, 152, 154, 168n, 170
eptagonus, 164, 380
equalis, 124–128, 152, 216n, 226, 294
equalitas, 122, 124, 326
equare, 152, 348
equatio, 276, 278
equedistans. See equidistans
equiangulus, 158n
equidistans, 132, 134, 138, 142, 144, 150, 152n, 154, 158, 178, 210n, 212n, 318, 326, 328, 390
equidistantia, 332
equilaterus, 122, 126, 128, 150, 152, 158, 160, 378
equinoctialis, 230–234, 240, 258, 260n, 266, 286, 288, 292
equinoctium, 232n, 288
equipollentia, 152
equivalere, 432
erigere, 226, 302
estivalis, 232, 240, 292
euclidens, 380
Euclides, 122, 124n, 134, 144, 338, 346, 378, 394n
exagonus, 158n, 162, 170, 380, 412
examinatio, 430
excessus, 134, 192, 196, 198, 202, 236, 330, 332, 348
exercitare, 380
exercitium, 372, 378
exigonius, 144n
extendere, 220
exterior, 128

extractio, 392, 406, 412, 456

fastigiosus, 206
ferire, 206, 244
fides oculata, 116, 390
figura, 130, 158n, 170, 378, 380,
    414, 418, 446
filum, 300, 308
finire, 242, 244
foramen, 114, 310
forma, 376, 406
formatio, 406
frons, 206, 216
fundus, 310, 348, 352

gemini, 254, 256, 260, 262, 284
genus, 398
genzahar, 290
geometria, 110, 112, 116, 122, 128,
    130, 192, 324n, 372, 384, 470
geometricus, 112, 130n, 382, 384,
    424, 452
Girbertus, 122
globositas, 330
gradus, 220, 230-234, 238, 240,
    244, 246, 250-254, 258, 260,
    264-270, 274, 278-284, 288-290,
    294, 334, 394, 398, 400, 404
granum, 384
granum ordei, 384
gravis, 272
grossus, 286

habitudo, 272n
hiemalis. See hyemalis
hodiernus, 282, 286, 288n
hora, 188, 242-246, 268-272, 286
hyemalis, 232

iacere, 302
ignorantia, 380
immobilis, 318
impar, 412n, 442, 446-450, 456
impariter, 448
imperfectus, 238
improbare, 130n
imus locus. See ymus locus
inaccessibilis, 298, 304
inaccessimetra, 298
inauratura, 182
indus, 286, 288
industria, 380
inequalis, 134, 152
inequalitas, 292

infigere, 322
infinitus, 108, 392
ingenium, 380
initium, 288
inpar. See impar
inperfectus. See imperfectus
inscribere, 140, 196, 198, 200n,
    202
insensibilis, 192
instrumentum, 322, 324
integer, 130, 286, 428. See also
    integrum
integraliter, 130, 386n
integritas, 392n
integrum, 384, 386, 390, 396, 398,
    406, 412, 424, 430, 448, 454-
    470. See also integer
intersecare, 186
intersecatio, 186n, 188n
intersectio, 186, 188
intervallum, 298, 300
intrare, 286
irregularis, 370-374, 376n
irregularitas, 372, 374, 380
isocheles, 132, 170n
isopleurus, 122, 124n, 126, 128,
    164, 378
iunctura, 286, 288

jupiter, 280, 282, 290

kalendae, 230
kathetus. See cathetus

lateratus, 328n
laterculus, 204, 216
latinus, 286
latitudo, 116, 152n, 204, 210n,
    212-218, 234, 238, 244, 270,
    274, 292, 314n, 316, 318, 366
latus, 110, 114-118, 122-140, 146,
    148, 152, 156, 158, 162-170,
    178, 190, 194, 200, 202, 206-
    210, 216, 296, 298, 302, 310,
    316-320, 330, 332, 364, 366,
    378, 380, 408-412
leo, 256, 262
levis, 272
libra, 230, 256, 260, 262, 292
lignum, 318
limes, 446, 448
linea, 114-118, 124, 126, 130, 154,
    156, 172, 176, 186, 212n, 226,
    228, 234, 236, 240, 264, 268,

274, 276, 286, 288, 302, 304,
316, 318, 328, 330, 342, 386,
390, 392
linea equinoctialis, 286, 288
linealiter, 384
linearis, 386, 388
locus, 264, 270, 276, 278, 282-286,
290, 298, 302
longitudo, 120, 204, 206, 210-218,
328, 352-358, 364-368, 384
luna, 280, 282, 334-340

magister, 110
magnitudo, 342, 344
manifestus, 182n
mars, 280, 282, 290
martius, 286, 288
maximus, 232, 236
mediclivium, 114, 298
medium, 206
medius motus, 276
memor, 110
mensis, 264
mensor, 304
mensura, 122, 130n, 146, 216n, 294,
358, 370
mensurabilis, 120n
mensurare, 110-116, 120, 130, 132,
300, 302, 306, 308, 318, 320,
324, 384, 434
mercurius, 280, 282, 290
meridianus, 228-232, 242, 258n,
264, 266, 274
meridies, 228, 242, 244, 268
meridionalis, 226, 228n, 236, 264,
268, 274, 316
metiens, 304
metiri, 124-128, 132-138, 144, 148,
150, 154-158, 162-170, 182, 194,
200, 210, 212, 220, 286, 294,
298, 304, 310, 316, 318, 334,
354-362, 366, 372-378
metra, 114n
miliarium, 334
minimus, 226, 228, 236
minutia, 112, 190, 382, 384, 392,
394, 398, 402, 406, 412, 424,
426, 448-456, 460, 462, 466.
See also minutum
minutia astronomica, 382, 392, 398,
448
minutia geometrica, 382, 384, 424,
452
minutia prima, 394, 398

minutia vulgaris, 392, 398
minutiatim, 466
minutimetra, 430
minutimetria, 382n, 470n
minutum, 232, 240, 244, 248-254,
258, 260, 278-282, 394-398, 402-
406, 448, 450, 456, 458, 462-
470. See also minutia
modernus, 288
modius, 206, 348-352, 384
monstrare, 228
morosus, 374
motus, 110, 276, 282
movere, 282, 290
multilaterus, 158n
multiplicatio, 170, 392
musica, 108

naturalis, 170
nodus, 274
nonogonus, 168
noticia, 210n
nox, 246, 272, 274, 292
numerare, 298, 326, 460-470
numerata, 352
numerus, 108, 388, 392, 414, 418
numerus cubus, 414
numerus quadratus, 388

occidens, 270, 282
occidentalis, 244, 270, 272, 316
october, 230
octogonius, 166
oculata fides. See fides oculata
oculus, 294, 302, 304
oppositus, 290, 310, 312
opus, 328n, 348n, 368n, 402
ordo, 170, 420
ordo naturalis, 170
oriens, 270, 282
orientalis, 244, 270, 272, 316
orizon, 246, 274, 292
orthogonalis, 212n
orthogonaliter, 180, 222, 224, 304
orthogonium, 178, 180, 224, 300,
302. See also orthogonius; orthogonus
orthogonius, 114-118, 138, 140,
154, 156, 176, 224, 316, 318.
See also orthogonium; orthogonus
orthogonus, 294. See also orthogonium; orthogonius
oxigonius, 144, 170n
oxigonus, 170n

palma, 384n
palmus, 130, 348, 384
par, 442, 446-450, 456, 460, 462, 466
paralellogramum. See parallelogramum
paralellus, 240
parallelogramum, 128, 156, 180
parisiensis, 230-234, 238, 258
parisius, 240
pars, 226, 246, 286, 312, 318, 322, 384, 386, 390, 392, 408, 424
pars aliquota, 384, 386, 390, 392, 424
pars bis, 408
pars tota, 384, 386, 392
partialiter, 130
partitio, 324
passivus, 196n
pavimentum, 158, 164, 214
pentagonus, 156-160, 170, 380, 412
percussio, 398, 400, 404
percutere, 334, 402, 464
perfectus, 238
perforare, 318
periferia, 172, 182
perpendicularis, 212n
perpendiculariter, 310
perpendiculum, 318, 320
pertica, 206, 208, 384
pes, 116, 118, 122-136, 152, 158, 164, 176-182, 186, 190, 194-198, 204, 206, 212, 216, 218, 294, 298, 302, 308, 312, 314, 318, 322, 326, 330, 348, 372, 374, 384-392, 426, 428
pes crassus, 130, 314
pes cubicus, 130
pes linearis, 386, 388
pes longus, 130, 386
pes planus, 124, 132n
pes quadratus, 130, 386, 390
petitio, 174
pinnula, 318
piramis, 348n
pisces, 256, 260, 288
planeta, 266, 270-278, 282, 290
planimetra, 114, 116, 298n, 386n
planimetria, 216n, 218, 386. See
   also planismetria
planismetria, 110. See also planimetria
planities, 304
planum, 110n, 294, 302

planus, 124, 226, 324
podismare, 114, 168, 172
podismus, 138, 144, 168
polus, 234-240, 292
ponderosus, 272n
ponere, 246, 322, 324
postica, 316
practica, 108, 110, 470
practica geometrie, 470
practicum geometrie, 324n
precisura, 136
premissum, 130
preostensus, 412
preponere, 460, 462
preterire, 244
probare, 110, 114, 122-126, 132, 134, 138-172, 176-184, 188, 190, 194, 198, 202, 204, 210, 212, 216-220, 224, 226, 234, 248, 254, 260, 276, 280, 282, 292, 294, 300-304, 308, 310, 324-328, 332-346, 350-368, 374, 378, 380, 386-392, 402, 404, 408, 418, 422-430, 440, 444, 446, 448n, 450, 452, 458, 470
probatio, 148, 160, 166, 168n, 170, 176, 180, 220, 242, 374, 380, 402, 404
profundimetra, 218, 310
profunditas, 310, 312
profundus, 310n
prolixitas, 292
proponere, 116, 128, 200, 328, 426
proportio, 108, 114-118, 222, 224, 278, 294, 296, 300-312, 338, 342, 394, 398, 400, 406n, 424n, 430
proportionalis, 394, 396, 400
proportionaliter, 278
propositio, 144, 148, 150, 160, 164, 176, 188, 220, 366, 368, 374n, 378, 380, 424, 430, 460
protendere, 150, 318
protractio, 392
protrahere, 124, 176, 178, 186, 226
Ptholomeus, 284-288
punctum, 186, 226, 316
pupilla, 224
puteus, 310-314, 370
pyramis, 346, 370, 376-380

quadra, 316
quadrangulum, 124-128, 132, 144-148, 178, 180

quadrangulus, 124n, 210, 348, 360, 362

quadrans, 120, 124n, 222, 224, 310, 318

quadrare, 190n, 366n, 368n

quadratum, 148, 158, 160, 164, 190, 194–198, 200n, 202, 298, 308n, 310, 316, 318, 368, 384–388, 406–422, 426–444, 452

quadratura, 190, 366

quadratus, 130, 190, 364, 368, 380, 386–390, 408, 432, 444, 446, 452, 470

quadruvialis, 108

quantitas, 110, 114, 118–122, 128, 152n, 158n, 168n, 180n, 202n, 222, 302, 310, 322, 326, 334–340, 346, 348, 352, 362, 366, 370, 430

quinquena, 238

radius visualis, 114–118, 304, 310

radix, 132, 136, 150, 156, 190, 194, 202, 220, 286n, 288n, 308, 366, 386, 392, 406–420, 424–434, 438–470

radix cubi, 412

radix cubica, 422

radix cubici, 424

ratio, 380, 412n. See also ratio numeri

ratio numeri, 182, 204, 210, 212, 216, 308, 350, 352, 356, 358, 362, 390, 392

rationalis, 342

recisio, 330

recissura, 130

rectangulus, 378

rectus, 114, 118, 126, 172, 176, 186, 224, 250–254, 304, 316, 318

redditum, 286

reducere, 400, 456

reduplicatio, 458n

regula, 116, 128, 132, 134n, 136, 138, 170, 216n, 222, 300, 316, 372, 374, 378, 380, 402, 408–412, 422, 452, 456, 458

regula generalis, 300

regularis, 314, 324, 348, 370, 376

reliquum, 258, 260, 290

remanere, 248

remotio, 230, 232, 266

remotio solis, 230, 232

rescindere, 330

residuum, 132, 136, 140, 162–170, 198, 220, 230, 234, 238, 240, 248, 318

resolutio, 454, 456, 460, 466

resolvere, 454, 462, 466

retrogradus, 272

rotundus, 218, 328n

rumbus, 150, 152

sagitta, 300, 308

sagittarius, 256, 262

saturnus, 280, 282, 290

scalenos, 134

scientia, 108, 152n, 158n

scientia de corda et arcu, 158n

scientia de arcu, 152n

scorpius, 256, 260, 262

sectio, 124, 186, 222, 318, 326, 332

secundare, 442

semiartifex, 108

semicathetus, 126

semicircumferentia, 314, 342

semidyameter, 176, 178, 218, 314, 322

semiplenus, 130, 216n

september, 230n

septemtrionalis, 266, 274, 316

sextarius, 206, 352, 384

signum, 186n, 244, 256, 258, 264, 278, 282, 288, 292

sinister, 292

sol, 220, 222, 226–232, 242–248, 264, 266, 278–286, 292, 294, 324, 334–344, 430

solaris, 224

soliditas, 330, 332, 342

solsticialis, 232, 240

solsticium, 232n, 264

solutio, 130

sonus, 108

spatium, 122, 126, 128, 186n, 294, 316

speculum, 300

spera, 250–254, 258, 330, 332, 338, 342

spericus, 330, 334, 338n

stare, 298, 302

statio, 298

stationarius, 272

statura, 114, 116, 294, 302, 304, 310, 312

stella, 224, 236, 246, 266–272
stella fixa, 268, 282, 286
studere, 430
subponere, 216n
subtendere, 212n
subtilis, 392
subtilitas, 380
subtractio, 170, 300
summitas, 158, 224, 294, 302, 304,
    308, 312
superagregare, 128
superare, 260, 262
superficialis, 122, 126
superficies, 110, 122–126, 130,
    132, 142, 144, 150, 170, 204,
    206, 212–216, 226, 322, 324
superponere, 172
surdus, 190, 444, 454, 466
synodoche, 130

tabula, 206, 384
tangere, 140, 186n
taurus, 252, 256, 260, 262
tempus, 282n, 284, 286n
terminus, 114, 118, 178, 212n, 310,
    318
terra, 258, 260, 334–344, 430
tetragonalis, 156
tetragonus, 406, 410
theoreuma, 380, 426, 470n
theoreumatice, 112
theorica, 108
tollere, 132
trabicus, 384
trabucum, 206n
trabuscus, 206, 384
tractatus, 110, 392
tracubus, 384
trapezitum, 154

triangulus, 114–118, 122–134, 144,
    154, 160, 176, 180, 216, 224,
    294, 300–304, 310, 348, 378
tribucus, 384n
trigonus, 128, 130, 158n, 406, 408,
    410n
turris, 294, 298

umbra, 220, 224–228, 248–254, 258,
    294, 316, 320–324
umbra plana, 220
urna, 370
utilitas, 392

variatio, 292
vas, 348, 350, 360, 362, 372, 374
venus, 280, 282, 290
veritas astrolabii, 270
vertex, 154
verus motus, 276
vicula, 242
vinum, 348, 352
virga, 224, 302, 312
virgo, 256, 262, 288

walzagora, 114, 238, 268, 316

yelizar. See genzahar
yenzar. See genzahar
ymago, 288
ymus locus, 276, 278
ypothenusa, 114–118, 128, 138, 140,
    296, 302
ypothesis, 174
ysocheles. See isocheles
ysopleurus. See isopleurus

zodiacus, 264, 270, 276

INDEX OF OLD FRENCH TECHNICAL TERMS

Most of the technical terms found in the <u>Pratike</u> are included in this index. Numbers in their various forms are usually excluded. Words from the variants are indexed only where they seem particularly interesting. Nouns are listed in the singular, in the subject case, with occasional cross-references from the object case. Adjectives are included in their masculine singular subject form. Verb forms are subsumed under their infinitives, except that participles used as adjectives are indexed as adjectives. A few abbreviations kept in the text have been spelled out for indexing; for example the final "x" standing for "us" has been indexed as "us," and "s." has been listed as "sous." A page number followed by "n" refers to a word occuring only in the variants of that page. If there is no "n," the word may occur in the text alone or in the text and variants.

abatre, 480
Achre. <u>See</u> Acre
Acre, 568, 574
Aicre. <u>See</u> Acre
aiouster, 480, 492
aire, 476–480, 482–487, 488n, 489–
   491, 493–496, 498–500, 505–508,
   511, 512, 515, 517–520, 522,
   532, 544, 546
aloiemens. <u>See</u> alouemens
alouemens, 566, 598
ambigoines, 482
angles, 524, 526
angoine, 570
angulers, 491
Antioches, 570
Aquilee, 574
arbre, 475
Ardane, 568
arismetike, 576
argens, 562, 564, 566, 568, 574,
   598, 600

arismetique. <u>See</u> arismetike
arpens, 502, 503, 546
art, 528, 576
artisien noef, 568
assambler, 481, 487, 488, 502–504,
   510
astrelabe. <u>See</u> astrelabre
astrelabre, 474, 475
astronomie, 473
aucerrois, 570
augorismes, 582
Auscorgne, 572
avoir de pois, 584

Babilonie, 572
barbarin, 570
barselevois, 572
basse, 474, 478–483, 487
besant, 572
Besencon, 570
besse. <u>See</u> basse
bialvoisien, 568

billons, 562, 566
blanc, 568
bloisois, 570
boison, 574
bordelois, 570
bouce. See bouge
bouge, 528, 542
brouselois, 568
burgalois, 570

cambrisien, 568
camp, 502, 503
cange, 550
cangier, 550
caorsin, 570
Cecille, 574
cercle, 481, 494, 496-500, 509
cerin, 572
certain, 570
chartain, 570n
cifre, 582
circonference, 492-494, 497, 506, 508, 513
circuler, 491
clé, 524
clermontois, 570
clignisien, 568
clochier, 475
clocier. See clochier
combe. See conbles
comber, 509
compas, 492, 493, 522, 524, 526
conbles, 485, 486, 507-510
cone, 526
Constantinoble, 574
contable, 562, 564, 566, 580, 584, 592
conte, 562, 564, 566, 568, 580, 584, 592
contemens, 598
contenance, 501, 504, 509-514, 516-520, 528, 536
contenir, 500
contenanche. See contenance
conter, 566, 578, 580, 584, 586, 588, 590, 592, 594, 598, 600
contraires, 486, 502-504
corne, 524
costes, 476, 478-480, 485, 486, 488-491, 496-498, 500, 502, 503, 505, 507, 516, 522, 524
coumbe. See conbles
courir, 570

crasse, 473, 507, 509
cuber, 509

Damas, 568
denier, 560, 562, 564, 566, 568, 570, 572, 574, 584, 586, 588, 590, 592, 594, 596, 598, 600
denominations, 479, 493, 494, 497, 502, 503, 505, 506
descendre, 554
deviser, 476, 479, 489, 490, 491, 493, 497, 501-506, 509, 511, 512, 517, 520, 532, 534, 536, 542, 550
devisions, 504, 509-512, 520, 536
diagonal. See dyagonal
dijounois. See diounois
diounois, 570
disaine, 578, 580
disisme, 576
divisions. See devisions
dobble, 477
doisien, 570
doubler, 578
draant, 568
droit, 474
dunoisien, 570
dyagonal, 474, 485, 488, 548
dyametres, 481, 492-500, 506, 508, 509, 511-514, 518, 519, 528, 532, 534, 536, 540
dyometrie. See geometrie

Egipte, 576
emperial, 570
ensegnier, 483, 491, 506
enseingnier. See ensegnier
ensengier. See ensegnier
entendables, 496
equilarere. See equilatere
equilatere, 476, 485, 486, 522
escange, 598, 600
espere, 509
escrire, 481, 498, 499
esse, 570
essengnier. See ensegnier
esterlins, 560, 562, 568, 586, 588, 590
estateur, 475
estievenent, 570
estrois, 546
eufosin, 572
ewe, 475

fause ligne, 481
ferton, 566n
firton, 566
figure, 483, 491, 522n, 524, 526, 582
fin, 474, 564, 568
Flandres, 562
flueve, 475
foible, 550
fons, 511, 512, 518, 528, 532, 538, 542
fors, 550, 568, 570
frons, 502
frizacois, 568

Gascoigne, 568
genevois, 570
geometrie, 473, 475, 528, 538
geter, 576, 580,
gimgembre, 584
grain, 562, 564, 566, 570, 594
gros, 528
grousenois, 568
gymgembre; gyngembre. See gimgembre
gyometrie. See geometrie

haut, 511, 515, 517, 520
hautece. See hautesche
hautesche, 473, 475, 477
hollandois, 568
huce, 515, 517-519

ieter. See geter
ieus. See oeul

kerré, 474
keure, 476, 478-480, 482-487, 489-491, 501-504, 506-508, 518, 519
keuvre, 568, 574

Laon, 568
laterculers, 501, 505
lé. See lei
léece. See léeche
léeche, 501
lei, 475, 501, 503, 504, 506, 516, 519, 524, 528, 538
leonnois, 572
libre, 554, 558, 560, 584, 586, 588, 590, 592, 594, 596
libre d'avoir de pois, 584
liegois, 568
ligne, 474, 475, 477, 481, 548

lisiere, 524, 548
livel. See liviaus
liviaus, 474, 476, 478-483, 485, 487, 489-491, 505, 524
livre. See libre
lonc, 474, 475, 477, 501-504, 506, 508, 512-514, 516, 518, 519, 524, 528, 532, 534, 536, 538, 540, 542, 544, 546
lonisien, 570
lons, 528, 538, 546, 548
Loon. See Laon
Losane, 568

maille, 566, 572, 594, 599
Manasces, 574
mansois, 568
marbotin, 570
marin, 572
marrois, 572
marros. See marrois
mars, 560, 562, 564, 566. 568, 586, 588, 590, 592, 594, 598, 600
mars d'argent, 562, 564
mars de billons, 562, 564
mars de fin argent, 564
mars de Flandres, 562
mars d'escange, 598, 600
mars d'or, 560
masconois, 568
mener, 506
messaig, 568
mesure, 473-475, 506, 507, 509, 522, 550
mesurer, 474, 475, 477-479, 511, 520, 522, 524, 546
minuce, 473, 496
missions, 596
moilekin, 572
molekin. See moilekin
molain, 568
monoie, 530, 566, 570, 598, 600
monteplier, 489n, 491n, 493n, 494n, 496n, 497n, 509n-512n, 528n, 534n
monter, 554
moritien, 568
Muce, 572
muis, 511, 512, 538, 542
muison, 528, 532
multeplier. See multiplier
multiplier, 479, 480, 485, 491, 492-494, 496, 497, 509-512, 515, 517, 520, 522, 528, 532, 534, 536, 540, 542, 550, 576, 582

nacois, 572
nantois, 570
Navers, 570
noiret, 570
noirois. See noiret
nombrer, 522
nombres, 479, 485, 488, 494–496,
   503–505, 507, 511, 514, 516–518,
   576, 598

obole, 554, 556, 560, 562, 568, 570,
   572, 586, 588, 590, 594n, 596
oeil, 474
oevre, 473
once, 560, 562, 564, 566, 568
onni. See ouni
ors, 560, 572, 574, 594
ordene, 483
orgone. See ortogone
orneure, 494
ortogoine. See ortogone
ortogone, 474, 480, 481, 483, 487,
   488
oster, 584
ouel, 476
ouni, 476, 478, 489–491, 496, 497,
   514, 518, 519
ounier, 486

papiois, 572
paresies, 570
paresis. See paresies
parcerain, 570
parmesain, 572
partie, 473, 479
partrieus, 474
pauc. See paus
paus, 528, 530, 532, 534, 536, 542
pauchiers, 534, 542, 544
pauls. See paus
paume, 477
pentagones, 489
perce, 502, 503, 546
perfondece. See perfondesce
perfondesce, 473, 477
perpe, 574
pertruis. See partrieus
peser, 560, 562, 564, 566, 570
piés, 474, 476–479, 487, 497, 501,
   504–506, 509–511, 522, 524, 526,
   528, 530, 538, 540, 544, 546
piés cras, 477
piés lons, 477

piés quarrés, 477, 540
piés reons, 532
piagones, 491
pilier, 508
plaine, 501
planece, 473–475, 477, 501, 504–
   506
planete. See planece
Poelge, 574
pois, 584
poitevine, 560, 562, 566, 568, 570,
   572, 574, 586, 590, 594, 596,
   598
poivre, 584
portigalois, 572
princois, 570
proportion, 474, 475
prover, 474–476, 478–491, 493–501,
   503–505, 507, 508, 510, 511,
   513–518
Provins, 570
pugoce, 550
pugoise, 550n, 554, 598
puis, 520

quadrangles, 475, 476, 478–480,
   482, 487, 488, 546
quadruple, 503
quarer. See quarrés
quarrés, 484, 496–500, 507, 509,
   510, 515, 516, 518, 522, 524,
   528, 530, 532, 534, 536, 540
quarrie, 544, 546, 548
quarte, 538, 540, 542
quarterons, 584
querré. See kerré
quintangles, 489
quinunois, 572

rachine, 478, 479, 485, 488, 496,
   497, 518
racine. See rachine
raencien, 570
raison, 475, 479. See also raison
   de nombre
raison de nombre, 494, 495, 503–
   505, 507, 511, 514, 516
ramnés borgois, 568
ravivés, 568n
recevance. See rechevance
rechevance, 515
remanance, 478, 479, 485, 499
reonde. See reons

reont. See reons
reons, 506, 508-510, 515, 522, 528,
    532, 534, 536, 540
rieule, 475, 478, 480, 528, 532,
    534, 538, 596
riule. See rieule
roedenois, 572
roial d'Acre, 568
roial de Marseilles, 568

Saint Quentin, 570
Saint Tirre, 570
sanserois, 570
saus, 550
scalonge, 568
scenaspois, 574
sen, 552
septangle, 491
Sesile. See Cecille
sestier, 511, 538, 540, 542
seus, 496
sevois, 572
sextagones, 490
sixteangle, 490
Soignie, 570
Soisons, 570
somme, 476, 478, 479, 502, 503,
    506-512, 515, 517, 520, 522,
    524, 532, 534, 542
sorcrois, 498-500, 509, 510, 522,
    542
sorcroistre, 509
sostraire. See soustraire
sous, 558, 562, 564, 566, 586, 588,
    590, 592, 594, 598, 600
soustraire, 478, 479, 481, 485,
    498, 499, 509
susain petit, 570
Suse, 568

terin, 574
Tollete, 572
tonel. See toniaus
toniaus, 512-514, 518, 519, 528,
    530, 532, 536, 538, 540, 542
toulousain, 568
Toulouse, 568
trabezire. See trapezire
trapezire, 487
triangeles. See triangles
triangle. See trianguler
triangler. See trianguler
triangles, 474, 476-479, 482, 483,
    522, 524, 526
trianguler, 505
trossien, 568
trover, 476, 478-482, 485-489 and
    passim
tunitsois, 568

vaisseau. See vaissels
vaissels, 515, 520. See also
    vassels
valencenois, 570
Valencienes, 568
valeur, 590
valoir, 556, 558, 584, 586, 588,
    590, 592, 594
valor, 560
vassels, 536, 538. See also vaissels
vendomois, 570
vendre, 584, 586, 588, 590, 594
Venice, 572
venicien, 568, 572
venisien petit, 572
vertis, 487
vienois, 568, 572
viés, 570
visée, 474

INDEX OF ASTRONOMICAL PARAMETERS

Significant astronomical parameters and some other numbers with quasi-astronomical use are found in this two-part index. The sexagesimal parameters are given first, in order of their digits' number value; the position of the sexagesimal point (;) is not considered in ordering the numbers. Zodiacal signs of longitudes are also disregarded, but the number of whole signs is given in parentheses to aid in identification.

The non-sexagesimal numbers in the second part are arranged in strict numerical order. For both parts, I have provided indications of the units of the parameters; dimensionless numbers are usually ratios. A page number without an "n" refers to the English text and perhaps the commentary; an "n" indicates an item in the notes but not in the body of the text of that page. The references to the notes of the Latin text mean that the parameter at issue is found in the variants on that page.

Sexagesimal Parameters

1,0 (parts)	221, 243, 249	6°	241
1,6°	233, 239	6;13°	280n
1;24	457n, 459	6;31°	281
1;24,48	463n, 465	(8^s) 10;47°	282n
1;24,50,24	469n, 471	(8^s) 10;49°	283, 285n
1,30;2°	255n	(4^s) 11°	285n
1,30;8°	255n	(8^s) 11°	285n
1;59°	281n	11;24°	281
1;59,8°	279, 281	11;25°	281n
1;59,10°	281n	(4^s) 11;48°	283, 285n
2,0 (parts)	335	12 (digits)	221–225, 249, 260n
2,20°	291n	13;4°	281
2;23°	281n	13;9°	281n
2;24°	280n, 281n	13;18 (digits)	259, 260n
2,30 (parts)	221n, 243n	14;17°	260n
3;1°	281	14;20°	263n
3;2°	281n	15 (°/hour)	243
;5°	284n	(2^s) 15°	285
5;15°	281	15;1°	263n

$15;36^{\circ}$	259, 263n	$28;8^{\circ}$	263n
$(2^{s})\ 17^{\circ}$	285n	29°	251
$;17,33$ (parts)	339n	$29;38$ (digits)	250n
$18;23^{\circ}$	263n	$29;54^{\circ}$	253, 255n, 257n
19°	233	$30;23^{\circ}$	253n, 255n
$20;15^{\circ}$	261, 263n	$31;44^{\circ}$	255n
$(5^{s})\ 21^{\circ}$	289n	$31;52^{\circ}$	263n
$(11^{s})\ 21^{\circ}$	289n	$32;16^{\circ}$	255n
$21;38$ (digits)	251	$32;34^{\circ}$	255, 257n
$21;39$ (digits)	251n	$33;33^{\circ}$	252n
$21;40$ (digits)	251n	$33;33,8^{\circ}$	240n
$(5^{s})\ 22^{\circ}$	289	34°	240n
$(11^{s})\ 22^{\circ}$	289	$35;11^{\circ}$	263n
$(5^{s})\ 22;58^{\circ}$	283, 285n	$(6^{s})\ 34;40^{\circ}$	282n
$23;33^{\circ}$	233	$35;43^{\circ}$	263n
$23;33$ (digits)	253	36°	241
$23;33,8^{\circ}$	241	$;0,36\ (^{\circ}/\text{year})$	285n
$23;51,20^{\circ}$	233n	$37;15^{\circ}$	263n
24°	241	$(3^{s})\ 37;45^{\circ}$	282n
$24;17^{\circ}$	261, 263n	$(5^{s})\ 37;54^{\circ}$	282n
$24;36$ (digits)	255	$(5^{s})\ 37;57^{\circ}$	282n
$24;49^{\circ}$	263n	$39;45^{\circ}$	263n
$(0^{s})\ 25^{\circ}$	285n	$41;20^{\circ}$	263n
$(6^{s})\ 25^{\circ}$	285n	$41;25^{\circ}$	263n
26°	255	42°	231, 235n
$(6^{s})\ 26;40^{\circ}$	283, 285n	$44;24^{\circ}$	263n
27°	253	$44;59^{\circ}$	263n
$27;17^{\circ}$	263n	46°	234n
$27;40^{\circ}$	251, 255n, 257n	48°	235, 263n
$(2^{s})\ 27;40^{\circ}$	283, 285n	$;0,51\ (^{\circ}/\text{year})$	285
$27;50^{\circ}$	251n, 255n	$;0,54\ (^{\circ}/\text{year})$	285n
$27;55^{\circ}$	251n, 255n		

Non-sexagesimal Parameters

$1/100\ (^{\circ}/\text{year})$	283	$70\ (\text{stadia}/^{\circ})$	335n
$1/70\ (^{\circ}/\text{year})$	285n	$87\ 1/2\ (\text{miles}/^{\circ})$	335n
$1/66\ (^{\circ}/\text{year})$	285n	$166\ 1/4 + 1/8$	339, 345, 431
$1/3$	259n, 261	170	339n
$12/16$	259	$365\ 1/4 - 1/30$	
$15/16$	259	(days/year)	287
$113/116$	259n	$365\ 1/4 - 1/300$	
$114/115$	259n	(days/year)	287n
$40/31$	251–255	$365\ 1/4$	
$3\ 2/5$	337, 339, 345, 431	(days/year)	287
$5\ 1/2$	335, 339, 343, 345, 431	$365\ 1/4 + 1/120$	
		(days/year)	287
$18\ 4/5$	339n, 341, 345, 431	1188 (A.D.)	282n
		1193 (A.D.)	283
$39\ 1/4$	339n	6644	339n, 341
$39\ 1/4 + 1/20$	339n, 341, 345, 431	$6644\ 1/2$	339n, 340n, 345
		$905142\ 1/3 + 1/21$	
39.304	339n	(miles/$^{\circ}$)?	334n
$60\ (\text{miles}/^{\circ})$	335		

GENERAL INDEX

The principal subjects treated in the introduction, text, and notes are covered in this index. I did not intend to list occurrences of every geometric concept implicit in a proposition. Thus, although one needs to know the areas of a square and a circle in order to find the difference between the areas of a square and its inscribed circle (ACC I, 28), the index makes no reference to the area procedures for that proposition.

The index applies to the English text and notes or commentary. References to topics in ACC and the Pratike are, therefore, to pages of the translations of those works (except where there is no facing translation in Pratike I and III). In other words, entries such as 259-263 apply to the English text and perhaps the notes on pages 259, 261 and 263; the Latin text related to that entry is found on pages 258, 260 and 262. A page number followed by an "n" indicates an item found only in the notes or commentary on that page. Latin and French terms are generally not indexed here since I have compiled separate indexes of Latin and Old French technical terms. The thorough searcher will look for references there as well as in this index.

abbreviations: in ACC, 101; in the Pratike, 103; editor's, in variants, 92
Abbott, T.K., 106
Abraham bar Hiyya ha-Nasi (Savasorda), 27. See also Curtze, Maximilian
ACC. See Artis cuiuslibet consummatio
accents in the Pratike, 103
Adam Marsh, 38
additions to ACC, marginal, 98-99
Adelard of Bath, 7, 12-14, 43-44, 47; his version of Euclid's Elements, 7, 43-44. See also Curtze, Maximilian; Willner, Hans

Adelbold. See Bubnov, Nicolaus
agere de arte, 3-4
agere ex arte, 4
agere per artem, 3-4
agrimensores (Roman), 5-6, 50, 64, 72, 83. See also surveyor
algebra, 27, 50
aliquot part: product of, with another aliquot part, 391, 393; product of, with an integer, 385-389
alloys, 55-56, 89; calculation of, 567, 599, 601
almucantars, number of incomplete, 239
Alonso Alonso, M., 11n

altimetry, 6, 15, 18, 21-22, 25,
    29-30, 57, 74, 80, 221-325
altitude: equinoctial solar, 24-25,
    231; line of meridional, 227;
    polar, 237; shadow of, 323, 325;
    shadow of solar, 249; meridian
    solar, 229; solar, 221; solsti-
    tial solar, 233; stellar, 225
angles of incidence and reflection,
    51
apogee, 24, 283, 285
Arabic learning, 7, 9
arc, 27, 48-49
Archimedes, 46, 50. See also
    Clagett, Marshall
architecture, 2, 58-59; biblical,
    33-37; practical geometry in,
    65-72
Archytas, 141, 481n
area, computation of, 76-77; divi-
    sion of, 77. See also plani-
    metry
argument from number, proof by, 39,
    44, 183n, 205, 211, 213, 217,
    351, 353, 357, 391
Aristotle, 3, 40
arithmetic, 16, 44, 74-75, 81-83;
    fractional, 90, 385-471, 553;
    of integers, 88, 577-583; intro-
    duction to, 577n; mercantile,
    26, 29, 88-91, 551, 555-575,
    585-601; sources of, in ACC, 86-
    87; treatises on practical, 86.
    See also fractions
ars extrinsecus, 8
ars intrinsecus, 8
ars mensoria, 15
ars mensurandi, 15-17
artifex, 13
Artis cuiuslibet consummatio (ACC),
    24-29, 70; commentary on, 105;
    contents of, 17, 63, 70, 74-87;
    date and place of composition,
    24-25; edition, translation and
    commentary, 107-471; figures in,
    104; later redaction of, 96;
    metrology in, 55n, 61; Old
    French version of, 27-29 (see
    also Pratike de geometrie); ori-
    ginality of, 87; parts of, 25-
    26, 74, 95; procedures in edi-
    tion of, 92-101; stemma of manu-
    scripts of, 93-101; sources of,

82-87; theory and practice in,
    44-47; translation of, 103
ascendent, 245
ascension: oblique, 259-263; right,
    251-257
astrolabe, 18, 30, 55n, 76n, 79,
    84, 115, 117, 121, 189, 317;
    treatises on, 7n, 84, 86
astronomical tables, 85-86, 257n
astronomy, 4, 14-15, 17, 24-26, 28,
    31, 49, 72, 74-75, 78-81, 83-86,
    221-293, 335-345; Arabic terms
    in, 85; sources of, in ACC, 84-
    86
Augustine, Saint, 32, 33n
Azarquiel. See Millás-Vallicrosa,
    José María

Bacon, Roger, 37-39, 41, 56-58
Bandinius, A.M., 106
Bandoun, John, 38
Baron, Roger, 3n, 7n, 19n, 33n, 111n
    ed. Hugh of St. Victor's Prac-
    tica geometriae, 3n-5n, 17n-19n,
    39n, 115n-119n, 295n-305n, 311n,
    317n, 335n
barrel: dimensions of, 529, 533,
    535, 539, 541, 543; octupling
    the volume of, 539; volume of,
    349, 351, 511, 531-543. See
    also tun
al-Battani. See Nallino, C.A.
Baur, L., 8n, 9n, 11n
Beard, Rick, xi
Beaujouan, Guy, xi, 54n
Beresford, Maurice, 62, 63n, 64n
Bible, geometry in. See exegesis,
    biblical
binomial, 51; squaring a, 581
al-Biruni, Abu'l Rayhan Muhammad
    ibn Ahmad, 221n, 227n
Blume, F., 5n
Boethius, 83, 87, 477n. See also
    Folkerts, Menso; Friedlein, G.
Boinet, A., 101
Boncompagni, Baldassarre: ed. John
    of Seville's Liber algorismi de
    practice arismetrice, 86n, 385n,
    393n, 399n, 403n, 405n, 433n,
    441n-449n, 453n-457n; ed. Leo-
    nardo Pisano's Liber abbaci, 419n;
    ed. Leonardo Pisano's Practica
    geometriae, 26n, 47n, 48n, 60n, 61n

Bowie, Theodore, 68n

Bridges, J.H., 37n

Britt, Nan Long, 19n, 21n, 22, 23n;
    ed. Geometrie due sunt partes
    principales, 19n-21n; ed. Trac-
    tatus quadrantis, 21n, 22n,
    119n, 121n, 231n, 235n, 237n,
    267n, 295n-301n, 305n, 311n,
    315n, 319n, 329n, 349n-353n,
    361n

brothers, problem of six, 553

Bubnov, Nicolaus, 5n; ed. Epaphro-
    ditus et Vitruvius Rufus, 129n,
    135n, 139n, 141n, 145n, 149n,
    151n, 155n-159n, 163n-171n,
    177n, 183n, 295n, 409n, 413n;
    ed. Epistola Adelboldi ad Sil-
    vestrum II papam, 173n, 177n,
    327n, 331n, 343n; ed. Epistola
    ad Adelboldum de area trigoni
    aequilateri, 123n, 127n, 129n;
    ed. Geometria incerti auctoris
    115n-119n, 129n, 133n, 135n,
    139n-151n, 155n-159n, 163n-169n,
    173n, 177n, 183n, 197n, 199n,
    203n-207n, 211n, 213n, 217n,
    219n, 227n, 295n-305n, 309n-
    315n, 329n, 331n, 335n, 373n,
    375n, 379n, 381n, 385n, 409n-
    413n; ed. Gerbert's Geometria,
    131n, 135n, 205n-209n, 385n,
    391n; ed. Gerbert's Liber de
    astrolabio, 221n, 247n, 317n;
    ed. Varro?, 409n, 413n

Burke, R.B., 37n

Busard, Hubert L.L., 29n, 52n, 61n

Buttimer, Charles Henry, 3n, 5n,
    6n, 18n, 32n

Bynum, Caroline Walker, xii

calendar: Egyptian, 287n; Julian,
    287n, 289n

Campanus, 105

Canones in triangulum pictagoricum
    de mensuris practica geometrie,
    39-41

Cantor, Moritz, 6n, 379n, 413n

capacity, 371n

chest: making a tun equal to, 369,
    519; making a, equal to a tun,
    367, 518; volume of a, 363, 365,
    516, 517

chord of an arc, 27

Cicero, 8n

circle: area of, 177-181, 219, 493,
    506, 535; circumference of, 173,
    175, 492; doubling, 70, 71; equi-
    lateral triangle inscribed in,
    523-527; Indian, 227n; inscribed
    in an equilateral triangle, 525;
    inscribed in a square, 197-203,
    498, 500, 523, 533n; ratio of
    areas of, 185; square inscribed
    in, 199-203, 499, 500, 523;
    squaring, 191, 193, 496; touching
    three points, 187, 189

city, area of a round, 219, 506

city planning, 58, 61-65, 72

Clagett, Marshall, xi; ed. Adelard
    of Bath's Preface to Euclid's
    Elements, 12n, 13n, 44n; Archi-
    medes in the Middle Ages, 49n,
    173n-179n, 183n, 195n

classification: of geometry, 12;
    of the sciences, 2-17, 31, 32,
    41-43, 56, 75

coinage: problems about, 29, 89,
    90, 551, 555-567, 587-601; title
    of, 89, 569-575

column: diameter of an engaged, 69,
    70; volume of, 537

commodities, sale of, 561-565, 585

"common usage" in geometry, 47-49,
    52

compass, geometric, 48

computation, 81; astronomical, 84,
    85; fractional, 26, 28, 86; geo-
    metric, 24, 30, 34; proportional,
    29, 31

construction, geometric, 70, 187n

Copeland, Henry, xi

cosmimetria, 6, 18, 19, 21, 23, 24.
    See also crassimetria

craft traditions, 7, 9, 13

craftsman, 4, 11, 54, 56, 73

crassimetria, 19, 25, 77. See also
    cosmimetria; volume, computation
    of

Craster, H.H.E., 105

cube (number), 425-431, 439

cube, volume of, 327, 507

cube root, 425-429; extraction of,
    415-425. See also root

Curtze, Maximilian: ed. Abraham bar
    Hiyya ha-Nasi's Liber embadorum,
    27n; ed. Adelard's ? Algorismum,
    435n, 441n, 445n-449n, 455n,
    457n, 461n, 467n; ed. Peter of

Dacia's In algorismum . . . commentarius, 419n; ed. al-Zarkali's Canones super tabulas Toletanas, 221n, 249n

curve, rectification of, 173n, 179n, 492n

cylinder, volume of, 329, 349, 351, 361, 508, 511, 515

Daly, John F., 106

day: hour of, 243; variation in length of, 293

decagon, area of, 171

decimal position, 447, 461–471

declination: planetary, 267; solar, 267; stellar, 267

Denholm-Young, N., 105

Deschamps, Paul, 63n

De Smet, Antoine, 59, 60n

diameters, ratios of, 185

difference, non-sensible, 193n

Dilke, O.A.W., 6n

direction, planetary, 273

Dominicus de Clavasio, 29, 30, 43, 51–53, 61

Dominicus Gundissalinus. See Gundissalinus, Dominicus

earth, size of, in relation to sun and moon, 335–345, 431

eccentricity, planetary, 279, 281

ecliptic, obliquity of, 233, 241

edition, procedures in, 92–104

education, practical geometry in, 31–42, 73

Epaphroditus et Vitruvius Rufus. See Bubnov, Nicolaus, ed.

equation of center, planetary, 277–281

equipollence, 153

Euclid, 45–47, 83, 381n; Elements, 5–7, 12, 39, 43, 46, 47, 50, 51, 70, 72, 78, 123n, 355n; Adelard of Bath's versions of, 7, 12, 43, 44; medieval versions of, 5, 44

example, numerical, 34

exchange rates, 89, 90, 551; table of, 555–559

exegesis, biblical, 32–37

Ezekiel's vision, geometry of, 35–36

Fanfani, A., 72n, 91

al-Farabi, 11, 12

Fibonacci. See Leonardo Pisano

field, area of, 60, 207, 211, 503

figurate numbers, 409–413

figures in ACC, 104

Folkerts, Menso, ed. "Boethius" Geometria II, 5n, 129n–135n, 139n–159n, 163n–171n, 177n, 385n, 409n

foot, round, 533

Foulet, Alfred, xii

fractions: arithmetic of, 28, 74, 81, 89–90; astronomical, see sexagesimal; common, 25, 81, 90, 385–393, 553; decimal, 82; geometrical, see common; sexagesimal, 25–26, 40, 49, 81–82, 90, 193n, 395–407, 449, 451, 455–465, 496

Frankl, Paul, 66

Friedlein, G. 129n, 131n

frustum, volume of, 373, 375

Gagné, Jean, 14n

Geometria incerti auctoris. See Bubnov, Nicolaus, ed.

Geometrie due sunt partes principales, 19–24, 75. See also Britt, Nan Long, ed.

Geometry: artificial, 19–21, 50, 75, 78; non-artificial, 19–21, 50, 75–78; practical, in education, 31–42; practical, nature of, 1–31; practical and theoretical, 3–7, 9–10, 42–53; sources of, in ACC, 83–84. See also practical geometry

Gerard of Cremona, 11

Gerbert, 5, 6, 83, 87. See also Bubnov, Nicolaus, ed.

ginger, sale of, 585

Gingerich, Owen, xii

gnomon, 18, 19, 55n, 76n, 79n, 323, 325

Gold, Donna, xii

gold, sale of, 561, 563, 595. See also coinage, title of

Goldstein, Bernard R., 243n, 245n

González Palencia, Angel, 11n

Gordanus. See Clagett, Marshall, Archimedes in the Middle Ages

Gundissalinus, Dominicus, 7-13, 56
Gunther, R.T., 247n, 265n-275n

Hahnloser, Hans R., 68n-70n
Haring, Nicholas, 7n
Hartner, W., 117n
Haskins, C.H., 7n
height, measurement of, 295-313.
    See also altimetry
Henry, Charles, 27n, 28, 101
heptagon, area of, 165, 491
hexagon, area of, 163, 490
horizon, depression of. See lati-
    tude, terrestrial
hours, unequal, 189n
Hugh of St. Victor, 3-7, 17-19, 21,
    23, 31-35, 39n; Didascalicon,
    3-7, 18n, 31-32, 34; De arca
    Noe morali, 33-34; De ponderi-
    bus et mensuris, 33n; De sacra-
    mentis, 32; Practica geometriae,
    3-7, 17-19, 34, 39n. See also
    Baron, Roger, ed.
Hunt, R.W., 3n, 4n, 8n
Hyde, J.K., 63n

Ibn al-Muthanna. See Goldstein,
    Bernard R.; Millás-Vendrell,
    Eduardo
Ibn 'Ezra. See Millás-Vallicrosa,
    José María, ed.
Indian astronomical methods, 251n,
    255n, 259n, 287, 289
instruments: in architecture, 65-
    67, 69-70; in astronomy, 16, 75-
    76, 113; in geometry, 4, 17-18,
    20, 23-25, 27, 30-31, 40, 51,
    53, 75; making, 54-58, 75-76,
    227, 317-325
instrumentum gnomonicum, 30
integer, product of, with aliquot
    part, 385-389
isoperimetric figures, 57-58

James, Montague Rhodes, 106
jauzahar. See node
Johannes Anglicus of Montpellier,
    21, 23-24. See also Britt, Nan
    Long, ed. Tractatus quadrantis
Johannes de Muris, 43, 49-51
John Bandoun, 38
John of Seville, 86. See also
    Boncompagni, Baldassarre, ed.
Judy, Albert G., 15n, 16n

Kennedy, E.S., 85n
al-Khwarizmi, 86. See also Neuge-
    bauer, O., ed.; Suter, H., ed.
Kibre, Pearl, 31n
Kilwardby, Robert, 8-12, 15-17
Kohler, Charles, 101

Lachmann, Karl, 5n
latitude: planetary, 275; terres-
    trial, 235, 293
length: computation of, 76; measure-
    ment of, 115-121. See also
    height
Leonardo Pisano, 26-27, 43, 47-50,
    52-53, 60-61, 64. See also
    Boncompagni, Baldassarre, ed.
Levey, Martin, 27n
Liber de astrolabio. See Bubnov,
    Nicolaus, ed.
liberal arts, 32, 41-43. See also
    quadrivium
Llobet de Barcelona. See Millás-
    Vallicrosa, José María, Assaig...
longitude: planetary, 271; solar,
    265; stellar, 267

Macrobius. See Stahl, R.
Madan, F., H.H.E. Craster, and N.
    Denholm-Young, 105
Magister 2, 68-72
magistri ymaginum, 289n
magnitudes: commensurable, 43; con-
    tinuous, 44, 78
manuscripts: of ACC, 93-101, 105-
    106; collation of, 101; of the
    Pratike, 101-102; sigla of, 101,
    105-106; stemma of, of ACC, 93-
    101
marginal notes, 101
Marsh, Adam, 38
Marx, J., 106
masonry, geometry in, 65-66, 68, 71
mathematics, restoration of, 38-40
measurement, 4-6, 10, 12-25, 28-31,
    50-52, 57, 75, 78; artificial,
    20, 23, 29-30, 48, 53, 55, 60,
    75, 78, 84; artificial, in as-
    tronomy, 79, 84; indirect, 66;
    non-artificial, 20, 23, 29-30,
    50, 53, 75-76, 78, 83-85; non-
    artificial, in astronomy, 79;
    practical, 43
measurer, 11-14, 52
mechanical arts, Arabic tradition
    of, 12

Mendelsohn, Everett, xi
mensor: geometrie, 52; laicus, 52
mercantile material, 55–56, 72, 91,
  551–601
meridional line, 227
Messahala. See Gunther, R.T.
metrology, 33–34, 43–44, 55, 61,
  65, 81, 131, 207, 209, 385, 477,
  529–533, 539–543
Migne, Jacques Paul, 5n, 32n, 33n,
  35n, 36n
Millás-Vallicrosa, José María:
  Assaig d'historia de les idees
  físiques i matemàtiques a la
  Catalunya medieval, 247n; Estudios sobre Azarquiel, 233n,
  237n, 243n, 245n, 251n, 259n,
  283n; ed. "Un nuevo tratado de
  astrolabio de R. Abraham ibn
  'Ezra," 235n, 237n, 265n, 267n,
  271n, 273n; ed. Abraham ibn
  'Ezra, El libro de los Funda-
  mentos de las Tablas astronómicas, 221n, 231n–235n, 251n,
  257n, 279n, 283n–291n
Millás-Vendrell, Eduardo, 221n,
  223n, 231n–237n, 243n, 249n,
  251n, 259n, 261n, 287n, 293n
minutes, astronomical. See fractions, sexagesimal
minutes, geometric. See fractions,
  common
minutimetria, 74, 81, 385–471
Molinier, A., 106
money: alloy content of, 567–575,
  599, 601; conversion of, 561–
  565, 597; problems about, 88–
  91; sale of, 587–595; title of
  gold, 573, 575; title of silver,
  569–573; minting, 567, 599, 601
Montgomery, Charles F., xi
moon, size of, in relation to earth
  and sun, 335–345, 431
Mortet, Victor, 28n, 63n, 68n, 89,
  101
Murdoch, John E., xi, 43–44, 54n

Nallino, C.A., 221n, 227n, 241n,
  245n, 251n–261n, 279n, 287n
Neugebauer, Otto, xii, 251n, 255n,
  257n; ed. al-Khwarizmi, 221n,
  231n, 235n, 237n, 249n, 257n–
  261n, 283n, 287n, 291n
Noah's ark, geometry of, 33–35

nodes, planetary, 291
nonagon, area of, 169
number: figurate, 82, 87; polygonal,
  77, 82, 161n–171n, 409–413; pyramidal, 78, 82, 379n, 381n;
  triangular, 129n, 409
number theory, 82, 87
numerical example. See proof, numerical example as

octagon, area of, 167
oral tradition of geometry, 67, 72

P. (author). See Canones in triangulum pictagoricum
Panofsky, Erwin, 66
Paris: equinoctial altitude at,
  25, 231, 233n; latitude of, 235;
  oblique ascensions for, 259–263
part: aliquot, 385–393; fractional,
  597; whole, 385–393
pentagon, area of, 159, 161, 489
pepper, sale of, 585
Peter of Dacia. See Curtze, Maximilian, ed.
Picard dialect, 27, 90, 102
Pirenne, Henri, 60n, 91
planet: apogee of, 283, 285; declination of, 267; direction of,
  273; equation of center of, 277–
  281; latitude of, 275; longitude
  of, 271; node of, 291
planimetry, 6, 15, 18, 21, 22, 25,
  29, 57, 75–78, 115–219, 474–506,
  523–527, 545–549
Plato of Tivoli, 27n
plot, house, 62–64, 213, 215, 219,
  504, 506
Podlaha, A., 106
pole, altitude of, 237
polygon, area of, 171
position, decimal, 447, 461–471
positional notation, 88, 581, 583
Poulle, Emmanuel, 22n
Practica geometriae. See Dominicus
  de Clavasio; Hugh of St. Victor;
  Leonardo Pisano
practica liberalis, 12, 13
practical geometry: agent of, 10–13;
  Islamic tradition of, 50; Italian,
  72n; nature of, 1–31; parts of,
  2, 6, 9, 10, 17–31, 57, 74, 111,
  113, 473; and practical concerns,
  53–73. See also geometry

practice, relation to theory. <u>See</u> theory and practice

Le <u>Pratike</u> <u>de</u> <u>geometrie</u>, 27–29, 72; arithmetical contents of, 88–91, 103, 553, 577–583; edition, translation and commentary, 472–601; geometric contents of, 87–88, 103–104, 473–549; manuscripts of, 101–102; mercantile contents of, 55n, 103, 551–601; metrology in, 61; order of propositions in, 103–104; procedures in edition of, 101–104; spelling in 102–103

precession, 283, 285, 289n

precisura, 137, 479

proof: techniques of, 5, 39, 44–51, 83; numerical example as, 39, 44, 183n. <u>See</u> <u>also</u> argument from number, proof by

proportion, 4, 18–22, 29–31, 50–51

Pseudo-Boethius, 5, 6. <u>See</u> <u>also</u> Folkerts, Menso, ed.

Ptolemy, 46, 221, 233, 249, 251n, 253n, 255, 263n, 279n, 281–289, 291n, 335, 337, 339n, 381n

pyramid: volume of an equilateral, 379; volume of a right, 347; volume of a square, 381

pyramidal number. <u>See</u> number, pyramidal

Pythagorean: theorem, 52, 309n; triple, 139n, 480

quadrangle, area of irregular, 153, 211, 486, 503

quadrangular vessel, volume of, 361, 515. <u>See</u> <u>also</u> chest

Quadrans vetus, 21–24, 26, 55n. <u>See</u> <u>also</u> Britt, Nan Long, ed.

quadrant (instrument), 18, 22, 30, 55, 76n, 115n, 117n, 121, 319, 321

quadrivium, 5, 32, 45, 74–75, 78

quantity: continuous, 16; discrete, 16

ratio, 18–21, 29–30, 51, 76–77, 83; of sizes of sun, moon, and earth, 335–345, 431

ratione numeri, 44. <u>See</u> <u>also</u> proof, numerical example as

reciprocal values, table of, 557, 559

rectangle: area of, 123–129, 147, 484; diagonal of, 157, 488; enlarging a, 545, 547; number of, in a known surface, 205, 213–219, 501

rectangular solid, volume of, 329n. <u>See</u> <u>also</u> chest

Reed, Roger, xii

rhombus: altitude of, 151, 485; area of, 151, 485

Richard de Escham, 62

Richard of St. Victor, 34–37

river, width of, 69

Robert Grosseteste of Lincoln, 38

Robert Kilwardby. <u>See</u> Kilwardby, Robert

Roberts, Susanne F., xii

Robertus Anglicus, 21n

Roger Bacon. <u>See</u> Bacon, Roger

root, 47–48, 81–82. <u>See</u> <u>also</u> cube root; square root

Rudorff, A., 5n

Saint Denis, abbey church of, 65–67

Schum, W., 106

science, middle. See <u>scientia</u> <u>media</u>

sciences, classification of the. <u>See</u> classification of the sciences

<u>scientia</u> <u>media</u>, 14–15

sexagesimal fractions. <u>See</u> fractions, sexagesimal

shadow: of altitude, 323; plane, 223

shadow square (instrument), 76n, 317–321

Sharp, D.E., 15n

Shelby, Lon R., xii, 65, 68n

silver, sale of, 563, 565. <u>See</u> <u>also</u> coinage, title of

Smalley, Beryl, 32n, 34n, 35n

solid, volume of a regular, 349, 351

space-filling figures, 57–58, 205, 213–219, 501, 504–506

sphere: area of, 183, 494; volume of, 331–337, 509; volume of, inscribed in a cube, 331, 333, 510

square: area of, 149, 523; circle inscribed in, 197–203, 498, 500, 523, 533n; circling of, 195, 497; doubling a, 549; inscribed in a circle, 199–203, 499–500, 523

square (instrument): carpenter's, 67; mason's, 70. <u>See</u> <u>also</u> shadow square

square number, 409–413, 433–441;
table of, 577–581. See also
number, figurate
square root, 447–455; extraction of,
443, 445, 457–471. See also root
Stahl, R., 241n
star: altitude of, 225; declination
of, 267; longitude of, 269
Steele, Robert, 37n–39n, 57n
stereometria (or steriometria), 15,
19, 21–24, 29, 57. See also
crassimetria
straight line, to measure, 115–121,
474–475
Suger, Abbot of Saint Denis, 65–67
sun: altitude of, 221; apogee of,
24n, 283–289; declination of,
267; equinoctial altitude of,
231; longitude of, 265; meridian altitude of, 229; shadow
of the altitude of, 249; size
of, in relation to earth and
moon, 335–345, 431; solstitial
altitude of, 233
superiometry, 57
surveying, 55–56, 58–59, 64–65, 72.
See also surveyor
surveyor, 11–12, 48–49, 53–54, 59–
61; Roman. See agrimensores
Suter, H., 221n, 231n, 235n, 237n,
249n–261n, 283n, 291n
Swerdlow, Noel, xii
Sylla, Edith Dudley, 54n
Sylvester II, Pope. See Gerbert;
Bubnov, Nicolaus, ed.

tables, astronomical, 85–86, 257n
Tannery, Paul, 21n, 23n
Taylor, Jerome, 3n, 5n–7n, 18n, 32n
temple of Solomon, geometry of, 35
theory and practice, relationship
of, 2–3, 45, 109, 111. See also
geometry, practical and theoretical
Thomas Aquinas, Saint, 14n
Thorndike, Lynn, 19n, 21n, 105, 111n
Thulin, C., 6n
time: of day, 243; of night, 247
Toledan tables. See Toomer, G.J.
Toomer, G.J., 279n–283n, 291n
tower: height of, 69, 295–301, 305,
307; height of an inaccessible,
299, 301, 305, 307
Trabut-Cussac, Jean-Paul, 62n

Tractatus quadrantis. See Britt,
Nan Long, ed.; Quadrans vetus
trapezoid: area of, 207, 213, 502;
area of a right, 155, 487
triangle: altitude of, 48, 52, 60,
133, 135, 137, 478, 479; area
of, 46, 60, 179, 181; area of
an acute, 145; area of an equilateral, 123–131, 476–477; area
of an isosceles, 133, 217, 478,
505; area of an obtuse, 143,
482; area of a right, 139, 480,
483; area of a scalene, 135,
479; circle inscribed in an
equilateral, 525; circle inscribed in a right, 141, 481;
equilateral, inscribed in a
circle, 523–527; right, 36;
sides and hypotenuse of a right,
139, 480
triangle (instrument), 18, 40
triangles: congruent, 78; similar,
51, 305n
triangular field, area of, 217,
505
trigonometry, 49, 80
Troebel, Johannes, 59
tun: chest equal to, 367, 518;
doubling the volume of, 357,
514; equalling the volume of,
359; making the volume of a,
equal to that of a chest, 369,
519; quadrupling the volume of,
355, 513; volume of, 353, 512.
See also barrel

units, conversion of, 89, 561n,
585–597

Van Egmond, Warren, 72n
Varro, 8n
Varro? See Bubnov, Nicolaus, ed.
Victor, Stephen, 49n
Victorinus, Marius, 8n
Villard de Honnecourt, 68–72
virga (measure), 64
Vitruvius Rufus. See Bubnov,
Nicolaus, ed.
Vogel, Kurt, 27n
volume: computation of, 77, 315,
327–383, 507–520, 529–543; dis-
tinguished from capacity, 371n
volumes of sun, moon, and earth,
ratios of the, 339–345
volumetry. See volume, computation of

Walterus de Kirchem, 59-61
walzagora, 117
Weisheipl, James A., 15n
well: capacity of, 315; depth of,
    311, 313; volume of, 371, 520
Willner, Hans, 295n, 305n, 313n
window, width of, 69
Witelo, 51

year: of composition of ACC, 24,
    283; sidereal, 287n; solar, 287,
    289; tropical, 287

zenith, distance of, 241
al-Zarkali. See Curtze, Maximilian, ed.
zij, 257n
Zuckerman, Charles, xii